# THE DOMAIN OF NATURAL SCIENCE

# The Domain of Natural Science

THE GIFFORD LECTURES DELIVERED
IN THE UNIVERSITY OF ABERDEEN
IN 1921 AND 1922

BY

E. W. HOBSON

DOVER PUBLICATIONS, INC.
NEW YORK

Published in Canada by General Publishing Company, Ltd., 30 Lesmill Road, Don Mills, Toronto, Ontario.

Published in the United Kingdom by Constable and Company, Ltd., 10 Orange Street, London WC 2.

This Dover edition, first published in 1968, is an unabridged republication of the work originally published in 1923, and is reprinted by permission of the Cambridge University Press.

*Library of Congress Catalog Card Number: 68-24653*

Manufactured in the United States of America
Dover Publications, Inc.
180 Varick Street
New York, N. Y. 10014

# PREFACE

THE prominence which that complex of knowledge and ideas which is denoted by the term Natural Science has attained in the thought and life of the modern world is undeniable, and is indeed fully recognized by all thinking persons. But the question as to the exact position which what is often termed the scientific view of the world should rightly occupy, in relation to the other factors of human experience with which Religion and Philosophy are concerned, is one which gives rise to very divergent opinions amongst earnest thinkers. This is a matter which causes grave perplexity to the minds of great numbers of men and women. The question has been, in various connections, a subject of controversy for centuries, and it has in our time become of inestimable importance, in view of the effects which any generally accepted answer to it may have, not only upon our theoretical views and mental life, but also upon our attitude towards existence in its more practical aspects. My main object in preparing the lectures which are here published was to provide a reasoned contribution towards the clarification of ideas in relation to this fundamentally important question. In order to attain my object it was necessary to undertake an examination, as close as the circumstances permitted, of the historical development, aims, and true characteristics, of various departments of Natural Science, with a view to the characterization of the proper position of Natural Science in relation to Thought in general.

Based upon this examination of the leading features

of scientific theories and laws I have advocated particular views of the essential characteristics of Natural Science, the acceptance of which would imply that, in our more general outlook upon the world, a position of much less dependence upon Natural Science may rightly be taken up than has been supposed by many men of Science to be admissible. What is sometimes spoken of as the "descriptive view" of the functions of Natural Science is, in its main outlines, far from new, and has received some measure of acceptance on the part of prominent men of Science and other thinkers. I have endeavoured to make this view more explicit and precise than in the forms, often fragmentary, in which it has previously been stated. That Natural Science is in no way concerned with questions as to the nature of reality, or with efficient causation, and that the edifice it has reared is independent of any special ontological assumptions, and in particular of that set of assumptions known as physical realism, are propositions which I have, throughout the lectures, illustrated and maintained. A position of detachment, or neutrality, as regards the ontological and other conceptions which divide various schools of Philosophy would entail as a consequence that the authority of Natural Science cannot properly be invoked, as of decisive weight, in favour of any assumptions which it does not need for its own purposes. The freedom which would thus accrue to Religion and Philosophy from any compelling influence due to Natural Science would be of course limited by the exigencies involved in the admission that all questions relating to the order of the world of physical percepts should be treated in accordance with the canons of Natural Science alone; a condition which has in the past by no means always been fulfilled. In the last two lectures I have given some consideration to the further question, what influence Natural Science may exert upon our wider outlook on the world, when it is supplemented by ontological assumptions which are extraneous to it.

Apart from slight emendations and a few short insertions, the lectures are here published as they were delivered in Aberdeen.

In the preparation of those lectures in which the histories of special departments of Natural Science are sketched I have been dependent upon information obtained from a large number of sources. Of these I here mention only those of which I have made most use. In the lecture on *Time and Space*, and in other lectures on departments of Physics, I have utilized the works of H. Poincaré, especially *La Science et l'Hypothèse*. In the lectures on *Corpuscular theories of matter*, on *Dynamics*, and on *The conservation of matter and energy*, I have made considerable use of the historical information contained in the work of E. Myerson entitled *Identité et Réalité*. In these and other lectures I have also utilized Lange's *Geschichte des Materialismus*. In the lecture on *Electricity, Magnetism, and Light* I have drawn much historical information from Prof. E. T. Whittaker's work entitled *History of the theories of Aether and Electricity*. In the lecture on *The constitution of matter* I have utilized the work by A. E. Garrett on *The periodic law*, and also Prof. Soddy's lectures in *Science and Life*. In the lecture on *Cosmical Theories* I have drawn historical information from Miss Agnes M. Clerke's *History of Astronomy in the nineteenth century*, and I have also utilized J. H. Jeans' work on *Problems of Cosmogony and Stellar Physics*.

In the part of the lecture on *Biological Science* which deals with a comparison of the living organism with a machine, as regards relations of Energy, I have utilized Prof. J. Johnstone's work on *The Philosophy of Biology*. In a portion of this lecture I have also made use of Merz's *History of European thought*. In the preparation of the lectures on *The living organism*, on *Heredity*, and on the *Evolution of Species*, I have utilized the works of Prof. J. Arthur Thomson, especially those entitled *Heredity* and *The Science of Life*. In these lectures I have also made much use of the historical information contained in E. S. Russell's work entitled *Form and Function*; and I have also drawn information from H. F. Osborn's work *From the Greeks to Darwin*.

Before I wrote the two final lectures I perused Prof. J. B. Baillie's *Studies in Human Nature* and was influenced in some points by his views.

To Dr F. H. A. Marshall, F.R.S., and the Rev. F. R. Tennant, D.D., each of whom read the type-scripts of some of the lectures, I am indebted for advice on special points. To my friend Prof. James Ward, F.B.A., I owe much. Without the stimulus received in the course of conversations carried on with him during many years I should probably never have ventured upon the task of writing upon a subject so much wider than my main subject of study.

E. W. H.

CHRIST'S COLLEGE, CAMBRIDGE
*March* 1923

# CONTENTS

# THE DOMAIN OF NATURAL SCIENCE

# I

# INTRODUCTION

THE Domain of Natural Science is the subject I have chosen for the course of Gifford Lectures which the Senatus Academicus have done me the honour of inviting me to deliver. Any attempt to define and delimit the domain of Natural Science must concern itself with the questions what Natural Science is, and what it is not; what its precise functions within the general domain of Thought may be; what the nature of its relations is with other parts of that wider domain; in what manner it has grown up, and what we may reasonably expect from it in the future; and to what kind of essential limitations it is subject.

Our general mental outlook upon the world contains a great mass of ideas and knowledge of which the origin is so remote in the development of our race that they may perhaps least inadequately be described as instinctive notions and knowledge. We possess also a heterogeneous mass of unsystematized ideas, in large part traditional in the society in which we have grown up. Besides these we possess a stock of ideas and knowledge of a more or less systematized character, drawn from the religious, philosophical, and scientific, thought of the past and the present. My choice of subject has been made in the hope of doing a little to promote clarity of view as regards the relation which that special kind of thought which, in accordance with established usage, we denote by the name Natural Science, has with that greater complex of ideas to which I have alluded, and which conditions our general mental attitude towards the psychical and physical world.

The denotation of the term Science is not fixed by usage with absolute precision. In the wider sense of the term, a critical and systematic study of any clearly marked out region of thought is described as a Science, provided that study attains, or at least aims at, a reasonable standard of rigour, in relation to classification and subsumption under general laws, in dealing with its special subject matter. Thus, for example, Philology, Ethics, Psychology, Economics, Anthropology, Textual Criticism, and even Heraldry, are spoken of as Sciences. On the other hand, Metaphysics is not usually spoken of as a Science, probably because, however systematic it may aim to be, its subject matter is too universal, embracing in fact the whole of experience and existence. The term Natural Science, with which I am here specially concerned, is generally restricted to denote the group of those special Sciences which concern themselves with the study of what we call physical phenomena, including the cases in which the phenomena are connected with living organisms. In the somewhat narrow sense in which I shall employ the term, Natural Science excludes any direct consideration of the mental or psychical facts in living organisms from its purview; although this restriction is not universally accepted in connection with the group of Biological Sciences. It may be objected against this avowedly narrow use of the term Natural Science that it implies an undue restriction of the term Nature, involving the relegation of the mental side of life to a place outside Nature. This objection has undeniable weight. My employment of the term Natural Science, in the meaning that it denotes the Science of the physical world, is a matter of convenience only, and is not intended to indicate the acceptance of the theory that there exists any ultimate barrier between physical Nature and the mental life of man.

The object I have indicated, of describing and delimiting the domain of Natural Science, can only be attained by undertaking an examination, from the inside, of the methods of procedure, the assumptions, and the course of development of various special parts of Natural Science. Anything like a complete examination of the history of the development, and an analysis of the methods, of each one of the group of the Natural Sciences would be a gigantic task far beyond the powers of any single individual. To do this would require the co-operation of an army of specialists intimately acquainted with the details of the present and past states of the very numerous departments of Natural Science. Happily, it does not seem altogether impossible to attain at least some measure of success in an attempt to define in general terms the nature of the contributions which the Natural Sciences are fitted to make to a general view of the world, by adopting the comparatively modest procedure of examining the methods and principles of some typical branches of Science, and especially of those branches which may be regarded, from a methodological point of view, as in a more advanced stage of development than the others. It might perhaps be regarded as the more logical method to commence at once this process of examination and analysis, and upon its completion to deduce the more general conclusions which could be extracted from the results obtained. But it is more practicable, if less logical, first to state, with as much precision as possible, the general conclusions which it is hoped to establish, and then later to support and illustrate these conclusions by means of an analysis of the history and general features of special branches of Science. Accordingly, in the first few lectures, a general account will be given of what may be regarded as the foundations of the method adopted in the various departments of Natural Science, and conclusions will be stated as to the nature of the knowledge of the physical world which organized Science affords. Moreover, some indications will be given of the relation of Science with other elements of thought, and especially with Philosophical thought and speculation. In particular an attempt will be made to trace the character of the limitations to which Scientific knowledge is subject. After these earlier lectures, the examination, in some detail, of various typical portions of Science, and of many scientific theories, will be undertaken. If, in this, the main part of the course, I devote what may seem a disproportionately large amount of time to scientific theories which have at the present time been abandoned, modified, or subsumed under wider theories, the explanation must be taken to be that my object is not to attempt to perform the impossible task of giving an accurate account of the present state of Natural Science, even in a few of its branches, but rather, by means of an historical retrospect, to disclose the essential characteristics of all the theories and laws which go to make up Natural Science. It is probably less difficult to discern the principles which underlie the essential methods of Science, and thus to obtain a grasp of their true character, and of the limitations to which they are subject, by examining and dissecting theories which have attained a crystallized form, than by attempting a complete analysis of the tentative and rapidly changing theories which actually direct, at the present time, the gigantic efforts that are being made to advance our knowledge of natural phenomena.

The state of various branches of Science, in our day, is such as to lead to a vivid appreciation of the unwisdom of considering any scientific theory as having attained finality. Recent experience has shown that a well-established and successful theory may be liable to fundamental change, owing to the discovery, by observation or experiment, of new facts incapable of reconciliation with the theory in the form in which it has for long been accepted. At the end of the course an attempt will be made to draw some general conclusions of a kind which may be regarded as having a bearing upon the great central problem, the unbiassed treatment of which was indicated by Lord Gifford in his Will as the object for which the Gifford Lectureships were to be founded. The difficulty of the task I have undertaken is such that, adequately to cope with it would strain the resources of anyone who had spent his life in the consideration of the Philosophy of Science. It can be undertaken only with much diffidence by one whose consideration of the weighty matters involved therein has been limited by the scant leisure available in a life in which attention has been concentrated, in the main, upon the absorbing technique of a single branch of Science.

During the last century and a half the life and work of multitudes of human beings have, in our own country, and in the civilized world in general, undergone a revolutionary change, due to the application of Science in all branches of Industry. Vast populations now exist which, at a former period, could neither have been fed nor provided with work. The use of the motive power of steam, introduced in the eighteenth century, its application to locomotion on land and sea, in the nineteenth century, the invention of the oil engine with its application to aviation, in the twentieth century; the electric telegraph, the telephone, wireless telegraphy and telephony, and other applications of Electromagnetism, are all traceable to the discoveries of Pure Science. The development of the Biological Sciences has led to applications, of the most far-reaching importance, in Agriculture, Hygiene, and in the healing Art. These applications of Science give indications of future extensions, in which the present limits may be indefinitely transcended. Other results of the application of Science are embodied in the great chemical industries, of which one of the most ominous results is apparent in the discovery of explosives of vast destructive power. These results of Natural Science, and many others too numerous for mention, in its persistent efforts to dominate physical nature, have furnished us with the mechanical means of securing an indefinite improvement in the welfare of mankind, if a wise use is made of the power with which they endow us. They have also provided our civilization with the material means of committing suicide, if the increased mechanical powers which they afford are not accompanied by a corresponding rise in the ethical standards which actuate nations in their dealings with one another. In conjunction with the revolutionary industrial and economic consequences which have arisen from modern applications of Science, vast changes have been produced in the mental outlook of multitudes of men, and these changes have given rise to social problems of the most far-reaching character, of which the solutions are not in sight. These problems are such that, even in the view of optimists, it may need centuries of unrest and strife before relative social stability is restored. These matters must however here be left on one side. I have referred to them because some of the effects of Science upon the thought of the world are of an indirect character, and are not simply logical consequences of the development of purely scientific thought, but are rather due to the stimulating effect on the imagination produced by the great increase in the mechanical appliances at our disposal, and to the enlargement of the mental horizon consequent upon the increased means of communication we possess having abolished the comparative isolation in which considerable aggregations of human beings formerly lived. However, our attention must be confined to the more direct relations between scientific thought and mental life in general.

Prolonged investigation of the ideas and beliefs of existing races in a primitive state of development, together with indications, obtained from historical and archaeological sources, of conditions in the earlier ages of mankind, have made it possible to reconstruct in broad outline the concurrent factors which made up the complex of ideas about natural phenomena among primitive peoples. The earliest conceptions of this kind may be roughly subsumed under three heads. First, there existed at all times a body of knowledge of the uniformities of Nature, some of which had an origin reaching back in the experience of the race to the simian, or to the still earlier, ancestry of man. Much of this knowledge involved no reflection, and it may perhaps not unfittingly be described as instinctive knowledge. A considerable part was obtained as the result of the experience of the individual, or was communicated to him by his fellows. Such knowledge, whether of the kind I have called instinctive, of racial origin, or of the kind dependent on individual experience, was due to the necessity for *action*, of men or their ancestors in their physical environment, for self-preservation, the obtaining of food and shelter, and generally for the maintenance of the life of the individual and the tribe. Such knowledge, of an unsystematic character, largely unconscious, and involving little or no reflection, we may describe as common knowledge. Its origin was empirical, and this common knowledge, acquired empirically, was the true parent of Science.

The second class of ideas of primitive peoples may be brought under the head of Animism. To primitive man, the distinctions between himself and his environment, and between the psychical and the physical, were blurred and indistinct. His own thoughts, feelings, appetites, and passions, he ejected outwards into the objects by which he found himself surrounded. Thus he peopled the world of objects with spirits and demons, of like nature with himself, under the influence of the same motives of action, the same passions and appetites.

For him, all Nature was alive; all objects were the residences of spiritual beings, whose actions might be influenced by supplications, persuasion, sacrifices, and threats. But the spirits which animated Nature were, like men themselves, highly capricious, and their actions often unintelligible. To their agency were attributed many of the calamities, such as diseases and storms, which fall upon men. It is to Animism and animistic conceptions that we may trace back many of the Philosophical and Religious problems which have perplexed mankind in later ages.

The third class of ideas which are found in primitive man may be brought under the head of Magic. Instead of supplications and sacrifices being employed to persuade external Nature to action of a kind favourable to the wishes and interests of man, a similar end was attained by a form of compulsion embodied in Magical ritual. The essence of Magic has been described by Carveth Read as follows[1]:

> Magic supposes constant connexions of events due to the agency, force, influence or virtue of charms, rites and spells; which connexions, however, are found only to be tendencies of some events to excite others, inasmuch as they may be frustrated by counteracting charms, rites or spells. Magic is entirely constituted by notions of force, sometimes violent, as in the discharge of an enchanted spear; sometimes subtle, like the efficacy of an opal; intangible, invisible, and operating at a distance through space and time, like a witch's spells that eclipse the sun or moon. These forces have only a one-sided relation to the workaday world; they meet with no resistance from what we take to be the "properties of matter," such as weight and impenetrability; but are themselves entirely exempt from natural law; what we call the "real world" has no hold upon them; they live in a world of their own. They are absolutely immeasurable; and hence the causation, which is certainly implied in the notion of their operation, is indefinite, and becomes vaguer and vaguer as the magical system develops; and all this is the opposite of what happens in the history of Science. In spite of having a

[1] *The Origin of Man and of his Superstitions*, pp. 326, 327.

necessary ground in the human mind, Magic and Science are contrasted from the first, in their development grow wider and wider apart, in their methods and ideas more and more opposed. If either can be said to precede the other, it is Science (at least, in its earliest and crudest form) that precedes Magic.

In the main the beginnings of scientific knowledge arose out of that element which we have called common knowledge, but throughout its development Science has not completely disentangled itself from notions drawn from the domains of Animism and Magic. Traces of Animism and Magic in their cruder forms are still to be found in the most civilized communities, and they are to be observed in undiminished strength in some of the lower races of mankind. Magic and Science have in common with one another the recognition of certain uniformities in the sequence of events, and in this point both of them are sharply distinguished from Animism; but this similarity is of small importance compared with the deep seated differences in their outlook and methods; and the history of the development of Science exhibits at all times the tension and conflict between the fundamentally divergent attitudes of mind which they represent. To a considerable extent this tension between Science and Magic and Animism was, in earlier times, held in check by the fact that what elements of genuine scientific knowledge existed were largely in the hands of the same persons who were adepts in Magic. The Wizard, or Medicine Man, or the Priest, had very frequently some knowledge of the course of natural events greater than that possessed by the ordinary members of the tribe or community. This real knowledge assisted him in bolstering up his power, and in increasing his reputation for possessing special means of influencing the course of natural phenomena. In conjunction with what we now regard as the fantastic cures for ailments that the Medicine Man prescribed, he also employed some modicum of empirical knowledge which can be regarded as rudimentary Science. The magical power, possessed by the rain-maker, of influencing the weather, was often eked out by some knowledge, obtained from observation, of the course of the seasons. At all times, moreover, there probably existed, here and there, individual members of the community, whose exceptional mentality or occupation impelled them to obtain some increased empirical knowledge of the properties of material objects, and of the course of natural events, and to leave aside, or subordinate, the methods of Magic. The comparatively slow progress of Natural Science in early times may with much plausibility be attributed to the fact that Magic, combined with the prevailing hypostatization of mental and physical qualities, gave almost unrestricted rein to the imagination, and led to a luxuriant growth which left far behind that steady progress of knowledge which could only be made by the exercise of the sober qualities requisite for the patient investigation and comparison of facts.

No doubt, Animism and Magic had some indirect effect in stimulating the acquisition of knowledge, by their encouragement of the practical arts, as adjuncts of their ceremonies, but in the main the attitude of mind which they encouraged was hostile to those habits of mind which favour the growth of scientific knowledge. But the principal driving force in the beginnings of Science was of the practical order; the urgent necessity for some kind of measurement, for the determination of position in navigation by means of observation of the stars, and for knowledge of the properties of materials used in building and in the arts. Slowly the beginnings of genuine scientific curiosity arose in some individuals, and under favourable conditions in particular places. Some account of Greek Science, and of the sharp contrarieties within it due to the contest between genuine scientific method and ideas of animistic and magical origin, will be given in later lectures. From the time of Leucippus and Democritus onwards there was in Greece a real rival to Animism; the physico-mechanical theory of the world. In medieval times the progress due to Greek thinkers was submerged by Scholasticism, with its belief in occult qualities; genuine Science was only kept alive by the Arabs, whose knowledge was partly drawn from Indian sources. After the renaissance, involving as it did the rediscovery of Greek thought, Science recommenced its path of progress which has continued with unabated vigour from the sixteenth century until our own time.

Both Animism and Magic, like the common knowledge of phenomena out of which Natural Science has grown, have their modern descendants. If the validity of the theories be admitted which trace back many of the features of the more highly developed mental life of the modern world to ancient animistic and magical ideas, we have no warrant for treating these features with indiscriminate disdain, on account of what appears to us the fantastic crudity of their ancestry. It would indeed appear that Animism, Magic, and common knowledge, in their earliest forms and in their later development through the ages, are all three natural growths representing persistent and normal forms of activity of the human mind under the influence of ineradicable impulses. The cruder forms of animistic conceptions still exist among uncivilized races to-day, and they have never completely disappeared even in the higher races. In the sixteenth and seventeenth centuries, in the age in which the foundations of modern Science were being laid by such men as Kepler, Galileo, and Newton, animistic notions took terrible shape in the employment of the stake and the gibbet for the purpose of combating the works of ubiquitous demons, as exhibited in witchcraft.

As the functions of the thinking part of the community became gradually more differentiated, the tension between Science and the other elements of thought attained increased sharpness. This tension took, at various times, the form of violent conflict; especially when scientific ideas arose which seemed to threaten the traditional ideas, and the higher or lower interests, of the great religious corporations. An ancient religious system carries with it through the centuries, in its written or oral traditions, a great deal of material which, either explicitly, or in its descriptive language, incorporates the views of natural phenomena which were current at the time when the traditions first took a relatively fixed form. These views about the natural world usually come to be regarded by most people as forming an essential part of the system, which cannot be changed or eradicated without entailing serious injury, or even destruction, upon the whole religious system. This attitude of mind is shared by the official representatives of the system in question, who are frequently disposed to use the influence of their corporation to combat any attempt to replace in the popular mind the older scientific or quasi-scientific conceptions by more modern ones due to the progress of scientific knowledge in later times. After a longer or shorter period of strife, the newer scientific view receives acceptance, or is at least acquiesced in; and it is ultimately discovered that the older elements, of a scientific complexion, which were embodied in tradition were not really an essential part of the system; and this even in cases in which the change to the newer scientific view involved some modification in the more strictly religious tradition.

I will here refer to three striking instances in which a violent shock has been given, consequent upon the rise of new scientific theories, to all those who had a traditional attachment to conceptions of the physical world which were in opposition to consequences of the new theories. The first, and perhaps the most striking, case to which I will refer was in the domain of Astronomy, and involved the substitution of what is known as the Copernican system for the older Ptolemaic scheme. The method of representing the motions of the bodies of the solar system which Ptolemy worked out in detail, and parts of which were due to Hipparchus and Apollonius, depended upon the idea that the earth is at rest, and the sun and planets in motion. Thus the Ptolemaic system of Astronomy was essentially geocentric. The earth being taken to be at rest, each of the planets was supposed to describe an epicycle, by a uniform revolution in a circle, the centre of which moved uniformly in a circle round the earth, which did not however occupy the centre of that circle. By proper adjustment of the radii of the circles, and of the velocities of the planet and of the centre of the epicycle, Ptolemy was able to give a fairly accurate representation of the apparent motion of the planet. In this manner he gave a systematic representation of the apparent motions of the sun and the planets; in particular the stationary points and the retrograde motions were represented. He was also able to represent the principal inequalities in the moon's motion. That the earth is a planet revolving round the sun as centre had been taught by Pythagoras to his disciples, but the definite overthrow of the Ptolemaic system was initiated by Copernicus, who represented the motion of the earth and planets as consisting of the uniform description of circular orbits with the sun at rest within each orbit, but not at its centre. It was only later, owing to the labours of Kepler, that the elliptic form of the orbits, with the sun in a focus, was established. So far, the whole change from the Ptolemaic to the Copernican system consisted, as we should now describe it, in the recognition of the fact that the relative motions of the sun and planets are much simplified by describing them as they would appear to an observer in the sun, rather than to a terrestrial observer; that is, by the adoption of a heliocentric description. It will be observed that, so far, the only change consisted in setting up a simpler description of one and the same set of facts. It was owing to the great work of Newton that the full importance of the change of descriptive scheme was made manifest; when the heliocentric description of the motions of the bodies of the solar system was seen to be consistent with a dynamical scheme, of great formal simplicity; and one to which the geocentric description did not lend itself. Only after a prolonged period of bitter strife did the substitution of the Copernican system, including the motion of the earth round the sun, and its diurnal rotation round its axis, for the Ptolemaic, meet with general acceptance. The invention of the telescope contributed much to the consolidation of the new view. The violent opposition to this change arose from the fact that the substitution of the sun for the earth as the body of reference in the solar system appeared to involve a fundamental change in the importance of the earth, its inhabitants and their affairs, relatively to the rest of the Universe. Our world was degraded physically to the position of a single satellite of a single star amongst the vast numbers of bodies in the stellar Universe; and this appeared to involve a corresponding moral and spiritual degradation which it was thought would seriously react upon the current theological conceptions, making it more difficult to maintain the importance assigned by those conceptions to man in his spiritual relations. The notion of the plurality of worlds, which arose in this connection, was one of the chief stumbling blocks; the teaching of Giordano Bruno, on this matter, formed one of the chief charges on which he was condemned to death.

The second instance to which I will refer, of a shock to traditional conceptions, was occasioned by the geological discoveries which assigned enormously greater antiquity to the earth than was admitted by the traditional biblical chronology, in accordance with which the age of the earth was estimated at not more than about six thousand years. After a shorter, and less violent, conflict than in the case of the Copernican system, the strength of the cumulative evidence adduced in favour of the great antiquity of the earth led to its general acceptance by educated persons.

Lastly I will refer to the controversies excited by the publication, in 1859, of Darwin's *Origin of Species*, and in 1871, of his *Descent of Man*. Although the idea of evolution of species, and even the suggestion that man was descended from a race of lower animals, were by no means novel, the special form which Darwin gave to these ideas, the weight of the evidence he produced, and the cogency of his reasoning, were such as to call the attention of the world to these views in so striking a manner as to produce a violent, though short, storm of opposition from those who feared the effect upon the traditional ideas of the spiritual nature of man, and his dignified position in the scale of living beings. The passionate repulsion which Darwin's theory called forth in some quarters is illustrated by the celebrated debate on the subject between Huxley and Bishop Wilberforce at the Oxford meeting of the British Association in 1860.

Belief in the descent of man from races of lower animals was in direct opposition to the current views about the Fall of man from a state of perfection, or at least of innocence; in fact it appeared to amount to a reversal of this idea, by substituting the conception of a slow ascent of man from indefinitely lower conditions of body and spirit. Here again the outcome has been much as in the first two cases I have referred to. The doctrine of evolution has been generally accepted by educated people, at least as regards the physical side of man, though not in all cases in accordance with the detailed views of Darwin himself. By the world at large it has been silently acquiesced in; at all events the world has survived the shock, and the time of violent and public opposition to the main conception of the evolution of man and other species belongs to the past.

The conflict of conceptions of which I have spoken has lasted, with varying degrees of intensity, into our own time. One of the maxima of this conflict occurred about the middle of the nineteenth century, at a time when the striking discoveries in various branches of Science, especially in Astronomy, Chemistry, and Biology, together with the far-reaching character of such generalizations as the law of the Conservation of Energy and Biological evolution, had filled men of Science with a feeling of confidence which sometimes took an aggressive form. At the present time, owing both to a more critical examination of scientific theories, and to the flood of newly ascertained facts which have led to modifications of theories which were formerly supposed to have attained final forms, and to the recognition of distinct limitations in their scope, the attitude of men of Science towards their theories has become much more cautious than it was in the nineteenth century. This change of attitude of men of Science, together with greater openness of mind on the part of the representatives of religious thought, has led to a marked diminution of the acrimony which has at times characterized the relations between those whose interests are in the main scientific and those for whom religious considerations are paramount. The tension of which I have spoken is essentially due to deep lying divergences of mental attitude towards the world, dependent, not only on differences due to varying types of education, tradition, and occupation, but also to fundamental temperamental divergences. This tension is often to be found within the complex mental make-up of one and the same individual, impelling him either to adopt some form of compromise between the contending conceptions and impulses in his mind, or else to set up a state of equilibrium in which the diverging sets of ideas are retained, so to speak, in separate compartments of his mind, and lie side by side without coming into direct conflict with one another. But this kind of mental dissociation is not possible for all persons; and thus in many minds a state of mental unrest, of uncertainty, of alternations of scepticism and belief, has been produced which is not conducive to that contentment of mind which is requisite for the highest efficiency of active agents in the work of the world. Even in our own time there are a considerable number of persons, some of them highly cultivated and intelligent, whose feeling towards the Natural Sciences is one of suspicion or repulsion, sometimes conscious, but perhaps more often unconscious or instinctive. This feeling of hostility or repulsion is partly temperamental, due to a distaste for the schematizing habits of mind of scientific thinkers, which are alien to those more intuitional modes of apprehension that are congenial to many minds. But this feeling is also largely, and perhaps in a preponderating degree, due to a fear that the Scientific view of the world leaves no room for the domain of freedom, spontaneity, and values, for teleological conceptions, or generally for the spiritual order of things.

The undeniable success of Natural Science, of the palpable order, exhibited in the mechanical inventions which have transformed modern life, while investing Natural Science with a certain glamour, has sometimes intensified this fear. That this fear is not *prima facie* groundless, or due to the merely conservative habits of timid minds, appears with clearness when the bold attempts are taken into account which were made by various representative men of Science, during the last century, to erect an all-embracing World-Philosophy on the basis of a Mechanical theory of Nature. It is instructive to consider, in this connection, some specimens of the pronouncements of prominent men of

Science. Perhaps the most famous utterance of this kind is that of Laplace in his essay on probability (1812), which exhibits in the most striking way the feeling of confidence produced by the triumphant success of Astronomers in applying the law of gravitation to the calculation of the motions of the bodies of the solar system. Laplace writes:

We ought then to regard the present state of the universe as the effect of its antecedent state and as the cause of the state that is to follow. An intelligence, who for a given instant should be acquainted with all the forces by which nature is animated and with the several positions of the beings composing it, if further his intellect were vast enough to submit those data to analysis, would include in one and the same formula the movements of the largest bodies in the universe and those of the lightest atom. Nothing would be uncertain for him; the future as well as the past would be present to his eyes. The human mind, in the perfection it has been able to give to astronomy, affords a feeble outline of such an intelligence. Its discoveries in mechanics and in geometry, joined to that of universal gravitation, have brought it within reach of comprehending in the same analytical expressions the past and future states of the system of the world. All its efforts in the search for truth tend to approximate it without limit to the intelligence we have just imagined.

As a remarkable example of confident scientific dogmatism the following quotation from the preface to Büchner's well-known work, *Kraft und Stoff*, published in 1855, is of much interest:

If these pages may venture to claim any merit or characteristic, it is that of representing a determination not to shrink with dismal horror from the simple if unavoidable consequences of an unprejudiced contemplation of nature from the standpoint of empirical philosophy, but to admit the truth regardless of what may follow. We cannot make things different from what they *are*, and nothing seems to us more preposterous than the attempts of some distinguished naturalists at introducing *orthodoxy* into natural science. We do not pretend to bring forward anything absolutely new or anything that had never been heard of before. Similar views and views cognate to ours have been taught in *all* ages, and some of them were laid down by the oldest Greek and Indian Philosophers, but their groundwork, which is necessarily empirical, could only be supplied by the progress of natural science in the present century. It is therefore obvious that these views, in their present clearness and consistency, are essentially a trophy of modern times, and closely related with the new and gigantic achievements of empirical science. Indeed, scholastic philosophy, ever riding the high, though from day to day more and more emaciated horse, lays the flattering unction to its soul that these views have long been disposed of, and would fain consign them to the limbo of oblivion, with which object it has labelled them "Materialism," "Sensationalism," "Determinism," and so on; nay, the gentlemen of that school go so far in their assumed supercilious superiority, as to talk of having given them "the historical quietus." But they themselves are going down day by day in the public estimation, and losing ground in their speculative hollowness before the rapid rise of the empirical sciences, which are making it daily more evident that both the macrocosmic and microcosmic worlds obey, at every stage of their genesis, existence and subsidence, the *mechanical* laws which lie in the very nature of things. Starting from the recognition of the indissoluble relation that exists between force and matter as an indestructible basis, the view of nature resting upon empirical philosophy must result in relegating every form of supernaturalism or idealism from what may be called the hermeneutics of natural facts, and in looking upon these facts as wholly independent of the influence of any external power dissociated from matter. There seems to us to be no doubt about the ultimate victory of this realistic philosophy over its antagonists. The strength of its proofs lies in *facts*, and not in unintelligible and meaningless phrases. But in the long run there is no contending against facts, it is useless to kick against the pricks.

Again, in his chapter on Thought, Büchner writes:

That thinking is and must be a mode of motion is not merely a postulate of logic, but a proposition which has of late been demonstrated experimentally.

Consciousness, like thought, is a performance or action or phenomenal activity of certain parts or tissues of the brain, and in that capacity it is subject to all the changes which take place in the condition, nutrition and growth of the brain. . . . How and in what way the atoms, the nerve cells, or, to speak generally, matter began to produce and bring forth sensation and consciousness, is quite unimportant for the purpose of our investigation, it is sufficient to know that such is the case. . . . The simple solution of the problem lies in the fact that not only *physical* but also *psychical* energies inhere in matter, and that the latter always becomes manifest wherever the necessary conditions are found, or that, wherever matter is arranged in a certain manner and moved in a certain way in the brain or the nervous system, the phenomena of sensation and thought are produced in similar fashion, as those of attraction and repulsion are under other conditions. . . . The development of mind from matter is indeed one of the latest, most difficult and most complicated triumphs of physical forces, and is the product of a protracted toil, rising from step to step, through countless centuries, till reaching the height of humanity. Nor can we say what shall be brought forth of similar fruit by the coming ages: we must confess that perhaps as yet we see only the incomplete, the imperfect, and that perchance we have no conception of what matter may yet be able to accomplish in its further evolution in mental phenomena and faculties, by further complications and yet more highly developed forms of motion.

The same idea of the primacy of matter over life and mind has been expressed by Tyndall:

Divorced from matter, where is life to be found? Whatever our *faith* may say, our knowledge shows them to be indissolubly joined. Every meal we eat, and every cup we drink, illustrates the mysterious control of Mind by Matter.

Huxley, in spite of the fact that he professed to hold idealistic opinions, expressed views somewhat similar to those of Büchner and Tyndall, as to the relation of matter and consciousness. He writes[1]:

There is every reason to believe that consciousness is a function of nervous matter, when that nervous matter has obtained a certain degree of organization, just as we know the other actions to which the nervous system ministers, such as reflex action and the like, to be. As I have ventured to state my views of the matter elsewhere "our thoughts are the expression of molecular changes in that matter of life which is the source of our other vital phenomena." . . . I really know nothing whatever, and never hope to know anything, of the steps by which the

[1] *Critiques and Addresses*, 1873, p. 283.

passage from molecular movement to states of consciousness is effected. . . . All that I have to say is that, in my belief, consciousness and molecular action are capable of being expressed by one another, just as heat and mechanical action are capable of being expressed in terms of one another.

Whatever we may think of the crude dogmatism of some of the pronouncements I have quoted, we must not be blind to the enormous effect which such ideas have produced, on the one hand, in inducing large numbers of persons, especially among the half-educated, to believe that these and similar views, expressed by prominent representatives of Science, embodied completely demonstrated results of Science, and on the other hand, in producing a repulsion to Science, often of an undiscriminating character, among many who feared its disintegrating influence upon cherished conceptions and beliefs. It is for this reason that I have dwelt at some length upon the extreme claims of complete supremacy made by scientific thinkers in the last century. It will be observed that these claims were made on behalf of the special mechanical theory of matter in vogue in the nineteenth century, so extended as to have become a complete theory of the physical and psychical domains. The effect of the later advances in Physics in the last thirty years, especially in the domain of Electromagnetism, has however been such as to produce a revolutionary change in many of the older conceptions of the properties of matter. The flood of new facts which have been discovered in connection with radiation and radioactivity has thrown the most serious doubts upon the range of applicability of the mechanical theories of the eighteenth and nineteenth centuries, at least in the forms in which they were then held. The bold extension of those theories into a World-Philosophy has, in consequence of our extended knowledge, lost much of what plausibility it ever possessed. Not only the electromagnetic theory of the constitution of atoms, and the quantum theory of energy, but also the latest theory, involving fundamental changes in our conceptions of the measurement of time and space, and of gravitation, have shaken to their foundations the notions upon which the older mechanistic theories were based. It must not, however, be assumed that, however great are the modifications which physical theories of material phenomena may undergo in consequence of increased knowledge of facts, the possibility has been for ever removed that some physical theory may again arise which may make claims of a far-reaching character similar to those which were made on behalf of the mechanical theory of matter in the nineteenth century. From the point of view of general thought, the really fundamental question in this connection is whether, or how far, it is possible to represent the physical world as a closed and independent system of deterministic type, uninfluenced by the psychical world. Whether such a closed scheme be what has been called mechanistic or not, is, from the point of view of those who are concerned primarily with the relations between scientific thought and the larger world of thought, a matter of secondary importance.

As a result of modern criticism of the foundations of Science, a view of the nature of scientific laws and theories has arisen, principally from the ranks of men of Science, which differs in some important respects from the opinions formerly held on such matters. This view, which I propose to explain in the next three lectures, has met with acceptance by many men of Science of eminence, and by some Philosophers. But it would be going too far to say that the scientific world in general has attained that degree of emancipation from certain ideas which have held sway for ages which would be implied in the complete acceptance of the view of the domain of Natural Science to which I refer.

# II

# SCIENTIFIC LAWS AND THEORIES

A GENERAL survey of the world, from the ordinary point of view which we call that of common sense, makes apparent to us the existence of two great complexes or domains; that which we call the physical or material, and that which we call the psychical or mental. This is the ordinary dualism of matter and mind; a dualism which we all accept for the common purposes of everyday life, when we are not philosophizing, and which was erected into a philosophical system by Descartes and his followers. We regard ourselves as sentient beings, endowed with consciousness, with the powers of feeling, thinking, willing, and remembering, as psychically active. We appear to receive from without, during our waking hours, a continuous stream of sense-impressions leading to what we call the perception of external objects and events, or grouped sense-impressions. These perceptions succeed one another in time, and change with greater or less rapidity. Although our percepts are in a continual state of flux, the changes in them are not entirely haphazard, but exhibit a considerable degree of regularity of sequence, of such a kind that we possess a certain amount of power of prediction of their future course. This power of prediction is an essential condition of the possibility of any actions on our part, because, in its absence, we should have no knowledge of what the effects of any of our actions might be. We find moreover that we possess certain powers of classifying and sorting out our percepts, of forming by abstraction from them permanent concepts, under which whole classes of our percepts are subsumed. Among the perceptual objects of which we are aware are the bodies of other individuals, and we believe, on indirect evidence (leaving aside such debatable matters as the existence of telepathy), that these individuals, like ourselves, have perceptions, and powers, similar to our own, of classifying these perceptions, of abstraction, of forming concepts, and of remembering. We also believe, as a matter of inference from experience, that the actions of other individuals are influenced by their perceptions in a manner similar to that which we find to be the case in ourselves. Further, we believe, also on inferential grounds, that the percepts of other individuals have a large measure of resemblance to our own; and that other persons discern regularities in the sequences of their perceptions which are closely related to those which we ourselves discern. We all alike have the apparent power of producing, by the exercise of our wills, changes in the perceptual domain or physical world. Such changes in the external environment, involving the use of our voices, or the motion of parts of our bodies, apparently produced by our wills, become percepts for other individuals, and thus form the medium of communication between one individual and another. The means of communication of ideas between individuals, thus conditioned, lead to the formation of a body of common knowledge of the perceptual world, which is regarded as accessible to all normal individuals. If we regard this picture as descriptive of what has been going on during an indefinite past, and if we take into consideration the cumulative effect of the process, combined with the memories of individuals and the traditions of the race, we have, from a non-philosophical point of view, an account, of some plausibility, of the way in which common or public knowledge of the physical world may be regarded as having grown up. It must, however, be observed that the experience of an individual is, in its fulness of detail, unique and incommunicable; it is only incompletely made intelligible to other individuals by means of language or other forms of symbolism. Thus language, in its very nature, involves abstraction in which the elements of actual individual experience are replaced by symbols which fail to represent with absolute completeness what they are designed to describe.

It is a part of that view of the nature of Natural Science, or the Science of physical percepts, which I propose to develop, that Science is essentially a purposive continuation of the formation of what I have called common knowledge, but carried out in a more systematic manner, and to a much higher degree of exactitude and refinement. The earliest stages in the formation of scientific knowledge are of the kind which may be described as classificatory. Physical objects are arranged in classes, in accordance with observed similarities in the objects assigned to any one class; physical events or sequences of events are classified in accordance with observed regularities or similarities in those events or sequences. There is, however, always a certain degree of arbitrariness involved in the selection of the precise similarities or regularities which form the basis of the classification; thus at the very commencement of the process of building up scientific knowledge, the analyzing and generalizing powers of the mind find scope for their activity, and are necessary factors. The result of this process expresses itself in the formation of abstractions or concepts, which are not identical with any of the perceptual objects or happenings which conditioned their formation, but which serve as a conceptual symbolization or representation of those aspects of the latter which were regarded as alone relevant for the purpose of classification. A rule, or law, which affords a conceptual description of a particular kind of sequence of physical events in the perceptual world is set up in the first instance in the manner I have indicated, on the basis of actual observation or experiment, dealing with a selected class of sequences of physical events. It is then extended hypothetically to the descriptive representation of sequences which may occur under conditions not in all respects identical with those in which those observations were first made which suggested the law. When this hypothetical extension is justified by further experiments or observations, and especially by its power of prediction of what will happen, or be observed under certain conditions, the rule or law becomes what is usually called either a scientific law or principle, or a law of nature. The term law of nature has too frequently been taken to imply that it meant a purely objective law, as it were inherent in the perceptual world, which natural phenomena must of necessity obey, whereas a law of nature is in reality a conceptual law set up by the activity of the mind of man, but conditioned as regards its validity by the perceptual world which must be taken as a datum. The law is constructed by the mind, but not purely arbitrarily; it has limitations other than, and besides those imposed by, the canons of thought. In the physical complex there is the essential element of fact, which must be taken as a given, and unalterable, datum required for the construction of the law. Thus the complex of phenomena or appearances which we call the physical world must be taken to be in fact such that the law must stand the test of applicability for the purpose of resuming certain tracts of uniformity in it. Natural Science need not however go beyond this recognition of the existence of this element of fact; it is unnecessary for its purposes to make the assumption that a single law has a precise correspondence with a single definite set of relations which actually subsist in Nature. Still less is it necessary for the purposes of Natural Science to assume that the law corresponds to a set of relations between real entities. The formation of precise views on these matters is the task of Philosophers; the man of Science need not go beyond the superficial view which contents itself with the recognition of the essential element of fact in the world of percepts.

The two opposite tendencies, either to regard natural laws as entirely constructible by the mind, in accordance with what were regarded as *a priori* necessities of thought, or on the other hand to regard them as something entirely external to, and only discerned by, the mind, have both been prominent in the history of Science. It may perhaps be the case that some of the writers who uphold the "descriptive" view of the character of Natural Science have so expressed themselves that the importance of the element of fact or "givenness" in the realm of percepts appears to be unduly minimized. Both the activity of the mind, and the data of perception, are factors in the genesis of a so-called law of nature. Natural laws have been characterized by Mach[1] as "abridged descriptions" and as "comprehensive and condensed reports about facts." It may be doubted whether such a characterization is quite adequate as applied to conceptual schemes in general. The laws always contain less than the facts, because the concepts which the laws employ in their statements are but symbolical representations which leave out of account those differences of individual detail which are to be found in perceptual objects or events which fall under a particular class. In the words[2] of Boltzmann:

> It has never been doubted that our ideas are merely images of the objects (or rather symbols for them) which have a certain relationship with the objects, and never completely correspond to them, but are related to them as letters to sounds or notes to musical tones. Also on account of the limitation of our intellect they are able only to depict a small part of the objects.

There is no absolute line of demarcation between scientific knowledge and common knowledge. The

[1] *Popular Scientific Lectures*, Trans. McCormack, p. 193.
[2] *Vorlesungen ueber die Principe der Mechanik*, Vol. 1, pp. 1–2.

knowledge which we have, that solid bodies, when unsupported, fall to the ground, would hardly be described as scientific knowledge, but it is not in essence distinct from other rules of the order of phenomena which we do dignify by the name scientific knowledge, on account of the fact that they rest upon experience which is less readily analyzed, or separated off from the general complex of physical experience, or which requires more refined methods of observation. But the law of falling bodies, in accordance with the more exact observations made by Galileo involving numerical determination, became a genuine scientific law. By the process of hypothetical generalization it was extended by Newton into the general law of gravitation, the power of which to represent to a great degree of accuracy the observed motions of the bodies of the solar system was brilliantly established by Newton himself and a number of eminent researchers in the field of Gravitational Astronomy.

A law, or rule, which refers to some particular kind of sequence of events, under certain conditions, and which is regarded as a scientific law, or principle, is very often constructed by means of artificial production of the set of conditions under which the sequence of events in question is observed to take place; the law is then said to have been discovered by experiment. In other cases, in which the conditions are not produced artificially, but occur without the intervention of the observer, the law is said to be discovered by observation. But in all cases, the discovery, or rather the construction, of a scientific law involves that synthetic activity of thought which manifests itself in a constructive process in which actual percepts are employed only as the raw material and starting point of the mental process. In the attempt to discover a scientific law, a selective process is requisite in regard to the percepts, some greater or lesser part of what is perceived must be ignored, as irrelevant to the purpose on hand; this selective procedure amounts to a process of abstraction, in which some elements of our actual percepts are removed, and not attended to. The degree of abstraction employed, and thus the degree in which the concepts differ from percepts, varies greatly in different departments of Science, and also varies greatly according to the stage of development which a particular department has reached. A scientific law is accordingly always, in some greater or lesser degree, abstract, in the sense that it represents only a part of what is in any individual case actually perceived; it describes a particular sequence of physical events which, in an actual case, is accompanied by other percepts or events in relation to which the law has no application. For example, when Kepler discovered the law that the earth and planets describe elliptical orbits with the sun in one focus, he abstracted from all the physical properties of the sun and planets, even from their sizes; all the infinite details of their physical constitution being irrelevant to his particular purpose of describing the main features of their relative motions.

A complex of physical facts, given as percepts, appears in the first instance to be confused and irregular, on account of its complexity; but the effect of further scrutiny is gradually to reduce it, in some degree at least, to order, when similarities and relatively permanent elements are discovered within the complex; and the ultimate result of this process is that we are enabled mentally to reconstruct groups of facts within the complex. It must, however, be observed that in Nature there are no precise repetitions in every detail, for conditions are never on two occasions absolutely identical. Absolute likeness of sequence exists only in our conceptual schemes of representation, not in the world of percepts. With the purely individual, Science cannot deal; it operates with the typical, and a type is an abstraction. Even in the preliminary work of classification of those groups of sense-impressions which we call physical objects, such as is undertaken in the classificatory stage of Zoology, or in systematic Botany, abstraction is made of individual differences which distinguish one member of a group from another. Thus two plants of the same order, genus, species, and variety, are never exactly alike in all respects; they are judged to be of the same variety because they are alike in certain particulars defined in accordance with the arbitrary rules which are employed for the purpose of classification. A plant of a given variety, or of a given species, is an abstract conception; the successive grades of classification represent different degrees of abstraction in the formation of the concepts that correspond to them. It appears then that, even in the earlier stages of scientific thinking, what we really work with are not the percepts themselves, but concepts which symbolize types of percepts, and in the formation of which concepts some of the elements of actual percepts are left out of account.

It is, however, only in the earlier stages of development of a branch of Science that the procedure consists mainly of the classification of objects by means of the substitution of concepts for the objects themselves, or of the construction of the kind of rules which form the simpler laws of sequence of events. As the Science advances, such laws are generalized into those more comprehensive laws which are often called scientific principles. A law, or principle, which has reached a high stage of generality, or a group of such laws considered as forming a single body of doctrine, is what is known as a scientific theory, and this forms a conceptual scheme under which a wide class of physical sequences is subsumed. But whether such a conceptual scheme is called a theory, a principle, or a law—and that is to some extent a matter of usage, or historical accident—its genesis always involves a more or less extensive constructive mental process, or synthesis. Such a conceptual scheme involves the employment of conceptual objects, the character of which, and the relations between which, are assigned as postulations of the scheme. These postulations must always be regarded as hypothetical or tentative. The possibility of setting up such a scheme has actual physical experience as its essential condition, but the constructive and generalizing work of thought is no less essential. The original function of such a scientific theory, or conceptual scheme, is to provide an ideal representation of some more or less restricted range of physical phenomena as actually observed, that is of certain sequences and regularities in percepts. But the functions of a conceptual scheme are much wider than those of merely describing symbolically what has actually been observed. The scheme is applied hypothetically to predict what will be observed in circumstances which differ in some degree, or in some characteristics, from those in which the experiments or observations which led up to the theory were made. The value and the range of validity of the particular conceptual scheme have to be estimated by its actual success in the fulfilment of this function of prediction. Thus a scientific theory, considered as an hypothesis, not as a dogma, must be judged by a comparison of its consequences with perceptual fact; and especially by its power of forecasting the occurrence of hitherto unknown or unobserved facts, when, by more minute observation, or by artificial production of experimental tests, the occasion for such comparison arises.

The different conceptual theories which are employed in the various branches of Natural Science vary greatly in their degrees of abstraction, and in their degrees of precision. In the most abstract and precise theories, the language employed is the most precise form of language which we possess, that of arithmetic and its generalization in Mathematical Analysis. As any branch of Science progresses to a higher stage of development, there are to be found at least portions of it in which it becomes possible to employ theories of this character. There exists a distinction between two species of ideal elements of a conceptual scheme which it is important to recognize. Some concepts have direct perceptual counterparts, such concepts having been formed by a direct process of abstraction, in which, what are, for the purpose of formation of the scheme, irrelevant characteristics of the perceptual objects or processes have been removed by the abstraction. Concepts of the other species have no such direct perceptual counterparts, or it is not assumed *a priori* that they have such; they are formed by an effort of constructive imagination, for the purposes of the representative scheme. Although concepts of this latter kind may be regarded as due to the creative activity of thought, it must always be remembered that they would never be formed apart from a basis of actual physical experience. In any scientific theories in which both kinds of concepts occur, those which have no immediate perceptual counterparts must be regarded, at least provisionally, as purely ideal elements of the scheme, in fact as auxiliaries necessary for the purpose of the formation of a self-contained conceptual scheme which shall serve its purpose of providing a sufficient mode of representation in thought of the particular domain of physical events and objects. In less advanced conceptual schemes, as for example in those of the purely classificatory order, all the concepts employed may be simply abstract forms, or types, of perceptual elements. Such concepts are usually however more precise in character than those employed in ordinary intercourse, the definition of their characters and relations having sharper outlines. The power which we possess of introducing into a scientific theory, as part of the edifice, ideal elements which do not directly correspond to clearly defined percepts, although essential to the more advanced work of Science, is one which has its obvious dangers, and these dangers have by no means invariably been avoided in practice. In a workable and satisfactory scientific theory such conceptual elements should be as few in number as possible, and of a simple and well-defined character; otherwise the theory has the grave defect of over-complication, or even of vagueness, and the fundamentally important requisites of a scientific theory which shall serve its purposes of representation and prediction, those of clarity and simplicity, are not satisfied. Such ideal concepts only serve their purpose provided they are subject to precise definition. In some cases the introduction of such elements into a theory has amounted to little more than the employment of new words, which, on account of the indistinctness of their denotation, have only served the purpose of labelling our ignorance, and have consisted of a futile attempt to disguise inability to set up a really adequate representative scheme. If, when each new difficulty arises, new concepts are invented for the purpose of shelving those difficulties, the theory becomes so overloaded that it ultimately perishes by its own weight.

It has perhaps been in the biological sciences, which have to deal with ranges of phenomena of the highest degree of complexity, that this kind of danger has been oftenest incurred; we may refer to the various vitalistic hypotheses to illustrate the point. Some such insufficiently defined concepts have often been introduced for the purpose of remedying, by means of an enlarged scheme, the recognized inadequacy of some older and more restricted descriptive scheme. In Physics, for example, the temptation to introduce a new ether, or a new substance with insufficiently defined properties, in order to remove each particular inadequacy of an older mechanical scheme, is one to which men of Science have frequently yielded. However, in scientific theories of the highest type, there may exist such ideal elements without direct perceptual counterparts, the danger of the kind I have indicated being successfully avoided. It is precisely in the proper selection and definition of such conceptual elements of a theory that the highest powers of the great men of Science, who are necessarily supreme Artists, have been exhibited. Descriptive schemes which employ concepts, the smallest in number, and the most definite in character, are the distinguishing mark of Science at its best.

In some cases it is preferable, for expository purposes, not to reduce the concepts of a scheme to an absolute minimum, but to retain some which are reducible to others, as auxiliary concepts. This does not, however, afford a dispensation from the duty of investigating what is the smallest number of irreducible concepts. A theory is sometimes capable of being stated in more forms than one, according to the choice made of the irreducible concepts, and thus some latitude arises as to the precise form of the theory. I may refer to the case of molar Dynamics as an illustration of this possibility. Those departments of Natural Science in which the theories, or conceptual descriptive schemes, are furthest removed from the region of the perceptual, and which are thus characterized by a higher degree of abstraction than has as yet been attained by other branches, are Mathematics, and those parts of Physical Science which have become in a considerable degree amenable to mathematical treatment. It is the relative simplicity of those aspects of the perceptual world with which Mathematics and large parts of those Sciences which concern themselves with non-living matter deal, as compared with the Biological Sciences, which accounts for their relatively advanced character. Number and Extension, with which Arithmetic, in the extended sense, and Geometry, are alone directly concerned, can be developed as sciences descriptive of only the most superficial aspects of the perceptual world, since they may leave aside almost all the properties and qualities of perceptual things as irrelevant for the purpose of their construction, so that in these departments the process of abstraction and idealization can take gigantic strides at a very early period in their genesis. Other branches of Science, in which the disentanglement from one another of great complexes of properties, qualities, and sequences is a much more difficult process, are on that account such that the stages of their growth and development are much slower, and much more difficult than in the case of Geometry and Mathematical Analysis.

Notwithstanding the very great difference, as regards the degrees of abstraction, in the conceptual schemes which belong to different branches of Science, the difference between them need not, I venture to think, be regarded as generic. This difference seems rather to be one of degree of advance in the scope and character of the abstract schemes of representation which have at the present time been devised for describing the various regularities of sequence with which the various branches have respectively to deal. On the assumption of the correctness of this view, the Science of Geometry, which on the perceptual side starts solely with the extensional relations of bodies, forms a kind of model, to which other branches of Science, in their further advance, will gradually conform. This conformity with the model may, in the cases of most branches of Science, not become complete in any time for which we may venture to make a forecast, but their advances will consist of progress in the direction of such conformity. Geometry, the Science of spatial relations, is essentially in its origin a physical science, but it has long ago become a deductive Science in the sense that a complete conceptual scheme has been constructed, of such a character that all ordinary special spatial properties are logically deducible, in their ideal form, from the postulations of the scheme itself. These deductions are then applicable for the determination, subject to inevitable errors of measurement, of the corresponding relations in perceptual space. In our actual Geometry, purely experimental determinations of special spatial properties are no longer necessary, but only serve for purposes of illustration. It is not inconceivable that the labours of Physicists, in a domain on which a flood of light has fallen in the last decades, may result in the production of a schematic representation of the composite nature of the atom, so precise that the various possibilities as regards the detailed structure of the atom may be worked out. If this were accomplished it might be possible to deduce from such a conceptual scheme the main characteristics of the different chemical elements, and the nature and character of their possible combinations. The Science of Chemistry would then have become deductive in the same sense as that in which Geometry is a deductive Science. However far from the attainment of this goal most branches of Science may be, the general character of progress in them consists of the discovery of ever larger tracts of phenomena that are capable of conceptual description by means of schemes which succeed in predicting further phenomena which had not been observed until such predictions were tentatively made.

This view of the nature of a scientific theory, the most general outline of which I have traced, that it essentially consists of a conceptual scheme, designed by the synthetic activity of the mind, working with the data of perception, for the purpose of representing particular classes of sequences and regularities in our percepts, has been powerfully advocated in recent times by Kirchhoff and Mach on the continent, and by K. Pearson in this country; some indications of a similar view are to be found in the writings of Auguste Comte and other earlier writers. I propose to follow it out in further detail, and to illustrate its application in various special departments of Science which I shall consider in later lectures. It will be found that the adoption of this point of view involves changes in the older traditional conceptions of the relations between scientific thought and general Philosophy. These changes will be seen to have the effect of according to Natural Science a more independent position in relation to Philosophical theories than it has often been supposed to occupy. But they will also have the effect of placing limits to the functions of Natural Science as one of the factors which contribute to the construction of our general view of the world; and also of assimilating in a large degree, though not wholly, the attitude of Natural Science towards the great speculations of Philosophy with the attitude of what is usually called common sense. The ambiguity of meaning of various expressions employed in their writings by men of Science often makes it difficult to be sure whether, or how far, they are in agreement with the descriptive view of the nature of Natural Science of which I here give an account.

We have taken as our starting point the fact that we all have streams of percepts which we call our physical experience. The reduction to some kind of order and regularity of particular kinds of such percepts, by means of a representation of them by conceptual schemes, is what I have maintained to be the function of scientific theories and laws. The important question, however, arises that, as a percept always involves a percipient, who is the percipient whose percepts are conceptually described in this manner?

We cannot take Science to deal merely with the private percepts of some one particular person. The very essence of scientific knowledge is that it shall be what we may call public knowledge, that is, not the partially incommunicable knowledge of some one individual. It is true that, at a particular time, a particular scientific theory may only be known to a single person, its discoverer; but it is essentially capable of being communicated to, and understood by, other persons whose training and previous knowledge fit them to receive it. In the form in which a scientific theory, or law, is usually stated it seems to describe facts or occurrences which are independent of any and every observer, but the presence of some percipient is always implied. This implied percipient must be regarded as placed in circumstances which enable him to have the perceptions, and these circumstances include the use of necessary instruments for extending the scope of his senses, such as telescopes, microscopes, etc.; he is assumed also to be endowed with normal powers of perception. The answer to the question, who is the implied percipient in a descriptive scientific theory, is then that the theory consists of the conceptual description of what would be observed by any normally constituted observer supposed to be placed in a proper position, and under suitable circumstances, and provided with the necessary appliances. Such an hypothetical observer is always implied in the statement of a theory or law, although he is usually not explicitly referred to. I should, however, observe that in the latest physical theory, that of general relativity, certain special circumstances relating to the observer are taken into account as part of the theory itself.

No scientific theory is designed to describe at once the whole of the perceptions which an observer would have under given circumstances, but only particular classes of perceptions arbitrarily separated from the rest, in accordance with the particular kinds of percepts and their sequences which the theory is designed to describe, that is, according to the particular range of phenomena with which the particular theory deals. Thus, for example, a dynamical theory of the motions of molar bodies will be a conceptual scheme designed to describe only the motions of actual bodies as they would appear under assigned circumstances to an observer who perceived those motions, and the theory would have no concern with the perceptions which such observer would have at the same time of the colours and other physical characteristics of the moving bodies, or of any sounds which they might emit; these would form the subject matter of other theories. Thus the actual perceptions of an observer are described as it were piece-meal by different theories belonging to different branches of Science. A vast part of our percepts has hitherto not been found to be amenable to abstract description of a precise character, by any conceptual scheme.

Considerable branches of Science are concerned with attempts to trace back into the remote past complexes of phenomena of certain special kinds, that is to give an historical retrospect in a particular domain. Some branches of Science also undertake to perform a similar function as regards the future. Geology, for example, undertakes the task of tracing the history of the earth's crust; Cosmical Astronomy attempts to depict the history of the evolution of the solar system and of stellar systems. Biological Science has made great efforts to construct a general account of the evolution of living organisms and their various species. It is obvious that no single percipient such as we are acquainted with can be regarded as having followed out the changes even in a single geological period, or the evolution of a single race of animals, let alone the evolution of the solar system.

I do not, however, think that the general scientific schemes which purport to describe what has happened in very long, but of course strictly finite, periods of time, need be regarded as falling under a different category from those which describe the short time processes observable by a single percipient. We may regard them all as hypothetical attempts to describe what would actually have been perceived by an observer at any time falling within one of the long periods in contemplation, had such an observer been present at that time. These historical accounts have been constructed on the basis of a knowledge of present conditions and of actually observed processes which take place during short periods. Their construction involves the piecing together of a large number of short time processes, and of utilizing our conceptual knowledge in such a way as to construct a conceptual scheme, of a strictly general character, which serves to give a consistent account of what we may hypothetically regard as having led up to the actual present conditions. It is of course clear that these histories of processes through long periods involve a large element of hypothesis, and a degree of uncertainty, and especially of indefiniteness as regards lengths of time, much greater than those which attach to conceptual accounts of short time processes which can be observed by a single percipient.

In those advanced scientific theories which are of the quantitative kind, that is which describe conceptually the results of actual measurements, there is always an element of approximation in the application of the theory, because all actual measurement is approximate only. Absolutely exact measurement is conceptual only, and resides in the conceptual scheme, not in the actual measurements that can be made in the perceptual domain. Thus, even in the case of a quantitative theory, there is an essential element of approximativeness in the power of a scientific theory to represent actual perceptions; much more is this so in the numerous cases in which the theory is not quantitative but qualitative only. It is in general only in the more advanced branches of Science that quantitative Mathematical representation is available. But in many branches, such as Chemistry and Physiology, there are, in ever increasing numbers, particular ranges of phenomena which have been made amenable to such treatment, although in some cases such ranges of phenomena form but a very small part of the whole complex of phenomena with which each such particular department of Science deals.

An essential characteristic of every scientific theory is that it only serves to give a conceptual description of a range of phenomena which is of a limited and circumscribed character, spatially and temporarily. It only describes what an observer would perceive in some portion of space which, however great, is limited by his powers of perception, even when those powers are increased by the employment of appropriate instruments. Further, it only describes what will happen in some limited lapse of time, however great that period may be. It also only describes those percepts which are of some more or less limited type. Thus the percepts described by a theory are bounded, spatially, temporarily and generically. There always remain possible percepts beyond any particular range, to describe which the theory is applied. As regards time, the theory is wholly inadequate to describe any absolute beginning or end of sequences of phenomena. The period of time to which it may be applied is capable of being extended in both directions, at least hypothetically, either within the scope of the particular theory, or by means of some more comprehensive theory applicable to phenomena which take place in a greater period of time, during which we can assign, at least tentatively, the relevant circumstances. Just as absolute origins and terminations of sequences of phenomena are beyond the reach of scientific theory, so also is unbounded perceptual space. Even the perceptions of the Astronomer are bounded by the limitations of the telescope and of the photographic plate; the space of his perceptions is strictly finite, however enormous its dimensions may be, as estimated in our ordinary systems of measurement. Spatial perceptions have also lower limits imposed by the limitations of the microscope.

Although conceptually we can regard portions of actual space or of time as capable of indefinite increase, at any actual stage of the indefinite regress in which we are thus involved, the portion attained is strictly finite. Thus we are not entitled to regard Science as dealing with the whole of time, or the whole of space, or with the whole Universe. It is indeed not clear that we are entitled to assert that either of these is a *whole* in the sense that, to such an hypothetical infinite whole, the same categories are applicable as in the case of a finite whole. These considerations would appear to indicate the existence of limits to the application of the methods of Natural Science. The existence of such limits has, however, frequently been disregarded, at least in appearance, in statements made by men of Science of high distinction. It has often been said that the whole energy of the Universe, or the whole mass of the Universe, is constant, or that the whole entropy of the Universe tends to increase. To such statements a meaning can only be assigned by assuming them to be elliptical expressions to denote, for example, that in any finite region which received no accession of mass from without, the total mass remains constant.

We have seen that the construction of rules, relating to sequences and regularities in Nature, the so-called laws of Nature, and the more comprehensive schemes which we call scientific theories, are the work of the human mind, utilizing the raw data of perception. We might imagine that the perceptual world could have been so intricate and irregular in character that it would have been impossible to set up such rules or schemes, under which phenomena could be subsumed. Science, as we know it, would then have been impossible, it could never have begun to exist. But life, as we experience it, would also have been impossible, because that common knowledge of sequences in Nature, upon which the possibility rests of forecasting the results of our actions, would have been absent. The possibility of the existence of Natural Science accordingly depends upon the fact that there is in our percepts, that is in what we call Nature, a considerable degree of regularity in the sequences of phenomena. This fact is often described as the Uniformity of Nature. What the scope of this uniformity may be, and whether it is subject to extraneous interferences, are questions which cannot be answered *a priori*; only experience is relevant in any attempt to answer them. It is not necessary for the purposes of Natural Science to assume that the complex of physical phenomena is such that it is, even theoretically, capable of description by a single all-embracing conceptual scheme. It is possible that the present state of Science, in which different conceptual schemes are employed for resuming different tracts of the perceptual domain, is the only possible state. Although the actual progress of Science has involved the gradual welding together of separate conceptual schemes into larger wholes which embrace tracts of phenomena that were earlier treated separately, it is not a necessary assumption for Natural Science that this process of unification can, even theoretically, become complete. In other words it is not necessary to assume that Nature can even theoretically be subsumed under a single interconnected rational scheme.

Throughout the history of Science we find that the most potent influence on investigation has been exercised by the notion of efficient causation. The principle of causation, or sometimes the more general principle of sufficient reason, has been regarded as a necessary principle to which Nature must conform, and the aim of Science has been regarded as that of discovering the precise modes in which the principle is realized in natural phenomena. But in Natural Science, conceived in the manner I have explained, such *a priori* principles are replaced by the postulate, or working hypothesis, that sequences can be found in Nature which are capable of being described by laws or sets of laws embodied in determinate ideal schemes, and that it is possible to an extent of which we do not know any limits to discover what these schemes are. That these schemes can be applied with absolute precision to concrete cases in Nature is an illusion dependent upon the imperfections of our senses, even when their power is extended by using instruments, which often prevents us from perceiving all those individual peculiarities of a concrete phenomenon which differentiate it from other phenomena of the same class, described by the same conceptual scheme. To construct laws we frequently isolate artificially a part of what happens from the rest; there are always however in a given case accessory phenomena. In a scientific laboratory elaborate precautions are frequently made in order to isolate phenomena, and to reduce disturbing influences as much as possible, but there remains some residual which cannot always be completely ignored when the experiment is regarded in relation to some simple law of an exact kind. Thus a scientific law states what would happen under ideal conditions, more or less imperfectly realized in actual cases.

We have seen that all conceptual schemes are applicable only to the description of what goes on in a portion of the physical domain limited in various ways. This involves the isolation, for the purpose of conceptual description, of particular domains in the perceptual world. The success of a conceptual scheme in discharging its descriptive function in relation to a particular perceptual complex, depends upon the fact that, for the purpose on hand, that particular perceptual complex may be treated as an isolated system, in which all perceptual elements that do not belong to it are ignored. Thus the existence of *approximately isolated systems* in the world of physical phenomena is a fact upon which the possibility of the existence of Natural Science, as we know it, depends. For example, a descriptive scheme for the motions of the bodies in the solar system would be impossible except for the fact that the solar system is approximately isolated, in the sense that the disturbing effects of gravitation due to the stars can be left out of account.

A conceptual scheme which successfully serves its purpose in the particular department of Science to which it belongs must satisfy certain conditions. First, it must have logical consistency or coherence, as a scheme which is not self-contradictory; the various conceptual elements which it employs, and the set of definitions and postulated relations contained in it, must form a consistent whole, in which there is a complete logical nexus between the parts. Secondly, the scheme must satisfy the test of providing an actually adequate description of sequences of phenomena of the class to which it is intended to apply; it must have applicability, in relation to physical phenomena. It is quite possible to set up conceptual schemes which satisfy the first of these conditions, that of logical coherence, but which either do not satisfy, or are not known to satisfy, the condition of applicability. A conceptual scheme which differs from that of Newton's law of gravitation, in taking the law to be that of the inverse cube, or any other power, of the distance, instead of Newton's law of the inverse square, could be worked out, in detail, and would satisfy the condition of logical coherence as fully as does Newton's law; it would however not be applicable for even the approximate representation of the motions of the bodies of the solar system. It would be in conflict with fact in the perceptual domain. A particular system of abstract Geometry may be unimpeachable as a conceptual scheme, and yet it may be inapplicable to the description of actual spatial percepts.

A third condition, that of relative simplicity, should be satisfied by a scientific theory, if it is to be used with effectiveness in the building up of Science. It may happen that two or more conceptual theories satisfy the conditions of logical coherence and of applicability. In that case that theory will be chosen amongst all those which are applicable to the same range of phenomena which is of the simplest character.

The most crucial test of applicability of a scientific theory is that it should enable us actually to predict the results of experiments or observations of sequences of phenomena, under conditions consistent with the scope of the theory, which are made subsequently to the constitution of that theory. If it fails to do this successfully, either the theory must be abandoned, and a search must be made for a more adequate one by which it can be replaced, or else the determination of disturbing elements must be sought in the conditions under which the new experiments were made, of such a kind that it can be shown that the complex of conditions in these experiments was not in accord with those postulated by the theory. In other words it must be shown that the observed complex of percepts did not, in these experiments, possess that feature of approximate isolation which the theory presupposes.

It has frequently happened in the history of Science that a theory which had a very considerable measure of success as a representative scheme failed to satisfy one of the requisites I have described. It has then been abandoned in favour of some other theory which appeared more nearly to satisfy the requisite conditions. In the first place the theory may be abandoned on account of internal defects in the conceptual scheme itself, in the matters of logical validity and clarity of definition of the conceptual elements. In some cases a clear statement, in a logical form, of a conceptual theory has only been given, as the result of critical investigation, long after the theory was first constructed and applied for its proper descriptive purposes. Also a theory, apart from defects in its own nature, may be superseded by a new theory of a more comprehensive character, one which is capable of representing a wider range of phenomena than the older one. When such supersession takes place, the older theory does not lose its logical consistency and whatever degree of applicability it ever possessed.

The truth, or the falsehood, of a scientific theory is often spoken of in a manner which obscures the real nature of such theories. It is the fact of applicability, or the degree of applicability of one logically coherent theory as compared with another, which is their distinguishing feature. Assuming that both of them are logically coherent conceptual schemes, both of which are hypothetical in their nature, that one will be preferred, for many purposes, which has the greater range of applicability, and hence the greater power of prediction. If the Einstein theory of gravitation, or some modification of it, finally supersedes that of Newton, as may very likely happen, it will be because the former is capable of representing some phenomena which the Newtonian theory fails to represent, and also because, as part of a wider theory which represents phenomena of a different class from the gravitational, it is better fitted than the Newtonian theory of gravitation to form part of a comprehensive scheme relating to a wide class of physical phenomena. But for the purpose of describing nearly all the relative motions of the bodies of the solar system the Newtonian scheme is demonstrably sufficient, and it will always be as adequate for that purpose as it has proved itself to be in the past. Thus it is in possessing greater range and accuracy in its power of description, and not in what should properly be described as its truth, unless truth be understood to mean degree of applicability, that the Einstein theory will be regarded as superior to that of Newton. For all we know, the Einstein theory itself may one day be superseded by some still more general and comprehensive scheme that is applicable to the description of a still greater range of phenomena.

In general the method of setting up hypothetical theories to be tested by their power of representing and predicting phenomena may be described as methodological pragmatism. That does not of course imply the acceptance of Pragmatism as a general philosophical system.

In some cases there may be diversity in the manner in which a conceptual scheme can be applied to the purpose of describing the phenomena for the representation of which it was constructed. This may especially be the case in a theory, expressed in a mathematical form, in which some only of the conceptual elements have direct perceptual correspondents, the remaining elements being required to ensure the coherence and validity of the theory itself. A principle of correspondence is then required which shall assign the nature of the correspondence between perceptual elements and conceptual elements of the abstract scheme, and such principle of correspondence is not necessarily unique. An example of this is found in the theory of the Electromagnetic field as formulated by Maxwell. There has been general agreement that Maxwell's equations do represent what can be observed in the field, but different modes have been suggested of placing into correspondence what can be observed, with the vectors which appear in the equations.

In the most abstract branch of Science, Pure Mathematics, the powers which the mind possesses, of idealization and generalization, have been so extensively employed that conceptual schemes have been erected which are not known to be applicable to the description of any physical phenomena, although such schemes could never have come into being without an ultimate starting point in physical perception. This tendency on the part of Pure Mathematics to outrun the exigencies of applications has sometimes been made a subject of reproach to Mathematicians, especially on the part of

Physicists. Mathematicians have been accused of wasting their intellectual efforts in a region of bare formalism which has cut itself off from all connection with the actual world. It has been urged that Mathematicians have an excessive tendency to occupy themselves with ideas that are too remote from the physical order in which Mathematics had its origin, and in which it should still find its proper applications. Experience has however shown, in many instances, that it is exceedingly hazardous to prophesy that a Mathematical theory, however remote it may apparently be from the physical order of things, will not sooner or later find an application within that order. It is quite true that, although Pure Mathematics is in its origin a part of Natural Science, having arisen, like every other branch of Science, in efforts to deal with certain aspects of the perceptual world, it has become by rapid stages a deductive and formal Science, in a sense and to a degree which is not the case at present in any other branch of Natural Science. Nevertheless, Mathematical thought, as its history very clearly indicates, is one of the normal forms of mental activity. If it be allowed to develop itself, on its own lines, free from all fetters imposed upon it by the supposed necessity of keeping itself in close relationship with other departments which can make use of it as a tool, it will not prove itself to be in default when new demands are made upon it by other branches of Science, as they gradually reach those stages in which they require the help of refined Mathematical methods. It would be easy to establish in detail that many of the most important applications which have been made of Mathematics were wholly unforeseen by those who developed the Mathematical ideas and methods so employed. Had not Mathematicians, during more than a century, investigated those systems of Geometry that are known as non-Euclidean, and which appeared to many Philosophers and Physicists to be an extreme instance of the addiction of Mathematicians to lines of thought which could lead nowhere, the recent theory of Einstein could not have been formulated, because the way would not have been prepared for the conception of the four dimensional manifold with a non-Euclidean metric which appears in the scheme as space-time, and the theory of tensors which Einstein has employed in the elaboration of his scheme would not have existed. The new Physics and Chemistry may at no distant time require weapons, forged by Mathematicians; the theories of functions, of differential equations, of groups; thus providing applications for the most refined methods which Pure Mathematics can devise. It will be seen in later lectures that this time has in fact already arrived.

# III

## NATURAL SCIENCE IN RELATION TO PHILOSOPHY

THE current views on general Philosophy have usually exercised a considerable influence upon the course of development of scientific thought; and this influence has been reciprocal. The strength of this connection was in former times increased by the fact that one and the same man not infrequently combined the functions of Philosopher and man of Science. The cases of Aristotle, Descartes, Bacon, and Leibniz may be cited as instances of this fact. Even Kant had strong claims to be considered as a man of Science. Philosophers such as Locke, Hume, and Berkeley were largely occupied with the discussion of matters which have a close connection with the epistemology of the Natural Sciences. Not only the more systematized metaphysics of professed Philosophers, but also the more popular and less systematized ideas on Metaphysics, current at a particular time, had a marked influence upon the forms in which the scientific conceptions of the time found expression, and exercised a considerable directive power upon the lines of scientific investigation. Popular thought is at all times permeated by metaphysical ideas, however detached and unsystematized their form may be, and however unconscious the majority of men may be of their presence and true character. I have already alluded to the strong tendency which Natural Science has exhibited during the last two centuries to extend its original scope and to transform itself into an all-embracing World-Philosophy.

With a view to the delimitation of the domain of Natural Science in relation to human experience as a whole, it is necessary to make an attempt to clear up ideas as regards the position of Natural Science relative to some of the various systems of Metaphysical Philosophy with which it has at various times found itself in contact. Fortunately, it is, I believe, possible to do this without considering in much detail the characteristics of the various metaphysical systems which have in different ages arisen, and some of which were held contemporaneously by different schools of philosophical thought. For our purpose it will suffice briefly to refer to a very few of the more fundamental points in which these systems, with their innumerable points of difference in less fundamental respects, differ from one another. In particular, it is requisite to attempt to draw some conclusion as regards the answer that should be given to the question what ontological assumptions are necessary for the special purposes of Natural Science. The history of Science shows that assumptions of this kind have in fact played a large part, sometimes in promoting, and sometimes in hindering, progress. But the most important question we have to consider is how far such assumptions are indispensable for the existence and efficiency of Natural Science, as an organized system devised for the systematic ordering of our physical percepts, in whatever degree experience may show that this ordering is possible. It is clear that, in accordance with sound Methodology, all unnecessary assumptions should be discarded as unessential to Natural Science, even when they have in point of historical fact been employed as part of the scaffolding used in the building up of the various departments of the domain. The existence and successful functioning of Natural Science cannot properly be invoked as a decisive reason for accepting any ontological theory or assumption which cannot be shown to be an indispensable element in scientific thought considered as a coherent body of doctrine.

In the subject-object relation which is fundamental

in all experience, although the phenomena with which Natural Science concerns itself are objects for some subject, the precise nature of the activity of the subject as a factor in the relation is irrelevant to Natural Science in the sense in which the term is here used. The investigation of this activity falls within the domain of Psychology which, for reasons I shall presently explain, is not part of Natural Science as here delimited. For Natural Science, the rôle of the subject in perception may provisionally, or methodologically, be regarded as purely passive, that of experiencing physical perceptions. Thus Natural Science takes these perceptions as simple data, and need not attempt to give any account of how far the activity of the subject is necessary for their existence, or how far they depend upon something foreign to the percipient. On such philosophical and psychological matters, Natural Science, as such, has nothing to say, and may take up a neutral attitude in relation to conflicting views concerning them. It should however be observed that the formation of concepts must be regarded as due to the psychical activity of subjects, although Natural Science is not concerned with any detailed investigation of the modes and laws of such psychical activity. This form of psychical activity is considered methodologically as distinct from any activity, or acts of attention, connected with the reception of percepts; but Natural Science is committed to no assumption that such separation of psychical function is valid in an ultimate sense, or that it is necessarily anything more than one of those kinds of separation which discursive thought is compelled to make, of elements which, from a more fundamental point of view, may come to be regarded as inseparable elements of an essential whole.

What we call a material object is *prima facie* a construct, built up by the synthesis of a group of actual sense-impressions, and of images of earlier sense-impressions stored up in the memory. An effect of the memories of sense-impressions is to enable us, in perceiving an object, to dispense with some of the actual sense-impressions which make up the construct. Thus, when we see a stone, we know that it is hard, without verifying the fact by touch in each instance; this is due to the memory of earlier sense-impressions, and to a belief in their unchangeableness. This view of the nature of a material object implies that there is a percipient for whom the object is a percept; and this involves the subject-object relation. The question now presents itself, can the object be properly regarded as existent when there is no percipient for whom it is a percept? In other words, can the object be separated out from the subject-object relation, and be regarded as having an existence independent of any percipient and of all percipients? Is there any definite meaning, and if so what, to be attached to the assertion of such independent existence? The kind of answer which has been given to such questions varies greatly in different philosophical systems. For us, the importance of the matter depends upon whether it makes any essential difference to Natural Science what answer is given to these questions and to other ontological questions relating to the existence of the percipient or subject. The answer which I suggest and advocate is that, for the special purposes of Natural Science, it is immaterial what answer is given to these questions; that in fact ontological hypotheses or theories on these points are irrelevant to Natural Science; that it is sufficient, for example, for the purposes of Natural Science to regard a perceptual object as a construct of sense-impressions, whatever else it may for the purposes of systematic Philosophy be regarded as being or implying. But this answer is one which would not have been given until recent times by any man of Science, and probably would not be given at the present time by the majority of men of Science. A material object has most frequently been regarded as having some kind of sub-stratum, not identical with the synthesis of sense-impressions, but as a thing in itself, a kind of bearer of the various properties or qualities such as extension, motion, hardness, colour, elasticity, etc., which are regarded as giving rise to the component sense-impressions. Thus material objects are regarded as having an entity, material substance, as their essential foundation; the various properties or qualities being regarded as inherent in the substance. The real object, or thing in itself, is not identified with any or all of the properties which it possesses, and with which our sense-impressions are related; it is not itself directly perceived, although its existence is supposed to be a necessary inference from the existence of the percept. This point of view was clearly expressed by Locke, who writes[1]:

> When we talk or think of any particular sort of corporeal substances, as horse, stone, etc., though the idea we have of them be but the complication or collection of those several simple ideas of sensible qualities...., yet because we cannot conceive how they should subsist alone, nor one in another, we suppose them existing in, and supported by, some common subject; which support we denote by the name *substance*.

This conception of substance, as a sub-stratum of matter, has been held by the adherents of some realistic systems of Philosophy, by many, or probably most, men of Science, and it appears to be held by those who are dominated by the set of notions usually described as common sense, or as naïve realism. The thing in itself has been regarded as having an existence independent of any and all percipients; thus forming an independent Real. The doctrine of the distinction between substance and its properties or accidents was an essential part of Scholasticism, pressed into the service of Ecclesiasticism. When it was pointed out by Locke and others that at

[1] *Essay concerning human understanding*, Bk II, Chap. XXIII, Sec. 4.

least some of the accidental properties of the thing in itself are clearly not independent of the percipient, a division was made of the qualities of the substance into two sets, called respectively primary and secondary qualities. The primary qualities, extension, and motion, and probably inertia, were regarded as inherent in the object itself, and thus, like the substance, independent of any percipient; but this independence was not asserted of the secondary qualities such as colour, sound, smell and temperature. These last were regarded as essentially dependent on the percipient, but also as conditioned by the primary qualities of the substance. The substance, or the sub-stratum of all material things, is incapable of being directly perceived, but is regarded as something which is of necessity conceived by the mind as existing independently of itself, as the bearer of qualities, just as an adjective requires a substantive which it qualifies. In Leibnizian monadism, and also in more modern forms of spiritualistic pluralism, matter is regarded as the manifestation of the activity of a plurality of psychical beings; and thus substance is of a psychical character.

The exact opposite of this notion of the necessity of the category of substance as the sub-stratum of the material world was advocated by Berkeley, the great English idealist philosopher. With him, what we have spoken of as percepts, and what he calls ideas, constitute the whole reality of material objects. He wrote[1]:

> It is indeed an opinion strangely prevailing amongst men, that houses, mountains, rivers, and in a word all sensible objects, have an existence, natural or real, distinct from their being perceived by the understanding.

The view expressed by Berkeley that there exists no sub-stratum, or substance, independent of perception, is characteristic of idealistic Philosophy, and is in sharp contrast with the opposed realistic view.

[1] *The principles of Human Knowledge*, Collyns Simon's edition, p. 19.

To decide between these opposed opinions is no part of the function of Natural Science; in fact the method which Science employs makes it incapable of doing so. *Ex hypothesi*, substance, things in themselves, if they exist, cannot be directly known as percepts, but are inferences from what is perceived; and thus they cannot form part of the perceptual order of things which it is the function of Natural Science to classify and describe. Whether they exist, or not, is consequently quite immaterial to Natural Science, because, even on the realistic assumption, Science is actually concerned only with the percepts to which things in themselves give rise, and not with the real world which the realistic Philosopher takes to exist behind phenomena. As between these two divergent philosophical views, the position of Natural Science may be taken to be that of a neutral. Natural Science is not compelled either to affirm or to deny the existence of real substance, or of monads as independent reals. It has no need to take account of the category of real substance. The history of Science is full of assumptions of the existence of ethers and substances of various kinds; but these may be taken to be concepts, not realities independent of the psychical world. They are parts of the conceptual scaffolding which go to build up scientific theories and laws.

In speaking of the function of Mathematical theories, Poincaré has given[1] a description of their function which might, I think, be extended to scientific theories in general. He writes:

> Mathematical theories have not as their object to reveal to us the real nature of things; that would be an unreasonable claim. Their sole aim is to coordinate the physical laws that experience reveals to us, but which, without the aid of Mathematics, we could not even enunciate.

Even to-day men of Science under the influence of philosophic Realism, or of the unsystematized Philosophy

[1] *Leçons sur la théorie mathématique de la lumière*, p. 1.

of common sense, are extremely reluctant to abandon the idea that the notion of real entities, something more than concepts, is essential to Natural Science. Thus, for example, Professor A. N. Whitehead writes[1]:

Another favourite solution, the most attenuated form which the bifurcation theory assumes, is to maintain that the molecules and ether of Science are purely conceptual. Thus there is but one nature, namely apparent nature, and atoms and ether are merely names for logical terms in conceptual formulae of calculation. But what is a formula of calculation? It is presumably a statement that something or other is true for natural occurrences. Take the simplest of all formulae. Two and two make four. This—so far as it applies to nature—asserts that if you take two natural entities, and then again two other natural entities, the combined class forms four natural entities. Such formulae which are true for any entities cannot result in the production of the concepts of atoms. Then again there are formulae which assert that there are entities in nature with such and such properties, say, for example, with the properties of the atoms of hydrogen. Now if there are no such entities, I fail to see how any statements about them can apply to nature....The current answer is that, though atoms are merely conceptual, yet they are an interesting and picturesque way of saying something else which is true of nature. But surely if it is something else that you mean, for heaven's sake say it. Do away with this elaborate machinery of a conceptual nature which consists of assertions about things which don't exist in order to convey truths about things which do exist. I am maintaining the obvious position that scientific laws, if they are true, are statements about entities which we obtain knowledge of as being in nature; and that, if the entities to which the statements refer are not to be found in nature, the statements about them have no relevance to any purely natural occurrences. Thus the molecules and electrons of scientific theory are, so far as science has correctly formulated its laws, each of them factors to be found in nature. The electrons are only hypothetical in so far as we are not quite certain that the electron theory is true. But their hypothetical character does not arise from the essential nature of the theory in itself after its truth has been granted.

[1] *The concept of Nature*, pp. 44 *et seq.*

If I understand this passage aright, the ideas of the nature and functions of a scientific theory expressed in it are in fundamental opposition to those which I here advocate. Whatever else they are, molecules and electrons are concepts, and even Dr Whitehead is not quite sure that they are anything else. They are concepts in scientific theories, and the only question about a scientific theory is not whether it is in Dr Whitehead's sense "true," but whether it is logically coherent and how far it is adequate for the purpose of representation; whether it will be superseded by some other theory which employs other concepts, because that theory is applicable to a greater range of phenomena than the theory which employs molecules and electrons. We do not perceive in Nature entities such as atoms and ether; we do not perceive entities at all, if an entity be taken to be anything more than a construct of a complex of sense-impressions present and past. What Dr Whitehead speaks of as an elaborate machinery of a conceptual nature which consists of making assertions about things which don't exist, is really a scheme involving things which do exist *as concepts*; and conceptual atoms have been employed because they have been found not merely "interesting and picturesque," but because, at all events for the time being, they have been found to be the best available means for representing what "is true in nature" in Dr Whitehead's phraseology; in the sense that they resume conceptually part of what we perceive as natural phenomena. Natural Science postulates, as a working hypothesis, only that the perceptual complex is such that tracts of it are capable of conceptual description by scientific schemes. It does not require any postulate as to detailed systems of relations or of entities within that perceptual complex, or within any supposed reality behind that complex, which shall account for the fact that the working hypothesis has proved successful.

In speaking of the new conceptual scheme of space-time which is associated with the name of Einstein, Sir Oliver Lodge writes[1]:

> In such a system there is no need for Reality: only phenomena can be observed or verified: absolute fact is inaccessible. We have no criterion for truth; all appearances are equally valid; physical explanations are neither forthcoming nor required; there need be no electrical or any other theory of the constitution of matter. Matter is, indeed, a mentally constructed illusion generated by local peculiarities of space.

And again he writes concerning the same theory:

> But, notwithstanding any temptation to idolatry, a physicist is bound in the long run to return to his right mind; he must cease to be influenced unduly by superficial appearances, impracticable measurements, geometrical devices, and weirdly ingenious modes of expression; and must remember that his real aim and object is absolute truth, however difficult of attainment that may be, that his function is to discover rather than to create, and that beneath and above and around all Appearances there exists a universe of full-bodied, concrete, absolute, Reality.

The notion indicated by Sir Oliver Lodge that it is the function of Natural Science to search for reality behind physical phenomena still receives support from the ranks of Philosophers. Thus, for example, Professor H. Wildon Carr writes[2]:

> The modern era of philosophy from Descartes onwards has been dominated by the insistence of the scientific problem—that is, the problem of the ultimate nature of the reality we study in physical science by the experimental method.

These examples may suffice to illustrate the persistence of this view, which I believe to be erroneous, even at the present time.

It is interesting to observe that some men of Science who avowedly regard Metaphysics with distrust and aversion are most insistent in claiming "reality" for material objects, and for molecules, atoms and ether; oblivious of the fact that they are, as it would seem

[1] *Nature*, CVI, pp. 796 *et seq.* [2] *Nature*, CVI, p. 809.

unnecessarily, by using the term "reality," illustrating the fact of the extreme persistence in thought of metaphysical conceptions of the theoretical and of the common sense order. The term "reality" has an intensely metaphysical complexion; and its shades of meaning in different philosophical schemes vary considerably. Many thinkers, like Plato, have held that the domain of concepts has a claim to "reality" superior to that of the perceptual world.

I have already spoken of the dualism which recognizes in the world two distinct domains, the psychical and the physical; or as is sometimes said with less accuracy, mind and matter. The ultimate relation of these two domains is for the Philosopher to elucidate. For the man of Science this dualism should be regarded only as provisional and methodological, and need not be taken to involve any final assumption of the fundamental disparateness or separation of the two domains. Mind and Matter have frequently been regarded as ultimately reducible to one of the two, or to some third basal existent more fundamental than either of them; but Natural Science, so long as it confines itself to its own original functions, does not stand or fall with any hypothesis of this character. A monistic system of Philosophy, whether it be of the kind called materialistic, or of the kind described as spiritualistic, or whether it be described as neutral, will no doubt make the attempt to interpret scientific theories in terms appropriate to the particular system. Pluralistic schemes of Philosophy, such as the Monadology of Leibniz, or other more modern forms of Pluralism, must also be left to find their own modes of interpretation of Natural Science. But scientific thought, leading to what is known as positive knowledge, needs none of the ontological assumptions peculiar to any one of the many rival philosophical theories of the nature of reality.

The naïve realist, and some philosophers, regard the

objects of perception, such as we perceive in our normal life when awake, as still existing very much as we perceive them, whatever that may mean, even when no sentient being is actually perceiving them. This is also the view of common sense; but scientific knowledge, in accordance with the view of its nature that I am here propounding, is independent of the acceptance or rejection of this opinion. Scientific knowledge is also independent of the assumptions made in any other form of realistic Philosophy. The pure phenomenalist holds that perceptual objects exist only when, and so long as, they are being perceived as objects by a percipient; and thus that the only realities are mental states and the objective constructs actually perceived. Modified statements hold good for intermediate philosophical theories, as for example the kind of phenomenalism which admits the existence of *νοούμενα* behind phenomena. The descriptive view of the functions of scientific theories and laws, which I here adopt, has sometimes been characterized as phenomenalism; but this view does not necessarily involve the acceptance of any philosophical doctrine of phenomenalism. The phenomenalism of Natural Science is methodological only.

That the distinctions which we make in order to meet the necessities of our discursive methods of apprehension necessarily correspond to similar distinctions in an objective reality is an assumption, of which, though it has been persistent through the whole history of speculative thought, Natural Science has no need, whatever its merits may be in the view of Philosophers of various schools.

Accordingly, the view of the function and limitations of scientific knowledge which has arisen as the result of recent criticism, has as one of its implications the assumption that Natural Science is independent of purely ontological theories, as applied either to percepts or to percipients. If, in the sense employed by any particular philosophical system, the existence of real entities be asserted in any particular connection, Natural Science can make no use of such assertion, since all scientific theories can be stated in a form which is independent of such ontological assumptions. There is a considerable practical advantage in the rejection, as unnecessary for the purposes of Natural Science, of all superfluous ontological hypotheses. So far as Science can take up the position of neutrality, or of detachment, with respect to the ever-conflicting and ever-varying ontological views of Philosophers, it does not share in the vulnerability which may be held to attach to any of the metaphysical theories that are involved in the conflict. In view of the instability of all definite metaphysical schemes, an instability which does not show much sign of disappearing, the greater stability with which this position of neutrality endows Natural Science has great advantages.

In making these remarks as to the relation of Natural Science with Metaphysics, I have no desire to associate myself with the views of those men of Science who decry Metaphysics as a barren study which can lead to nothing but fruitless sophistries. On the contrary, I fully recognize that, in all ages, Philosophical thought has been indispensable, in ministering to one of the deepest and most ineradicable impulses of the human spirit; the imperative impulse to penetrate in some degree into the recesses of the great mystery of life and existence, with which we are all in relation. Although no generally accepted solution of even the fringes of that mystery may be in sight, yet Philosophy has done much to fix and clarify the form in which fundamental questions can be properly stated, and has also rendered great services in definitely removing, by its criticisms, many irrelevant accretions which had gathered round the main philosophical questions. The attempt to penetrate to a reality in which we may find something permanent and stable in, behind, or above, the impermanent and unstable elements of physical and psychical phenomena, is one which will continue to be made as long as the human spirit resembles what it has been in the past, and is in the present; although it is no part of the functions of Natural Science to take part in this attempt. The impulse from which this attempt arises will never for long be stifled by merely negative criticism. But progress on the path to reality cannot be made by the simple expedient of transmuting such concepts as those of matter, ether, atoms, electrons, into real substances or entities.

Philosophy has in recent centuries rendered valuable services to Natural Science in assisting it to remove various injurious accretions derived from medievalism, of supposed *a priori* knowledge, and of unnecessary and irrelevant ontological assumptions. The attention of scientific thinkers has thus been concentrated on phenomena themselves, and on the mental representation of them. Science has no longer been hampered by medieval prejudices relating to occult properties of supposed sub-strata behind phenomena, or by conceptions of Nature as necessarily conforming to ideas constructed by *a priori* thought. The advance of Natural Science involves two main factors, first the discovery, by observation and experiment, of facts; and secondly the procedure of reflective thought in classifying groups of facts, devising rules or laws to which the groups of facts conform, and constructing general theories which symbolize phenomena, and under which groups of laws can be subsumed. Both these factors are essential in all genuine Science, but the emphasis placed on the two factors has varied at different periods and with different thinkers. At some periods, and with some persons, there has been a tendency unduly to subordinate one or other of these factors. Among the Greek thinkers, Aristotle can be distinguished by the great importance he attached to the observation of facts, but he did not thoroughly grasp the nature of the slow and painful process by which empirical results are obtained and used to build up genuine scientific theories. A mere collector of facts is far from being a real man of Science, but there is room for real division of labour in contributing to the growth of Science, and this implies that the work of various grades of investigators can be of value in the construction of the edifice. Among these grades those workers who undertake the arduous duty, under suitable direction, of collecting precise facts, in some given experimental or observational domain, have an honourable place. But, in the trenchant words[1] of Poincaré: "Science is built up of facts, as a house is built up of stones; but an accumulation of facts is no more a science than a heap of stones is a house." The really important general theories are always the work of men endowed with very exceptional powers of imagination, insight, and generalization, which enable them to seize instinctively upon relations of similarity between things that have little or no superficial resemblance with one another. These great pioneers have formed but a small band in the history of Science.

Even after medieval conceptions of Science had waned, the Rationalistic school, which preceded Kant and the critical conceptions he initiated, attempted to construct nature out of pure Thought. The critical movement led by Hume and Kant established the impotence of Thought alone, devoid of the essential foundation of actual physical experience, to formulate any genuinely scientific view of the world of physical phenomena.

In what I have said hitherto about the functions of Natural Science, the conceptual description of the physical world alone has been referred to, but it is clear that no treatment of the general aspects of the subject would be adequate without some reference to the

[1] *Science and Hypothesis*, p. 141.

psychical order of things. That the physical and psychical domains have, in appearance at least, very close relations with one another is obvious to the most casual observation. *Prima facie* they are not independent of one another, but closely interconnected.

Modern Physiology, on the one side, and Modern Psychology, on the other side, have confirmed this fact of observation in minute detail. There is evidence, of overwhelming strength, that psychical events or processes in the mind are accompanied by physical events or processes in the bodily organism; that at least some kinds of physical events or processes in the body are accompanied by psychical happenings in the mind is also a fact not open to doubt. The question of the nature of the relations between such physical and psychical processes, between body and mind, or of how such relations are to be represented, is one of the most difficult and intractable problems of Philosophy. An important and influential school of Physiologists has advocated the view that physical processes in the brain and nervous system of the bodily organism form a closed sequence entirely independent of the psychical processes, or happenings, which accompany these physical processes; that, despite appearances, the sequences of events in the bodily organism are entirely unaffected by concomitant psychical sequences. In accordance with this view, if we possessed a sufficiently advanced knowledge of Physiology, a complete account, involving only the categories of Physics and Chemistry, could be given of all the actions of a man, of all his responses to external stimuli, without taking into account his consciousness or will, or any of the motives to which he himself attributes his actions.

His cognition, feelings, and conation are, in this theory, regarded as belonging to a domain which has no influence upon the world of physical phenomena, including all the physical happenings in his own body; the former are regarded as epiphenomena, or Begleiterscheinungen, which accompany, but have no influence upon the latter. The man is a conscious automaton, a machine endowed with consciousness, but not with the power to influence his own actions; although he is under the delusion that he has this power. The closeness of the correspondence between the sequences of events in the two domains has been formulated in the theory known as Psycho-physical Parallelism. Whatever be the value of this theory, it certainly makes colossal demands upon our powers of imagination, when we attempt to represent to ourselves how it works, and what it implies, in an individual case. On the 7th of May, 1915, multitudes of people read telegrams conveying the news that the *Lusitania* had been torpedoed. The reading of the telegram was followed, in a vast number of cases, by bodily feelings indicating very marked disturbances in the nervous system. That these nervous tremors were due to a conscious apprehension of the terrible meaning of the news is the ordinary view of common sense. But a believer in the theory of psycho-physical parallelism is bound to assert that the meaning of the telegram, only interpretable in psychical terms, had nothing whatever to do with these nervous disturbances. The image of the print of the telegram on the retina of the eye, the subsequent neural currents to the brain, certain changes in the smallest part of the grey matter of the brain, and neural currents from the brain to other parts of the organism, formed a sequence which could theoretically be accounted for as a sequence of purely physical phenomena; the only relevant factors in the determination of this sequence were the image on the retina, and the detailed physical constitution and condition of the individual organism. A slight change in the words of the telegram, such as for example the insertion of the word "not," with the corresponding slight change in the image on the retina, might have led to the absence of all the nervous disturbances in the body of the reader of the telegram. A similar absence of nervous disturbance would have been observed in the case of a person who did not understand the language in which the telegram was written.

The change of meaning of the telegram would have been of vital importance, but believers in the theory of psycho-physical parallelism must assert that the slight change in the image on the retina was the only factor which could account for the absence, instead of the presence, of what were, in many cases, very marked disturbances in the nervous system. This illustration has been given solely with the intention of making clear some of the implications of the theory of psycho-physical parallelism. It is not put forward as a refutation of the theory; a much more complete discussion of the theory would be required for such a purpose, and it is no part of my programme to put forward even a tentative solution of the problem to which the theory is related.

Another attempt to overcome the difficulties connected with the relations between physical and psychical phenomena in the living organism is embodied in the theory of interaction. In accordance with this theory, neither the physical nor the psychical phenomena form a closed system, but each group is affected by action from the other. So long as the two domains are regarded as completely disparate, the difficulty of this theory arises from inability to represent the precise nature of such interaction, or to discover the exact points in a physical sequence at which the effect of the psychical domain makes itself apparent as a disturbing factor, introduced extraneously into the physical sequence.

It has sometimes been assumed that, regarding the physical phenomena in the living organism as forming a dynamical system, the action of the psychical side of the organism is of such a character that no breach of the conservation of energy in the dynamical system takes place. If it could be established that the action of the psychical side is such that it does no mechanical work in the dynamical system, an interesting fact would have been discovered, but the main principle would not be thereby affected that the dynamical system ceases to be an independent system, the working of which can be described simply in accordance with the laws of Dynamics. It must be remembered that the principle of conservation of energy itself leads only to one of the equations which describe the motions of the parts of a material system possessing more than one degree of freedom. The effect produced by the psychical factor on the physical system could only be represented by an action to which there corresponds no reaction on the physical side. Accordingly the physical part of the organism could only be treated as a dynamical system if we had a knowledge of the character and numerical measures of the mechanical forces, acting on the physical system from without, by which the action of the psychical part of the organism on the physical system could be represented.

There is much force in the contention that the whole problem of the nature of the relations between the physical and psychical domains in the living organism is essentially insoluble because the problem is a purely artificial one, having arisen from the original assumption made that the physical and psychical sides are disparate, without an underlying unity. If it be held that we have, in treating body and mind as belonging to separate domains, set up a distinction which does not correspond to any really fundamental difference, this may be held to account for our inability to formulate any satisfactory and coherent theory of the relations between the two artificially separated domains. Having made an arbitrary separation into two domains supposed to be disparate in their natures, we are unable, without contradiction, to undo our work by recombining them into one system. In default of a purely monistic Science which should take up into itself all the phenomena which we now call physical, and also those which we call psychical, and which should be sufficiently advanced as to absorb and unify all our present knowledge on both sides, we are compelled to retain the methodological procedure of considering physical and psychical phenomena separately. Unless we are prepared to adopt, as a general principle of unlimited scope, the theory of psycho-physical parallelism, that the physical domain is entirely independent of the psychical, we are compelled to restrict ourselves, on the side of Natural Science, to the procedure of tracing out such sequences of physical events as we find by actual experience to be capable, at least to a practically sufficient degree of approximation, of conceptual description in which the concepts arise from the physical domain alone. That this is possible, even in the biological departments of Science, to some very considerable extent, the limits of which we do not know, has been abundantly established by modern Physiology. Nevertheless, it is impossible to regard the physical side of the living organism as a completely independent system, all the happenings in which are capable of being described conceptually, in accordance with the canons of purely physical Science, and to an extent without theoretical limits, unless we are prepared to adopt the unproved assumption that the physical organism is independent of the psychical side of the living being. The amount of partial independence is just what experience shows it to be, but no number of successes in subsuming particular chains of processes which occur in the living organism under conceptual schemes such as we employ in Natural Science, will warrant us in asserting the principle of the absolute independence of the physical side of the organism to be more than a surmise.

We are thus led to what must be regarded as a limitation upon the claims of Natural Science, in the sense of the term which has been here adopted, to the power of theoretically extending itself so as to become a complete Philosophy of Physical Nature, independent of all psychical factors. Physiology is completely justified in assuming this independence as a methodological principle, and experience alone can decide how far it will be able to extend its present far-reaching results in accordance with that principle. But the existence of Psychology, the Science of the normal individual mind, with the borderland domain of Physiological Psychology, indicates that the possibility of representing the phenomena of the bodily organism by means of conceptual schemes of the kind which we class as belonging to Natural Science may have limits which cannot be passed, however far Physiological investigation may at present be from having reached those limits.

Hitherto I have spoken only of Natural Science as the Science of physical phenomena, but in the wider sense, the term Science is employed in relation to the study not only of physical, but also of psychical, phenomena; and it is also used in the case of the study of complexes involving both physical and psychical phenomena. The question arises how far the methods of procedure which Natural Science adopts are applicable where psychical phenomena are involved. The Science of Psychology occupies itself with the conceptual description of sequences of psychical happenings in the normal human mind; the idiosyncrasies of particular minds being disregarded, just as, in the conceptual schemes of Natural Science, the irrelevant peculiarities of individual physical objects are disregarded. There are, however, important respects in which Psychology differs from a department of Natural Science. The only mental phenomena, of which an observer can directly take cognisance, are those which occur in his own mind; what occurs in the minds of other persons he can only ascertain indirectly through physical manifestations. Thus introspection, and inference from physical events, assumed to afford sufficient indications of corresponding psychical events, are the two sources of the facts with which the Psychologist has to deal in ascertaining laws and in building up his conceptual schemes of representation of psychical sequences. Moreover, psychical processes and states are not of a quantitative character such as are accessible to the methods of measurement employed in the physical sciences. These measurements of quantity are dependent upon that property of extension which has no direct correlative in the psychical domain. Psychical states and processes may lend themselves to notions of magnitude, but not to that of extensive magnitude; an intensive magnitude not being made up of units, and therefore not being capable of numerical representation as is an extensive magnitude. Such measurements as are made by experimental Psychologists are measurements of the physical concomitants of psychical phenomena, and not of those phenomena themselves. It has moreover been pointed out by Professor James Ward that Psychology cannot be defined by reference to a special subject matter, as in a department of physical science, since it deals in some sense with the whole of experience. Not only has Psychology to concern itself with a more complex subject matter than a department of Natural Science, but the position of the observer, or percipient, relative to the phenomena is less simple in the former case than in the latter; and it is thus difficult to assert that psychical events and processes are phenomena, or factors in the subject-object relation, in quite the same sense in Psychology as in the Natural Sciences. The Sciences of the Politico-social group, so far as they rise above the more superficial classification of special kinds of phenomena, and their statistical setting, are all dependent on psychological knowledge, and share in the peculiarities I have mentioned which differentiate Psychology from the Natural Sciences.

The methods of Psychology and of the Sciences of the Politico-social group must be similar in character and spirit to those pursued in the case of the Natural Sciences, so far as the nature of their respective subject matter allows; and there exist in all the former Sciences tracts in which the results and methods of Natural Science are directly applicable, and may render great services. Some departments of Biological Science, especially evolutionary Biology and the theory of Heredity, have made use of psychical categories in such a manner that they cannot, unless such psychical elements can be eliminated, be regarded as falling wholly under the denomination Natural Science, in the restricted sense in which I have employed the term. For example, Darwin, in his *Origin of Species*, occupies himself to a considerable extent with mental factors, such as sexual selection, as contributing to the causes of the natural selection which determines the evolution of species. Purposiveness, and all teleological conceptions, are foreign to Natural Science; and all departments of Biology, so far as they make use of such conceptions, must be considered as mixed Sciences, in the sense that they make use of concepts which represent not only physical percepts but also psychical elements.

The psycho-physical parallelist may maintain that the psychical elements in Biology can be ultimately eliminated, and the whole be reduced to a purely physical Science; but such assertion is of a highly speculative and contentious character. In view of what has been said earlier in this lecture, it cannot be assumed as established that such complete elimination of the psychical factor is possible, even in theory.

# IV

# CAUSATION AND DETERMINISTIC SYSTEMS

WHEN a physical event takes place it is usually regarded by common sense as determined by preceding events or processes which are deemed to have caused it to take place. Very frequently some one preceding event is singled out as the cause of the event in question. It is assumed that the particular event, the effect, would not have taken place in the absence of the cause; and that cause is regarded as affording an explanation of the occurrence of the effect. A somewhat less summary explanation of the event recognizes a plurality of preceding events or processes, the absence of any one of which would have entailed the non-occurrence of the effect. The ordinary notion of causation attributes a certain contingency to what is regarded as the cause of an event; thus an event or sequence $A$ is in general only considered as the cause of an event $B$ when the absence of $A$ can be easily imagined; when this is not the case, the invariable succession of $B$ after $A$ does not give rise to the idea that $A$ is the cause of $B$. For example, we do not consider the night as caused by the preceding day, although we regard both day and night as belonging to a fixed sequence; in this case the element of contingency is not regarded as present. In scientific thought, the notion of causation is expanded so as to embrace a whole complex of conditions, some preceding in time, and others simultaneous with, the particular event in question. That every event has a cause, formulates the conception of the determination of the event by a complex of preceding and present conditions. Taking into account the fact of the existence of approximately isolated systems in the physical world, the cause of an event is restricted to the notion that it consists of preceding and present conditions forming a limited complex of relevant circumstances; all other circumstances being regarded as irrelevant in regard to the explanation of the event. When the event under consideration and the relevant circumstances are of such a character that they are measurable, or in other words capable of being correlated with numbers, the event is frequently said to be a function of the relevant conditions. This notion of functionality may be regarded as a more precise form of that of causation. But, in the view of common sense, a view which has until comparatively recently been shared by men of Science, the notion of causation contains something more than the bare idea that the so-called effect has an invariable relation to the cause, in point of fact. It includes the notion that the cause, some particular event or process, or some sub-complex within that cause, can be regarded as an active agent which compels the effect, regarded as passive in the transaction. When this idea of compulsion by an active agent is an essential component of the notion of causation, the effect being regarded as forced by the compelling action of the cause, that cause is said to be an efficient cause. It is not difficult to recognize the essentially anthropomorphic character of this notion of efficient causation. When, by an exercise of his will, a man moves a body, or one of his own limbs, he regards himself as an active agent who, by the power of his will, compels the body, or limb, to move; the moved body being looked upon as purely passive in the occurrence. The organic feeling of exertion which accompanies the action is to him the sign of his activity. In all our interventions in the physical world this feeling of activity, dependent upon the consciousness of motion and contraction of muscles, is present in greater or lesser degree.

By ejection of this notion of activity into the physical world in general, we tend to attribute to the cause of an event the rôle of the active agent which we feel ourselves to be when we regard an event as caused by ourselves. We do not however usually suppose that consciousness, or any psychical factor, is necessarily present in a cause when that cause is not a living organism; in this respect the original animistic conceptions which were probably concerned in the ejection of the notion of efficient causation into the external world have taken a modified form. Hume, who in his *Enquiry concerning human understanding* has treated the subject of efficient causation with great clearness and fullness, writes[1]:

> No animal can put external bodies in motion without the sentiment of a *nisus* or endeavour; and every animal has a sentiment or feeling from the stroke or blow of an external object, that is in motion. These sensations, which are merely animal, and from which we can *à priori* draw no inference, we are apt to transfer to inanimate objects, and to suppose, that they have some such feelings whenever they transfer or receive motion.

The indirect character of the effect of the causation presumed to exist when a man moves a limb has also been pointed out by Hume, who wrote[2]:

> We learn from anatomy, that the immediate object of power in voluntary motion, is not the member itself which is moved, but certain muscles, and nerves, and animal spirits, and, perhaps something still more minute and more unknown, through which the motion is successively propagated, ere it reach the member itself whose motion is the immediate object of volition. Can there be a more certain proof, that the power, by which the whole operation is performed, so far from being directly and fully known by an inward sentiment or consciousness, is, to the last degree, mysterious and unintelligible?

The notion that causation in the physical world is similar to the efficient causation which we feel to be present when we act affords an explanation of the

[1] *Enquiry*, Sect. VII, Part II, footnote. [2] *Ibid.* Sect. VII, Part I.

ancient belief, one which has for ages been persistent in scientific thought, that all physical action involves the contact of material bodies, because our own mode of directly intervening in the physical world involves the placing of some part of our bodies in contact with external matter. The principle known as the "principle of causation" has frequently been stated in the concise form that every natural event has a cause; and this has usually been regarded as a necessary or axiomatic principle of thought to which Nature must conform. It should be observed that the principle leads unavoidably to an endless regress. For, if an event have necessarily a cause, that cause, being considered to be a preceding event or a plurality of such preceding events, must itself have a cause, and consequently we have a sequence of events in which there can be no initial, or first, cause.

A result of modern criticism of the notions which underlie scientific thought has been that the conception of efficient causation has been discarded, as unnecessary and useless, for the purposes of Natural Science. Hume's analysis of the nature of perception made it abundantly clear that efficient causation is not to be discovered in natural phenomena. He writes[1]:

> When we look about us towards external objects, and consider the operation of causes, we are never able, in a single instance, to discover any power or necessary connexion; any quality, which binds the effect to the cause, and renders the one an infallible consequence of the other. We only find, that the one does actually, in fact, follow the other.

It should be observed that the acceptance of Hume's position in this matter does not imply an acceptance of his views in this connection as regards the association of ideas. It is at the present time often thought desirable no longer to use the terms cause or causation, in connection with physical phenomena, on account of the fact that these terms have usually in the past been taken

[1] *Enquiry*, Sect. VII, Part I.

to imply the notion of efficiency. When the term cause is retained it is taken to be synonymous with the totality of antecedent conditions. The principle of causation, taken in the only sense in which it can now be retained in Natural Science, is not a logically necessary principle, but merely the postulation, or rather working hypothesis, that it is possible to predict the happening of particular events when certain complexes of antecedent conditions are known. The mode in which this prediction is made, in any particular case, is by the employment of conceptual laws or schemes which have reference to events of a class to which the particular predicted event belongs. Thus Helmholtz has written[1]: "The principle of causality is nothing else than the hypothesis that all the phenomena of nature are submitted to law." I shall however presently show that Helmholtz's statement is capable of a too general interpretation; and that in fact the principle of causality, as employed in Natural Science, is of a more stringent character than would appear from Helmholtz's formulation of it. The extent to which the postulate of causation actually holds good, when applied to natural phenomena, we can only find out by experience.

In the more superficial notion of causation of which I spoke at the beginning of the lecture, an event $A$, antecedent to an event $B$, is under certain circumstances regarded as the cause of $B$, some interval of time elapsing between $A$ and $B$. But our sense-impressions are no longer regarded by Psychologists as atomic, in the sense of consisting of a set of separate sense-impressions with intervals of time, however small, between consecutive ones, but rather as forming a continuous stream; so that we now have to contemplate, instead of two separate events $A$ and $B$, a continuous process to which both $A$ and $B$ belong. An advanced scientific theory does not take account only of the bare fact of the invariability

[1] *Ueber die Erhaltung der Kraft*, Wissensch. Abh. p. 68.

with which $B$ follows $A$, but should give a conceptual account of the process to which $A$ and $B$ belong. In some departments of Science, especially those which have become amenable to representation by a Mechanical theory, this has in many cases been accomplished. In other departments the efforts to trace continuous processes, to which both the so-called cause and the so-called effect belong, have met only with partial and fragmentary success. Sometimes, what has been accomplished, after minute scrutiny, amounts to the insertion, between $A$ and $B$, of other intermediate observed events, between each two consecutive ones of which a hiatus exists, of essentially the same character as the original hiatus between $A$ and $B$ themselves; and thus the nexus between $A$ and $B$ has not been successfully represented as a continuous process. Moreover it must be observed that an event at an instant of time can only be regarded as a conceptual limit, not as an actual percept, because an actually perceived event always has some duration, however small that duration may be; and thus the actual events $A$ and $B$ can themselves be regarded only as processes; finite parts of a single process which includes them both.

Although, as we have seen, the search for efficient causes in natural phenomena is chimerical, the belief in the existence and possible discovery of such efficient causes has been of great importance in the whole history of Science. That belief has often been advantageous in stimulating investigators to discover a few stages in the endless regress to which the *a priori* principle of causation leads; and this has often led to the establishment of the connection between different phenomena, although such connections have never given any lasting satisfaction to the desire to realize the notion of efficient causation.

Just as efficient causation cannot be discovered within physical phenomena, so also the search for any logical necessity, binding together the phenomena of a sequence, is doomed to inevitable failure. Logical necessity has its habitat in the domain of thought alone. It binds together the parts of a fully worked out conceptual scheme which represents a physical sequence, but no such logical necessity connects the successive stages of the physical sequence itself.

In a scientific theory in which conceptual elements are linked together by a scheme of postulations which fixes the relations between the different elements, the consequences which are deducible from the structure of the scheme follow by logical necessity from the postulations and definitions contained in the theory. That logical necessity is of the character in accordance with which the conclusion in a syllogism is a necessary inference from the major and minor premisses. But, when the scientific theory is employed for the purpose of describing a complex of physical phenomena, there is no justification for a transference of the logical necessity from the conceptual theory to the perceptual phenomena. For there exists no logical necessity that the theory should be applicable to the description of the phenomena; such applicability, when it exists, is a fact of experience only, not a necessity of thought. For example, we possess a conceptual scheme which assigns, in accordance with the Newtonian dynamical scheme and the law of gravitation, the relative accelerations of the bodies of the solar system, represented conceptually by points with definite mass-coefficients; and this scheme is employed to deduce the orbits of the points which represent the earth and planets about the point which represents the sun. The forms of the orbits follow by logical necessity from the postulations of the scheme. Yet there is no purely logical necessity that the earth should continue next year to describe, even approximately, the actual orbit round the sun which it has described during the past year, in approximate accordance with the orbit as deduced from the conceptual scheme. For there is no logical necessity that in the future year the conceptual scheme will continue to serve its descriptive purpose. The expectation which we have that it will do so is of overwhelming strength, based upon our past experience of the adequacy of the conceptual scheme. This expectation amounts to a subjective certainty, the existence of which in our minds is a fact for an explanation of which reference must be made to the Psychologist and the Epistemologist, not to the Logician. Thus there is no purely logical necessity that the seasons will recur as they have done in the past; there is only an expectation of such strength that it may be called a practical certainty, but not a theoretical or absolute certainty. Without such practical certainty, of this character, human action would be an impossibility; the great survival value which this subjective feeling of certainty as regards the continuance of a certain uniformity in Nature must have possessed during the stages of evolution may be an important factor in the explanation of its existence. The fact that this kind of expectation, which Natural Science accepts as a working hypothesis, is found by general experience to be satisfied is one of which Metaphysicians will take account in their theories of the nature and structure of reality; but for the purposes of Natural Science no such inferences of a philosophical kind are necessary.

The very common idea that it is the function of Natural Science to explain physical phenomena cannot be accepted as true unless the word "explain" is used in a very limited sense. The notions of efficient causation, and of logical necessity, not being applicable to the world of physical phenomena, the function of Natural Science is to describe conceptually the sequences of events which are to be observed in Nature; but Natural Science cannot account for the existence of such sequences, and therefore cannot explain the phenomena in the physical world, in the strictest sense in which the term explanation can be used. Thus Natural Science describes, so far as it can, *how*, or in accordance with what rules, phenomena happen, but it is wholly incompetent to answer the question *why* they happen. When a sequence of phenomena can be imitatively represented by means of some other sequence of a more familiar type, this latter is frequently said to afford an explanation of the former. Thus Lord Kelvin said, "when I have made a mechanical model I understand a process." This dictum might conceivably be understood in either of two senses. In the first, that in which, as the context shows, Lord Kelvin understood it himself, the model is taken to be a concrete model in which actual material bodies are employed to constitute its parts. Such a model affords an explanation of the phenomena which it pictures, in the more limited sense to which I have alluded; it makes plain by means of a reduction of the relatively unfamiliar to the more familiar. Such employment of a model is really only a preliminary, but often a very important, stage in the process of gradually arriving at a genuine scientific theory of the class of phenomena in question. The next stage commences with an attempt to explain the model, that is, to proceed further, with the model as starting point; unless indeed the working of the model has already been shown to be representable by a conceptual scheme, in which case the phenomena in question have, by an indirect process, been subsumed under a scientific theory. The second sense in which Lord Kelvin's dictum might be interpreted is that the model is of a conceptual character, of which the parts are conceptual objects, of defined characteristics, and related in assigned modes with one another. In this case the subsumption of the particular phenomena under a scientific scheme has been accomplished; to achieve this result is of the very essence of scientific method. The predilection which has been shown during centuries, for mechanistic modes of representing all physical phenomena, that is for schemes based upon the Mechanics of molar bodies, is to be accounted for, in large measure at least, by the fact that our most familiar experiences of natural phenomena are related to the motions, and interaction by contact, of gross bodies.

It will have been noticed that, not only in the more popular notion of causation, but also in that more refined and clarified form of the conception which we call scientific causation, the cause which conditions the effect precedes in time, or is at least not subsequent to, that effect; and this even when the notion of efficiency has been completely extruded. Thus, in the sequences with which Natural Science deals, an earlier part of a sequence is regarded as determining a part which is subsequent in time. The relation between what we call the cause and what we call the effect is not a symmetric relation; we do not in Natural Science regard an effect as conditioned or determined by a cause in the future. For the practical purposes of Science, those of prediction, it is essential that such an asymmetric determination should be the object of search. But this mode of determination is not the most general which we can take into account.

A deterministic scheme of the most general type is one for which we possess, or at least are assured of the existence of, a principle or set of rules which is sufficient to determine any single part of the sequence; although a knowledge of that part of the whole sequence which precedes in time the part to be determined may not suffice for such determination. In such a case the whole sequence is subject to a law, or formula, which makes it possible, at least in theory, to calculate the particular characteristics of any assigned part of the sequence. Thus all parts of the sequence are mutually related with one another, and with the sequence as a whole, in such a way that a knowledge of the principle of the whole sequence is requisite for the purpose of calculating the detailed nature of any part of it. In such a sequence there is no asymmetric relationship of cause and effect. The conceptual sequences with which Natural Science has to do, and the search for which is essential for the particular purposes of Natural Science, are deterministic, but not of the most general type. When I said that Helmholtz's definition of causation in Science is too general I meant that it appears to refer to a deterministic scheme of the general type, and not to the particular species with which Natural Science is concerned. The methodological assumption of Natural Science is that there are to be found, in natural phenomena, not merely physical sequences which are describable by means of deterministic schemes, but sequences capable of description by that particular species of such schemes as permits of the determination of the present from a knowledge of the past, without taking into account the future. In other words, those deterministic schemes which Natural Science employs are such that they involve the asymmetric relation of causation.

It is possible to imagine that, to an unlimited extent, natural phenomena could be representable by deterministic schemes of the general type, and yet that, only to a more limited extent, were the phenomena representable by such deterministic schemes as Natural Science, taken in the restricted sense, is capable of employing. Determinism in this general sense does not necessarily imply predictability of the future by means of a knowledge of the past or the present. To make such prediction, a certain knowledge about the whole complex, past, present and future, would be necessary. The principle of sufficient reason, or of Ground and Consequence, may be considered as more general than the principle of Causation. The Ground may be embodied in the principle which underlies the whole scheme;

whereas when the principle of Causation is applicable there must be a restriction of, or specialization in, the nature of the Ground, of such a character that asymmetric causation is realized in the scheme. A deterministic sequence of the general type does not necessarily involve the same kind of uniformity that exists in a scheme for which the causal relation holds. This is a fact which admits of a simple illustration. Let us suppose that a particle had been moving, during the indefinite long past, uniformly in a straight line; let us further suppose that, at a certain time, it suddenly proceeds to describe with uniform velocity a semi-circle, returning to the original straight line at the further extremity of the diameter of the semi-circle, and that then it proceeds to move for ever afterwards in the continuation of the original straight line in which it was moving before the semi-circular deviation took place. This is representable as a deterministic system, in which a single formula can actually be given, by means of which the position of the particle at any assigned time can be calculated. The law, or principle, embodied in this formula, makes provision for the single deviation from uniformity of the motion of the particle, which occurs during the semi-circular motion. Yet no observation of the earlier part of the motion would have given any means of predicting the excursion of the particle along the semi-circle; the asymmetric law of causation would be inapplicable in such a case.

It has sometimes been suggested that teleological conceptions may be taken into account in deterministic schemes of the general type. A process in which the teleological factor is involved, whether it be completely deterministic or not, is one in which the present cannot be regarded as independent of the future parts of the process. Whether this suggestion is of value in general Philosophy is a difficult question which lies outside the boundaries of the present discussion, and which I consequently must refrain from discussing. It is quite clear that, in that particular type of deterministic schemes which Natural Science employs, there is no possibility of any direct representation of the effects of purposiveness, regarded as introducing the element of contingency.

Although, as we have seen, the notion of efficient causation, implying the activity of an agent that compels or necessitates an effect, is one which forms no part of the stock of conceptions which Natural Science can utilize, it does not in the least follow that the conception of such activity is illusory and must be denied all validity in the more general domain of Thought. The consciousness of activity which we all have, whenever we act, cannot be shown to be illusory, unless it can be conclusively shown that it can be brought under some other category. The mere fact that the notion of efficiency is of anthropomorphic origin is quite insufficient to establish that this conception is illusory, or that it is necessarily possible to dispense with it in the account which Thought attempts to give of all the factors in experience. All that has been maintained in the earlier part of this lecture is that Natural Science, as distinct from a complete Philosophy of Nature, in the method it adopts of describing conceptually that part of our experience which we call physical, has no need of the conception of efficient causation, and can make no use of it, for its own special purposes. It may be a debatable point amongst Psychologists, whether the notion of efficient causation is necessary in Psychological Science. In any case the reasons which have led to its extrusion from Natural Science cannot, without much further examination, be held to be sufficient to justify its exclusion from the categories of Psychology.

I have already referred to the fact that, in Biology, considerable use has been made of mental factors in relation to Evolution, and that teleological conceptions have not always been dispensed with. Unless the theory of psycho-physical parallelism could be raised to something more than a surmise, or at most a postulation of which the range of applicability is unknown, it is exceedingly difficult to conceive that Natural Science, in the narrow sense, when applied to living organisms, can ever succeed in giving accounts of sequences of physical phenomena in such organisms which shall attain a degree of completeness as great as has been reached in some of the departments of Science which leave living organisms out of account. Neither can it be regarded as certain, perhaps not even as probable, that the future progress of biological Science will ever enable it, in all its departments, completely to dispense with the notions of purpose and activity, although indefinitely larger tracts of phenomena than at present may be found capable of the descriptive treatment which Natural Science applies. I have already pointed out that, in the strictest sense of the term explanation, Natural Science explains nothing; it only describes conceptually. It is just the absence of the conceptions of efficient causation and of logical necessity in the kind of account which Natural Science is able to give of portions of the world of physical phenomena that prevents us from regarding such account as explanatory in the strict sense of the term. Philosophy has always occupied itself with attempts to find explanations in the complete sense; and thus to condemn *a priori*, as necessarily illusory, the notions of efficient causation, activity, and purposiveness, amounts to a dogmatic denial of all validity, and of all possibility of success, to metaphysical thought. Some men of Science are apparently prepared to take up this dogmatic attitude towards Metaphysical Philosophy. Their position however admits of no justification by arguments of demonstrative, or even very cogent, force. It is in opposition to what is at least *prima facie* the most immediate knowledge we possess, that derived from the experience we have when we will and act.

Natural Science deals simply with what is observed to happen in the physical world. By minute examination and comparison of observations it discovers the existence of sequences of sufficient similarity with one another, to be capable of being described by rules, which are often called Laws of Nature. It synthesizes these rules into conceptual schemes, called general theories. It employs these rules and theories tentatively to predict what will happen when a portion of a sequence which appears to fall under one of these laws or theories is observed; the prediction refers to that part of the sequence which has not yet been observed. There is no logical necessity that the prediction should be successful; but in case the rule or theory has been abundantly verified by past observations, there is an expectation of success, the disappointment of which, by age-long habit, we do not usually contemplate. This expectation is often so strong that, when we are so disappointed in the verification of our predictions, we attribute the fact to a mistake in our belief that the sequence concerned is actually sufficiently similar in character to those to which the law or theory is known to be applicable. In fact we search, usually with ultimate success, for disturbing factors in the conditions, which remove the particular sequence in question from that particular type which alone the law or theory is designed to describe. As the result of such further investigations, either the law or theory is rejected as insufficiently established; or a wider theory is set up which takes account of what we at first regard as disturbing factors preventing the older theory from being applicable to the new observed sequence, so that the wider and amended theory suffices for application to the description of a wider class of sequences than did the older theory.

It is no part of my programme to make a contribution to the interminable discussions, with which Philosophers and Theologians have for centuries occupied themselves, on the subject of Free Will and Determinism; whether the world can be regarded as a deterministic system, or whether it is subject to interventions, either continually or sporadically, the character and amount of which are, even in theory, incalculable. There is however one question, relevant to these discussions, which is in the direct line of thought that concerns itself with the true character of Natural Science. I have already spoken of deterministic schemes as employed in Natural Science; but what *precise* meaning can be attached to the term deterministic scheme? This question seems to require a somewhat closer examination than I have already given to it, especially in view of the fact that it has frequently been assumed, without critical examination, both by Determinists and by Voluntaryists, that the conception of a deterministic scheme is perfectly definite.

Let us consider a finite system of ideal objects, endowed with a set of properties or qualities, and subject to a defined set of relations with one another. We regard this system as subject to change, as a certain variable, usually taken to represent time, takes up a continuous set of values. At any one time we regard the system as being in a certain state. By a known state, we mean that we have precise information as regards the positions, motions, and all other relevant facts relating to the various properties of and relations between the ideal bodies, as they are at the time at which the state is known. In what kind of language is it possible to express this knowledge of the state of the system at a particular time? The only language that we possess which would appear to be adequate for this purpose is that of Arithmetic, in which numbers are employed. We therefore assume that the state of the ideal system can be completely specified by a finite set of numbers. This amounts to the assumption that all the positions, properties, qualities, and mutual relations which constitute the state of the system can be completely specified by means of a finite set of numbers. This again involves the assumption that all these factors of the state are capable of exact measurement; that is that each one of them can be correlated with one number or with several numbers. It is necessary to consider in some detail what this correlation involves. In the perceptual domain there are two kinds of magnitude, known as extensive and intensive respectively. Only extensive magnitudes are capable of direct measurement, that is, of correlation, to a degree of approximation depending on the fineness of our instruments, with the system of numbers. The possibility of such correlation depends upon the fact that an extensive magnitude can be regarded as consisting of units, all of which are identical in respect of magnitude; this is the equivalent to the property that two extensive magnitudes may be added together, their sum forming another magnitude of the same kind. The sizes, spatial positions, velocities, and accelerations of perceptual bodies are all extensive magnitudes. On the other hand, such qualities as temperature and colour are not extensive magnitudes; they are not directly additive as are extensive magnitudes, although they may be regarded as having greater or less intensity. But nevertheless some qualities which are not extensive magnitudes may be correlated with such magnitudes, and in that case they are indirectly measurable, and the intensities can then be placed in correspondence with numbers. The correlation of different kinds of intensive magnitude with the number scale has actually been effected, as in the case of temperature, or in that of colour which has been correlated with several numbers. Unless this correlation has been accomplished for the case of any particular intensive magnitude, the quality which it represents cannot be specified numerically as regards amount. All actual measurements, direct or indirect, are more or less inexact, being dependent on the defects of our senses and of the instruments which we may employ to extend the scope of our senses. In a conceptual scheme, these practical measurements are replaced by ideally exact measurements, not subject to errors of observation; thus the correlation of magnitudes with numbers in a conceptual scheme becomes ideally exact.

We now assume that, in our finite conceptual system, all the elements which characterize the state of the system are capable of exact correlation, each with one number or with several numbers. On this assumption, the complete state of the system at a given time is specified by a certain set of numbers. In order to represent changing states of the system, we have to consider a set of variables whose values depend on the time-variable; and which for a particular value of the time-variable coincide with the numbers that represent the state of the system at that particular time. If we consider the ideal system during a particular interval of the time-variable, we may conceive the variables of the set, by the specification of values of which a state of the system is defined, to be functions of the time-variable. If the forms of all these functions are known, we have the means of calculating, to any assigned degree of approximation, the state of the system at any time falling in the time-interval for which the functions are defined; thus the state of the system at every point of that time-interval is determinate, and theoretically calculable. In the case of a particular conceptual system, such as we are here considering, the postulations and definitions which we make as regards the relations between the component parts of the system, and the laws governing the changes in that system, may suffice actually to determine these functions. When these functions are all known to us, the conceptual scheme is not only deterministic, but is actually determined, in the sense that we have such a complete knowledge of the character of the system as a whole, that we can by simple calculation determine in detail its state at any assigned time falling in the interval for which it is a deterministic system.

If we have not actually determined the forms of the functions, but have convinced ourselves by sufficiently cogent reasoning that these functions exist, as implicitly defined by the general laws which hold for the system, we still regard the system as deterministic, although only theoretically determined. For values of the time-variable outside the range for which the system is known to be deterministic, the functions employed may become meaningless, or in any case may have no application to the system. Without any essential difference in the definition, it is possible to contemplate a system which is deterministic for indefinitely great ranges of value of the time-variable, of either sign. A conceptual system which is deterministic, in accordance with this definition of the meaning of the term, is deterministic in the more general sense; there is no distinction between the relation of past or present with future states different from the relation of future with past or present states. The whole aggregate of states of the system for all the time during which it is deterministic are bound together by one common nexus. But a deterministic system of the particular type which Natural Science can employ must be such that the forms of the functions for the whole time can be inferred from a knowledge of some or all of the states of the system prior to a time falling within the time-interval for which the functions validly represent the system. In this case, future states of the system are determined by past states, on the assumption of the actual, or of the theoretical, existence of the functions. This may be stated in the form that functions determining the states of the system up to some particular time can be continued beyond that particular time, so as to continue to represent the states of the system for some finite, or indefinitely great, interval of time subsequent to the particular time considered. In this manner we can represent to ourselves a deterministic scheme in which the asymmetric causal relation (not efficient causation) exists. Having determined the forms of the functions for all the times up to a particular time, a knowledge that the system is deterministic in this sense, involves the postulation that it is possible to determine, by some kind of continuation, the forms of the functions for future values of the time-variable, and thus by employment of these functions to determine states of the system at future times. This involves the postulation that the laws which regulate the system continue to be valid in the future.

Is it possible to extend the definition of a deterministic system so as to apply to the case in which the states of the system are only representable by means of a non-finite set of functions? It is impossible exhaustively to exhibit in any manner an infinite set of functions, or even an infinite set of numbers. If such an infinite set of functions is to be regarded as known, or determinate, we must possess the means, when some finite part of the set is known, of determining any one of the others by means of some finite set of rules. Thus the states of the system are virtually determined by a finite set of functions together with a finite set of rules for the determination of any other of the functions when the former have already been determined. This amounts to the virtual reduction of the system to a finite one, and it is only on the assumption that this reduction is possible that the conception of a deterministic system whose states are specified by an infinite set of variables has any meaning.

We are undoubtedly able to define particular conceptual schemes of a deterministic type, and in particular such as include the causal relation, in the sense in which that term is employed without involving the notion of efficient causation. Does there exist in the perceptual world anything that corresponds to such an ideal deterministic scheme? The success of physical Science depends in a considerable degree upon the fact that we are able to mark off in the perceptual world, approximately isolated domains in which selected kinds of sequences or events can be described with a very considerable degree of approximation by means of ideal deterministic systems to which such sequences are correlated. It must however be observed that such correlation is confined in each case to a limited set of the relations between the physical objects of the perceptual domain, and never embraces the whole of their properties or relations. For one class of properties or relations we may require one ideal deterministic scheme, and for another class we may require another such scheme. Science cannot be said to have succeeded, even in the case of a strictly limited, approximately isolated group of percepts, in correlating all their observed relations and all their changes with a single ideal deterministic scheme. The notion of correlation of isolated systems with particular deterministic ideal systems may have its scope widened, so as to take account of the detailed differences between one perceptual system and another one of a similar character, by supposing that the functional relations in the ideal scheme involve a certain number of variable parameters, to which particular values have to be assigned when the ideal scheme is to be applied to describe what happens in one particular perceptual system belonging to the given class.

In those cases in which we are able to correlate the phenomena, of a certain class, in an approximately isolated group of percepts, with an ideal deterministic scheme which serves as a description of those phenomena, the correlation can never be taken to be valid beyond some limited period of time, of greater or less duration. That this is the case depends, at least partially, upon the fact that the perceptual system is only approximately isolated; and that outside some limited period of time the relations of the system with other systems external to it may produce an effect upon the system which is no longer negligible; the deviations of the system from the conceptual scheme which represents it may be cumulative in amount, and may produce after a sufficiently long time an effect which makes the correlation no longer even approximately valid. It thus appears that a group of phenomena can only be shown to be representable by means of an ideal deterministic scheme, to a certain degree of approximation, and only for some limited time which may however in some cases be very great, according to our ordinary measures of time. The description is never applicable to represent in one scheme all the phenomenal aspects of the perceptual objects which have to do with the phenomena described by the scheme. It must also be remembered that many phenomena have hitherto proved too intractable to admit of complete representation by any kind of conceptual scheme, least of all by a scheme which we can describe as deterministic in the precise sense that it is expressible by relations of magnitudes.

The assertion seems to be certainly true that Natural Science has not succeeded in showing that all that happens in a group of objects, however small that group may be, is deterministic, in the sense that it can be completely represented by an ideal deterministic scheme of the precise kind that I have depicted. Attempts, which I shall describe in later lectures, have been made to show that all the happenings in a group of perceptual objects, or even in the whole perceptual world, could be represented by certain kinds of conceptual schemes, especially of the kind called an atomic theory, in which everything is regarded as dependent upon the interactions of suppositive elementary particles. All such schemes have proved utterly inadequate to give a representation of more than a part of what is observed to happen. Even the theoretical possibility of the existence of a deterministic scheme applicable to the whole physical world and to all that happens in it, is open to many and grave objections; partly dependent upon the difficulty to which I have before alluded, of extending to the world as a whole, conceptions which can only be regarded as significant when applied to a finite part of the world, and partly dependent upon our inability to estimate the cumulative effects of those individual peculiarities of perceptual objects which have always to be disregarded by Natural Science. Even Laplace, when he contemplated the scheme which I referred to in my first lecture, had to postulate the existence of a kind of superhuman calculator; and the complex of electrical and sub-molecular phenomena which have been discovered since Laplace's time has indefinitely increased the difficulty of imagining what even the main outlines of such a scheme would have to be.

Moreover the possibility of regarding, even theoretically, the physical world, or a finite part of it, as capable of complete description by means of a deterministic scheme, cannot be contemplated without some consideration of the relations of the physical world to the psychical world. If indeed the theory of psycho-physical parallelism could be established completely, the psychical world could be completely ignored, and no special difficulty would arise from its existence as a system which exercises no influence whatever upon physical phenomena. If there exists complete detailed parallelism between the two domains, it would appear to follow as a consequence of the assumption that the physical world is completely describable by a rigidly deterministic scheme, that the same assertion could be made as regards the psychical world. On the other hand, on an interactionist theory of the relation between the two domains, the influence of the psychical world upon the physical must be in some way included in the deterministic scheme which *ex hypothesi* is applicable to the physical world. If this could not be done, the physical world would cease to be representable with absolute completeness by means of the deterministic scheme; the influence of the psychical factor would then be of a character which could only be described as an interference with the physical order of things. If the view, to which I have referred in the third lecture, were accepted, that the physical and the psychical worlds are in reality only parts, or aspects, of one single fundamental system, the assumption of the deterministic character of the physical world would appear to involve the assumption that the whole fundamental system, which would embrace both what we call physical and what we call psychical, is in some sense deterministic. At all events those processes or events in it which appear to us to belong to the physical world would be regarded as forming a deterministic system, and possibly other events or processes in it might be regarded as free, or not determined; but these latter would then have to be considered to have no influence on the former, otherwise we should be faced with the same difficulty as in the case of the theory of interaction between two disparate domains.

It is difficult, if not impossible, to attach a precise meaning to the conception of a deterministic psychical scheme, of a kind similar to the meaning which we attach to a deterministic physical scheme. For psychical qualities, processes, and events are not *prima facie* measurable in a manner similar to that in which physical qualities, processes, and events may be measurable. What Psychologists measure are not psychical events or processes, but their physical concomitants; and these measurements could only be taken to represent psychical measurements on the unproved assumption of some complete and exact correlation between the two domains, extending even to the transference of numerical measurements from the physical to the psychical domain. This transference may be, and probably is, sufficient, and of value, for certain purposes; but the assumption that all psychical states, processes, and events are completely and exactly measurable, by means of a transference of the measurements of physical concomitants, is at least a very large assumption, and one extremely difficult to justify.

The conclusion of the whole matter seems to be that the conception that the whole world of physical phenomena, or that a finite part of that world, is theoretically capable of being represented by a unified deterministic scheme is unproved and unprovable. All that Natural Science has established is that tracts of phenomena can be found which are sufficiently represented for certain purposes by means of deterministic schemes. A very large part of Natural Science has not yet reached the stage in which deterministic schemes, of the kind which involve relations of number, are applicable; much of it is in the stage in which only classification in abstract types can be employed, and to which precise measurement is not yet applicable. But it is a working hypothesis, employed in all the more advanced departments and stages of scientific thought, that tracts of phenomena can be discovered, to which deterministic schemes can be applied for the purposes of precise description and of prediction. The justification for this postulation is to be found in the past successes of such advanced parts of Natural Science, and we are not acquainted with barriers which will prevent ever larger tracts of phenomena from being correlated with deterministic descriptive schemes.

# V

# NUMBER AND ITS DEVELOPMENTS

ARITHMETIC, the Science of Number, taken in the general sense of the term now employed by Mathematicians, is the most advanced and also the most purely abstract department of Natural Science. Abstract Arithmetic is usually spoken of as a formal Science, like Logic; and it may therefore perhaps occasion some surprise that I should speak of it as a department of Natural Science. However, it resembles other branches of Natural Science in the fact that it was in connection with physical experience that it took its origin. Its earliest development as a Science consisted in setting up a conceptual scheme for the representation of certain aspects of the physical world. In this respect it does not differ generically from other branches of Natural Science; although the stages by which it became formal and deductive were much more rapid than in those departments which deal with what may perhaps be described as more complex, and less superficial, aspects of the world of physical phenomena. I have already suggested that every branch of Natural Science, when it reaches a sufficiently advanced stage of development, tends to become purely formal and deductive. The Science of Geometry has reached this stage of development, and the same thing may be said of Mechanics. It is unnecessary, and would be outside the scope of these lectures, to enter into any discussion of the questions whether, or how far, what are known as the truths of Arithmetic embody *a priori* knowledge, and may thus be regarded as logically necessary truths, presupposed as a precondition of experience, and whether this accounts for the kind of apodictic certitude which we attach to the knowledge of simple arithmetical relations. Whatever views be held as regards the much debated subject of the philosophy of Arithmetic, the fact remains that the knowledge of the individual has its origin, as had that of the race, in physical experience; although it may of course be maintained that the function of physical experience was simply to awaken, and make explicit, conceptions already present in the mind in a latent form. In any case the part played by physical experience in the development of arithmetical concepts in an explicit form affords, I think, sufficient justification for reckoning Arithmetic amongst the Natural Sciences. I have spoken of Arithmetic, including the higher Mathematical Analysis which is a development of more elementary Arithmetic, as the most advanced branch of Natural Science; this must be taken to mean that it has more completely than in the case of any other department, with the possible exception of Geometry, which I shall consider in the next lecture, reached the stage in which it consists of the detailed development of the implications of a purely conceptual scheme; no further recourse being required to observation or experiment in order to test its range of applicability to describe certain aspects of the physical world.

Whilst Arithmetic, in the wide sense of the term, is the most advanced and the most abstract branch of Science, its rudimentary parts are far more popularly understood, and used in applications, than is the case with any other branch of Natural Science. The grocer when he weighs out and sells his goods makes use of conceptions which were developed only by a long process of evolution; when he enters his receipts in his books, the notation he uses is a warrant of the great importance of notation in a formal Science, and embodies one of the triumphs of our race, as a mode of economizing thought. The concepts of Arithmetic in their more elementary form, or in the higher developments to which they attain in Mathematical Analysis, pervade all departments of Natural Science and all the Mechanical arts. The Philosopher, in his reflections on spatial and temporal relations, on number and quantity, on matter and motion, is in a region of thought in which the boundary between his own domain and that of the Mathematical Analyst is difficult to delimit with precision. The Epistemologist has always been accustomed to consider Mathematical knowledge as a kind of touchstone on which to test his theories of the nature of knowledge. The dominant views in some departments of philosophical thinking have been notably influenced by the results of recent Mathematical research, and may not improbably be in future further modified from the same quarter. The universality of Arithmetic in the Natural Sciences consists of the fact that numbers, or variables which are interpreted numerically, enter into every conceptual scientific scheme that has reached a stage characterized by extreme precision of statement. For Arithmetic, in the developed form of Mathematical Analysis, provides the very language in which the precise descriptions contained in such schemes are clothed.

The concepts of unity, and of number, or degree of plurality, were in their formation occasioned by physical perceptions. The precise mode in which these concepts were formed is a matter for psychologists to discuss and determine. As formal categories they lie at the base of the Science of Arithmetic, and consequently of Mathematical Analysis, which is now regarded as essentially no more than abstract Arithmetic carried to a higher stage of development by the help of certain postulations concerned with the domain of the infinite, or transfinite. The notion of unity, the product of mental activity in relation with an environment, is the form under which an object is subsumed when it is the object of attention.

Thus unification is the result of an act of attention which involves a differentiation of the presentational continuum. A physical object, brought under this category of unity, may for all other purposes be recognized as possessing any degree of complexity. It is sufficient, in order that the object may be subsumed under the form of unity, that is, regarded as a single object, that it be so far distinct, within the presentational continuum, as to be recognized as discrete and identifiable. What external marks are necessary, that an object may be so recognized as discrete, is a matter for the judgment of the mind which performs the act of unification. There is a large degree of arbitrariness, limited only by the powers of perception of the individual mind, in this act of unification, or of apprehending an object under the form of unity; no special, or uniquely definable, physical characteristics of the object are essential for this purpose, but only some sufficient degree of differentiation of the object from its environment.

The notion of plurality is involved when attention is paid either successively or simultaneously to objects, each of which is subsumed under the form of unity. This notion, at first indefinite, takes the form of definite plurality when a *collection* or *aggregate* of objects is attended to. Such a group or collection is then regarded both as a single whole to which unity is attributed, and also as a definite plurality, consisting of a set of objects each of which is regarded for the particular purpose as one. The single objects which compose the collection need not possess any parity as regards size, weight, or any other special quality, but may be of the most diverse characters, although in practice they usually have some similarity of nature which forms the ground of their being treated as a collection. In any case a certain logical parity is ascribed to them in virtue of the fact that each one of them is subsumed under the form of unity. The fact has recently been pointed out and illustrated by Prof. James Ward[1] that the earliest conception of number, as a degree of plurality, probably arose as the result of immediate intuition of the differing qualitative characters of very small groups of objects. Thus a pair of objects can be intuitively recognized as qualitatively different from a group of three objects, without recourse to the process of counting. This immediate intuition of the number of objects in a group is facilitated when the objects are arranged in some recognizable pattern. It is probable that not only human beings, but also higher animals, possess the power of discerning intuitively this qualitative characteristic of a very small group of objects, and of recognizing that something is changed in a group originally, say of three objects, when one of them has disappeared. It is clear however, that this avenue to the concept of number is of very limited scope. All further development of the concept was made in connection with the process of counting, or tallying. For this process, the two notions of *order* and *correspondence* are requisite.

In virtue of the notion of order, relative rank is assigned to each object in a collection, so that the collection becomes an ordered aggregate. In actual counting the order is usually assigned to the objects during the process itself, as an order in time, and this may be done in an arbitrary manner. The order of the elements in an aggregate may however be assigned in a manner dependent on their sizes, weights, or other qualities, or in a manner dependent on their relative spatial positions. Order may however be regarded as an abstract conception, independent of any particular mode of ordering; for an aggregate to be an ordered one, it is necessary and sufficient that each object of the aggregate be recognized to possess a certain rank, in virtue of which it is definite, as regards any two of the objects which may be selected, which of the two has the lower, and

[1] *Mind*, Vol. XXIX, p. 137.

which the higher, rank. The notion of correspondence underlies the process of tallying, or that of counting on the fingers. The objects of one aggregate are regarded as standing in a logical relation with those of another aggregate, of such a character that a definite element of one aggregate is regarded as corresponding to a definite element of the other aggregate. The correspondence between two aggregates is complete, or (1, 1), when, to each object of either of the aggregates, there corresponds one object, and one only, of the other aggregate[1]. The number, or degree of plurality, of an aggregate can then be defined as the concept of the quality of plurality which the aggregate has in common with all aggregates with which it can be placed in complete correspondence. Thus a number is the concept of the quality which the members of a family of similar aggregates have in common. The number 1, although not characteristic of a plurality, is still in formal Arithmetic regarded as one of the integral numbers. If, in counting an aggregate, the process is stopped before the aggregate is completely counted, we may regard the part counted as a section of the aggregate; such a section can then be taken to be an aggregate having a particular number. The sections of an aggregate, together with the aggregate itself, have an order assigned to them, the same as that of the objects of the aggregate itself. Thus the numbers of the sections are themselves ordered, and we thus conceive the numbers 1, 2, 3, etc., to be arranged in a definite order, usually called their natural order.

A number which specifies the degree of plurality of an aggregate is called a cardinal number, but a number which specifies the rank of a particular object in an aggregate is called an ordinal number. Cardinal and ordinal numbers are distinguished from one another in that their descriptive functions are different. It has been maintained by some writers on the foundations of

[1] The aggregates are then said to be similar to one another.

Arithmetic that the notion of an ordinal number is logically prior to that of a cardinal number. This does not, however, seem to be necessarily the case; either concept can be employed in a systematic treatment of the subject as fundamental, the other being then regarded as derivative.

Before Arithmetic can be considered to be a developed Science the further step must be made of the introduction of a scheme of relations between the numbers dependent on the operations of addition and multiplication, with the inverse operations of subtraction and division. There has been a very prolonged discussion amongst philosophers as to whether the judgment expressed by such a proposition as that $7 + 5 = 12$ is an analytic judgment, in the sense that the truth of the proposition can be deduced from an analysis of the concepts of the three numbers, or whether that judgment is synthetic, in the sense that some further extraneous knowledge is required to warrant the judgment. There can however be no doubt that, historically, and in the individual, the explicit knowledge of such relations, in simple cases, was empirical; being derived from actual counting of the combined aggregate when two aggregates are amalgamated into one. The general conceptual scheme of relations involving operations arose as a generalization of knowledge obtained from such physical experience. It seems certain that the fundamental notions that I have specified must have been possessed by primitive man, in an implicit form, long before the notion of abstract number reached an explicit and developed form. But the earliest records we possess of ancient peoples, those of the Egyptians, the Babylonians, and the Chinese, show that they possessed arithmetical knowledge that had already attained a very considerable degree of development.

The origin of fractional numbers is doubtless to be ascribed to the necessities arising in connection with measurement. The division of an object into equal parts, and the representation of one or more of such equal parts, was the empirical origin of the notion of a fraction. But in our present theories of Arithmetic, the concept of number, both integral and fractional, is taken to be independent of any conceptions relating to measurement. That operation is now regarded as requiring the application of Arithmetic, but as not connected with the foundations of the subject, the necessary empirical basis of which rests exclusively upon the operation of counting. A fraction, as for example $\frac{3}{5}$, may be regarded as representing the operation of counting three objects each of which belongs to an aggregate of five objects, no assumption of equality of the objects in respect of size or other quality being requisite. On this basis the theory of the Arithmetical operations involving fractional numbers can be made to rest. The more purely formal theory of fractional numbers, as usually expounded, regards each one as defined by a pair of integers forming a single object, a couple; formal laws are then postulated as to the relations between such couples, forming the basis of the scheme of operations involving them.

It is very important to remark that the operation of counting conceptual objects in which the integral numbers are employed, and the extension of that process, in which fractional numbers are employed, are both free from that element of approximativeness which appertains to every operation of actual measurement. The number of an aggregate of objects is an exact description of that aggregate in a certain respect, but number as applied to the measurement of extensive magnitude represents the measure only subject to the inexactitude inherent in our sensuous perception, even when refined instruments are employed. This unique peculiarity of the application of the conceptual scheme of Arithmetic to the perceptual domain depends upon the fact that the operation of unification is independent of any special physical properties of the object to which it is applied, and involves only the mental operation of general differentiation of the object from the environment. Measurement on the other hand is directly concerned with one at least of such physical properties. The application of the conceptual scheme of Arithmetic for descriptive purposes in the physical domain, so long as we confine that application to the original purpose for which it exists, has a certain absolute exactitude which does not appertain to any other conceptual scheme in its application to the concrete, and which no longer appertains to Arithmetic when it is applied to represent measures of physical magnitudes.

In all schemes of a symbolic character, such as written language, symbolic logic, or Arithmetic, the facility with which the scheme works depends very largely upon the choice of a suitable and simple notation. In the case of Arithmetic, the notation which we employ, and which is of Indian origin, represents perhaps the greatest labour-saving invention that has ever been made. The notation for numbers is systematic, in the sense that all integral numbers can be denoted, on a uniform plan, by the employment, in the decimal scale which we use, of ten distinct symbols. This number of symbols could be reduced to two, if we employed the dyad scale. The principle that the place which a digit occupies, in the set of digits used to represent a number, indicates the mode in which it is interpreted to represent a multiple of a power of ten, is the crucial point in the principle upon which the notation is founded. It is a remarkable fact that neither the Greeks nor the Romans were in possession of a systematic notation for numbers. An attempt to carry out even a simple addition or multiplication, in which the numbers are represented by Roman numerals, is the simplest path to a conviction of the vast importance of the great Indian invention which renders arithmetical operations practicable in accordance with uniform rules. There is an important invention, included in our system of notation, which appears to have been introduced later than other parts of the system. This is the use of a symbol denoting zero, which is employed to indicate the absence of a particular power of ten in the representation of a number as the sum of multiples of powers of ten, together with a digit which is one of the first nine numbers, or is itself zero.

The most ancient account of the Arithmetic of the Egyptians is contained in the Papyrus of Ahmes, and is entitled *Direction for attaining a knowledge of all secret things*. In this arithmetic both integral and fractional numbers appear, but the notation is not a systematic one. Only fractions with unity as numerator are employed, such a fraction being denoted by the integer which represents the denominator, but with a dot placed over it. The only exception is the fraction $\frac{2}{3}$, for which a special sign was employed. Other fractions are expressed as the sums of fractions all of which have unity for their numerators, and a table is given in the Papyrus for expressing fractions in this manner. Ahmes also dealt with some problems involving arithmetical and geometrical progressions. A Babylonian table has been discovered in which the first sixty square numbers are given, and also some cube numbers. In this table a sexagesimal system of notation is employed, in which the place in order represents a power of 60. This was in essence a systematic notation, and so far an anticipation of the invention of a later age, but the Babylonians do not seem to have employed a zero to represent vacant units. Although the Greeks possessed no systematic arithmetical notation and no sign for zero, they managed to perform arithmetical operations of some difficulty. For example, the greatest Greek Mathematician, Archimedes, in his discussion of the quadrature of the circle, inscribed a regular polygon of seventy-two sides in a circle, and obtained a good approximation to the ratio of a side to the diameter of the circle. To do this he had to extract the square roots of several large numbers, to a sufficient degree of accuracy for his purpose. The nature of the method he employed in performing these operations has been a subject of considerable discussion in our time. It has frequently been said that the ancient Greeks were great Geometers, but poor calculators. It is certain that their command of Arithmetic was seriously hampered by the unsystematic character of the notation they employed. The Greek Mathematicians, even the arithmetician Diophantus, did not take the important step of introducing negative numbers, in order to remove the impossibility of carrying out the operation of subtraction of a greater number from a lesser. This step was, however, taken by the Indians; the Indian Astronomer Aryabhata employed distinct names for positive and negative numbers, which denote respectively possession and debt. Even the representation of positive and negative numbers by segments of a straight line in opposite directions was known to the Indians.

As a result of the introduction of difference of sign, we possess what is known as the ordered aggregate of rational numbers, a conceptual arithmetic scheme in which the operation of subtraction is in all cases a possible one. We have here an example of the extension of the domain of number, in order to extend the scope of a formal operation, and thereby to increase the utility of the conceptual scheme in regard to its applications to the description of magnitudes. As we shall see, this extension is but one example of several extensions of the domain of number, made in accordance with the same general principle. The inadequacy of rational numbers for the complete representation of all ideal magnitudes was discovered by the Greeks. A rigorous theory of the ratios of incommensurable magnitudes, the discovery of the existence of which is attributed to

Pythagoras, was given by Euclid in the tenth book of his treatise on Geometry. A valid proof was given by the Pythagoreans that the ratio of the length of the diagonal of a square to that of a side cannot be exactly represented by any rational number. On the formal side, the restrictions, which hold within the domain of rational numbers, upon the possibility of operations such as the extraction of roots of numbers, pointed in the direction of an extension of the conception of number, of such a character that, in an extended domain of number, such operations would no longer be impossible, when both rational and irrational numbers are recognized as falling within the extended domain.

The modern theory of the aggregate of real numbers arose out of the exigency of the requirement of a complete theory of ideal magnitudes, and out of the limitations which exist in the domain of rational numbers as regards arithmetical operations within that domain. The term "real number" is a somewhat unfortunate one, but it arose historically from an attempt to distinguish between these numbers and the so-called imaginary numbers of which I shall presently speak. The theory of the aggregate of real numbers, as developed in a complete form by Cantor, Dedekind, and Weierstrass, involves a postulation related to the conception of the infinite. The integral numbers 1, 2, 3, ... form an ordered sequence which has a first term, the number 1, but no last term. The very principle of the sequence is that any particular term is succeeded by another term, and thus the existence of a last term would be in contradiction with this principle. The sequence is thus an example of the unending, or indefinitely great. The postulation is made that nevertheless the sequence can be regarded as a single object for thought, having definite properties. Considered as an aggregate of distinct objects, that of the integral numbers, this single object, the aggregate, is infinite, or transfinite. In the form given to this postulation by Cantor, the sequence is regarded as defining a new ordinal number $\omega$, which is not identical with any of the finite ordinal numbers 1, 2, 3, ... but is of higher rank than any of them, and is not *immediately* preceded by any one of them. This new number $\omega$ is called the first transfinite ordinal number, and is made by Cantor the starting point of a new series of transfinite ordinal numbers. The more general form of the postulate to which I have referred is that, if any sequence of objects $P_1, P_2, P_3, \ldots$ is so defined, by means of a finite set of rules, that a definite object of the sequence corresponds to each integer of the sequence 1, 2, 3, ..., then the unending sequence $P_1, P_2, P_3, \ldots$ may be regarded as a single definite object with definite properties. The justification for making this postulation is two-fold. In the first place it must not lead to contradiction, when the logical consequences of the scheme based upon it are scrutinized, and in the second place the scheme which involves this postulation must be of utility as a conceptual scheme for the application of number in a general theory of magnitude.

For the purpose of defining a real number which is not rational, that particular kind of unending sequence is employed which is known as a convergent sequence of rational numbers. If we carry out, for example, the steps of the process of extracting the square root of the number 2, we obtain successive sections of a convergent sequence. The rational numbers which form the elements of the sequence cannot of course be exhibited exhaustively, but we possess a set of rules which suffice for the calculation of any one of the numbers by means of a definite procedure involving in each case a finite number of applications of the rules. In this sense the sequence is regarded, in accordance with the postulate, as a single definite object, defined by a norm, or set of rules, although it cannot be completely exhibited as in the case of a finite set of objects. We observe that, when we have carried out a large number of steps of the process, we have determined a rational number represented by a decimal with a large number of digits, and this differs from any of the later rational numbers which may be determined by a decimal in which a corresponding large number of places are occupied by zero. This is a particular example of a convergent sequence, the rational numbers successively obtained representing a set of continually closer approximations to the object which we regard as defined by the sequence, namely the irrational number $\sqrt{2}$. In the general case, every convergent sequence of rational numbers is taken to define a real number, and, as in the case of $\sqrt{2}$, this real number is not necessarily a rational number. The establishment of the complete theory of real numbers involves a detailed investigation which fixes the precise theory of identity and of relative order of the real numbers so defined. It also establishes the essential fact that the arithmetical operations which express relations between rational numbers can all be extended, with enlarged scope, to the case of real numbers.

The result of this investigation is that we have before us an aggregate, that of the real numbers, positive, zero, and negative, in which any two of the real numbers have a definite relation of rank, specifying their relative order in the aggregate. This aggregate has two properties of capital importance. In the first place every convergent sequence of which the elements are real numbers itself defines a real number, known as the limit of the sequence. This is expressed by the statement that the domain of real numbers is *closed*. The domain of rational numbers is not closed, in this sense of the term, since a convergent sequence of rational numbers does not necessarily define, or have as its limit, a rational number. The second property is that every real number can be exhibited, in an indefinite number of ways, as a sequence of real numbers. This property may be described as that of connexity, and also appertains to the aggregate, or domain, of the rational numbers. Both these properties, in one of which the domain of rational numbers is lacking, are essential to the fitness of the aggregate of real numbers for the purpose of the complete conceptual representation of linear magnitudes.

The question has frequently been asked, how can Number, which is essentially discrete, be employed for the complete representation of magnitude, which is essentially continuous? In considering the answer which should be given to this question it is necessary to scrutinize the meaning which can be assigned to the expression "continuous." As in all other such cases we must consider both the perceptual and the conceptual meanings which may be attached to the term continuum. On the perceptual side, in which we are considering actual measurement of a magnitude, the notion of continuity, or absence of gaps, implies that between any two magnitudes, of the kind considered, there exist other magnitudes, and that the process of continual contemplation of magnitudes filling up the gaps between any pair can be carried out to an extent limited only by our powers of perception, even when instruments of the utmost precision are employed for the purpose. The notion of continuity, regarded in this way, gives rise to what may be described as the *sensible* continuum of magnitudes, and containing, as it does, an inherent element of indefiniteness, dependent on the approximative character of our sensuous perceptions, it can only be raised into the position of a precise conceptual scheme by a process of abstraction and idealization. For the representation of all actual measurements the rational numbers are sufficient, and can be applied to represent any actual magnitude to any required degree of approximation, the degree of approximation which it is worth while to aim at being dependent on an estimate of the inevitable errors which the mode of observation entails. The exigencies of our method of representing aspects of the perceptual world by ideally exact conceptual schemes necessitate however the development of a theory of measurement in which the characteristic properties of a *conceptual* continuum are assigned by definitions and postulations. That the employment of rational numbers only is insufficient for the purposes of such a conceptual scheme was known, as we have before remarked, as early as the time of the Greek Mathematicians. The modern theory of the aggregate of real numbers, or of the arithmetic continuum, as it is now called, has been devised as a scheme sufficient for the purpose of denoting the magnitudes in the conceptual continuum, which is taken to be the idealization of the notion of the sensible continuum of perception. The aggregate of rational numbers, since it is not closed, is not a conceptual continuum in the sense in which the term is applied to the aggregate of real numbers. It has been charged against Mathematicians that, in setting up such a scheme as the arithmetic continuum, they have introduced an unnecessary complication, in view of the fact that rational numbers suffice for the representation, to any required degree of approximation, of all sensibly continuous actual magnitudes; that in fact an instrument has been created of an unnecessary degree of fineness for the purposes to which it is to be applied. The answer to the charge is that Mathematical Analysis, which is based upon the arithmetic continuum, and essentially consists of operations involving numbers, would become unworkable as a conceptual scheme, or at least much more cumbrous, if the conception of irrational numbers were excluded from it. The results of operations involving rational numbers constantly lead to irrational numbers, without which the operations would be impossible if their effects are to be regarded as definite. But in order to appreciate the full weight of this answer it is necessary to consider the great generalization of Arithmetic which is made when variables are introduced which denote unspecified numbers. The passage is then made from the primitive form of Arithmetic to Algebra, in which the formal operations of Arithmetic are represented as relations between sets of unspecified numbers represented by non-numerical symbols. The result of an algebraic operation, expressed by general formulae, such for instance as the simple case of the solution of a quadratic equation, would not always be interpretable when special numerical values are assigned to the symbols, if the only admissible numbers were rational ones.

Without the employment of the conception of irrational numbers the function of Mathematical Analysis would be degraded to that of determining only approximate results of the operations it employs, and in consequence its technique would have indefinitely greater complication, of such a character that, at least in its more abstruse operations, it would break down, or lead to results which contained a margin of error difficult to estimate.

I have described in general terms the gradual extension of the concept of number, commencing with integral numbers, proceeding to fractional numbers, then to negative numbers, and thus attaining to the aggregate of rational numbers, and lastly, by the employment of an ontological postulate, the extension to the arithmetic continuum, which is capable of describing adequately relations of magnitude in the conceptual domain of continuous magnitude. There is however one further extension of the conception of number, essential for the purposes of general Mathematical Analysis, which has very frequently been a stumbling-block for non-Mathematicians, who have founded upon it the charge that it consists of a species of jugglery with symbols that from their very nature are meaningless. I allude to the introduction of complex, or so-called imaginary, numbers. The operation of determining a square root of a negative number is not a possible one, within the domain of real number, since the square of every real number is a positive number. On a principle similar to that by which the domain of numbers was extended by the introduction of positive and negative signs, in order that the operation of subtraction might become always a possible one in the enlarged domain, the domain of real number is further extended so as to become one in which every number, whether positive or negative, has a square root. The new domain is then that of what is known as complex number. The existence, subject to the law of contradiction, of a new number whose square is $-1$ is postulated; this is usually denoted by $\iota$. It can then be shown that, in a new domain, in which each number is the sum of a real number and of $\iota$ multiplied by a real number, where either of the real numbers may be any number belonging to the arithmetic continuum, it is possible to postulate a consistent scheme of relations of operations. This scheme is of such a character that the laws of operations are in formal agreement with the laws which hold for operations which involve only real numbers. This extension of the domain of real numbers to that of complex numbers has the advantage that it involves a notably enlargement of the scope of algebraical processes. For example, every quadratic equation has solutions, which would not be the case if the existence of real numbers only be admitted, and thus a much greater degree of generality is introduced into Arithmetic and Mathematical Analysis generally by the extension. Moreover, the boldness of the Mathematicians who ventured upon this extension has its reward in the fact that the new set of numbers, the complex numbers, are applicable to the specification of the positions of points in a plane, as is illustrated by the well-known Argand diagram. The theory of functions of a complex variable has become a most important branch of Analysis, indispensable for many purposes, among which are applications to abstract, or Mathematical, Physics. The popular prejudice against the use of the number $\iota$, or $\sqrt{-1}$, and of the whole system of complex numbers is based on the ground that $\sqrt{-1}$ represents an impossible operation. When the matter is regarded aright, there is no justification for this prejudice. What is a possible operation, or what is an impossible one, does not depend upon any absolute criterion of possibility, but upon the characteristics of the domain in which operations are carried out; the possibility or impossibility is in fact relative to a particular domain. So long as the domain was that of signless number, the operation of subtraction was not always a possible one, for example $3 - 5$ represented an impossible operation, and could only be taken to represent an "imaginary" number, in relation to the domain. Similarly, the operation of extracting a square root of a negative number is only impossible within the domain of real number; it becomes a possible operation within the enlarged domain of complex number. The so-called imaginary numbers have just the same conceptual reality as the so-called real numbers, provided the domain of such numbers has been defined in accordance with a precise scheme of definitions and postulations subject to the law of contradiction. The validity of the scheme having been justified, its utility is the justification for its actual construction.

In the time of the Greek Mathematicians, and also during the progressive period of Mathematical thought which commenced in the sixteenth century and has lasted to the present time, the conceptions relating to the *infinite* and *infinitesimal*, together with the related conception of a *limit*, have almost continuously occupied the attention of Mathematicians and Philosophers. Much of the confusion of ideas on these matters which lasted from the time of the origin of the Differential Calculus, created by Newton and Leibniz, was due to a failure to distinguish with sufficient clearness of outline between the conceptual and the perceptual sides of the measurement of magnitudes, and to the uncritical acceptance of notions derived from sensuous intuition as sufficient for the basis of a rigorous conceptual scheme. On the subject of the infinite and the infinitesimal the views expressed have frequently presented a diversity akin to that which has been exhibited in relation to general Philosophy. There have been all shades of believers, sceptics, pragmatists, and finitists. Although it is a popular belief that Mathematics is of such a character as to leave no room for differences of opinion, it is a fact that, even at the present time, there exist differences of opinion amongst Mathematicians about the foundations of the Science, and more especially about matters in which the notion of the infinite is involved. In the earlier presentations of the Differential and Integral Calculus, frequently called the Infinitesimal Calculus, theories concerning infinitesimal magnitudes played a large part. The notion of an infinitesimal was frequently complicated with ideas about motion, derived uncritically from spatial and temporal intuition. Resulting from the clarification of the conceptions of the foundations of Arithmetic and Mathematical Analysis, resting upon a purely arithmetical basis, which has taken place during the last half century, the conception of infinitesimal numbers has been excluded from ordinary Arithmetical Analysis, as an unnecessary conception in the scheme. Every number regarded, for the purposes of the Calculus, as existent, belongs to the arithmetic continuum, and is therefore finite, if it be not the single number zero. The term *infinitesimal* is no longer used to denote a number, or a magnitude; when it is used at all it is employed to describe a process of change, and even for that purpose it is better avoided, so as not to give rise to misunderstandings.

The fundamental conception of a limit, to which I have already referred in connection with the definition of an irrational number, emerged first amongst the Greeks, in a geometrical form, embodied in the method of exhaustions. It is remarkable that their conception of the nature of a limit attained a standard of rigour, greater than that which obtained amongst modern Mathematicians before our own time. This is exemplified in the proof given by Euclid, in the eleventh book of his Elements, that the circumference of a circle varies as its radius. Archimedes applied the method of limits in a rigorous manner to various problems of what we now call integration, such as the determination of areas and volumes. The advantages which the methods of Leibniz and of Newton had over that of Archimedes were of a practical kind, as they had a form which made them readily applicable to calculations, whereas the geometrical form in which the Greek method was clothed, together with the absence of a convenient arithmetical scheme, made the application of the Greek method to particular problems decidedly cumbersome. But in the matter of soundness of theory, the method of Archimedes was superior to those of Leibniz and Newton. The construction of the arithmetic continuum was an absolutely necessary requisite for a rigorous theory of limits, and thus for the foundation, on a logical basis, of the Differential and Integral Calculus. In default of such a construction, when the notion of magnitude, as directly given by sensuous intuition, was taken as part of the basis of the doctrine of limits, the frequent endeavours that were made to prove that every convergent sequence of numbers necessarily converges to a limit were doomed to inevitable failure.

That the contemplation of the infinite, in some form, is indispensable to Mathematicians arises from the fact that even comparatively simple problems, such as those of the determination of the length of a curvilinear arc, or of the area enclosed by a curve, can in no case be solved by employing a finite number of the operations of arithmetic, except when the curve consists of segments of straight lines. An approximation to the magnitude of an area can be found by division of the area into a sufficiently large number of rectangles, leaving a small undetermined part of the area out of account. By increasing the number of rectangles indefinitely in a suitable manner, the measure of the area required is exhibited as the limit of the sequence of numbers determined by the approximations. Thus the magnitude to be determined is only obtainable as the limit of a sequence involving an indefinitely great set of arithmetical operations. The Integral Calculus is concerned with methods of calculating the limits defined in such a manner. It is clear moreover that the very conception of an area or length, as having a definite magnitude, is dependent upon the concept of a limit, and that it is only defined by, and is dependent upon, the existence of a definite limit to a sequence. The notion of the gradient, or differential coefficient of one variable with respect to another, is one which is essentially dependent upon the conception of a limit, and the gradient only has a precise meaning when a rigorous theory of limits has been previously established. The older idea that a gradient may be regarded as the limiting ratio of vanishing quantities, and which was justly derided by Bishop Berkeley, affords a striking example of the hazy conceptions of fundamental matters with which Mathematicians for a long period of time contented themselves. The abstract conceptual schemes, of an advanced character, employed for the purpose of representing measurable physical processes, are very frequently expressed in the form of relations between the gradients of various magnitudes with respect to other magnitudes, or to the time-variable. These relations, known as differential equations, are of fundamental importance in connection with such abstract schemes of representation; and the mathematical theory concerned with the determination of the values of variables, under certain conditions, in terms of the time-variable, is consequently, for the purposes of theoretical physics, of great importance. In the theory of differential equations, we have one of the most important examples of the fact that the abstract theories of Pure Mathematics provide the means for utilizing conceptual scientific schemes for the purpose of representing in their quantitative aspect a great variety of physical phenomena.

The fact that Mathematical methods are, in a very large class of cases, unable to deal with objects, or with processes, except by breaking them up into parts, and increasing indefinitely the number of those parts, is a significant example of a limitation imposed upon us by what appears to be a definite characteristic of our modes of apprehension. We appear to be unable to grasp some of the relations of a whole, without breaking it up, as it were atomistically, and then proceeding to reconstruct the whole by a synthetic process which is confined to a continual approach to the whole along the path of an endless regress, which by its very nature, is such that the whole is never actually reached within the process, although a scrutiny of the laws of the regress may enable us to obtain a knowledge of the relations of the whole.

The mathematical theory of the numerically infinite, and especially the developments of Georg Cantor and his school in the theory of infinite aggregates of objects, have aroused considerable interest in the ranks of Philosophers. Some of them, such as, for example, Josiah Royce, have suggested that the results of the theory of transfinite numbers may throw light upon questions of General Philosophy, such as the fundamental problem of the One and the Many. I cannot enter here into a discussion of the extension of the theory of the transfinite into the more general region of thought, but I would suggest that extreme caution should be exercised in attempting to extend results of such a theory as that of transfinite aggregates to a domain wider than its original one. The theory has been created for a special purpose, that of dealing with certain aspects of the numerically infinite, and its constructions and results are all dependent upon a set of postulations and definitions which it has been the aim of investigators to make as precise in character as possible. The most careful scrutiny of the meaning to be attached to the terms employed in any extensions of the theory, such as those to which I have alluded, is of the last importance; otherwise there is a serious danger of falling into grave errors in setting up theories in which vague analogies, involving the surreptitious use of such terms as the infinite, in a sense different from that in which they are employed in the Mathematical theory, take the place which should be occupied by a critically explored foundation.

# VI

## TIME AND SPACE

THE physical experience of each individual contains as elements what are called his intuitions of time and of space. These elements are only separable by abstraction out of the unity of his actual experience; and when the separation is made, the elements represent his own private temporal and spatial intuitions. This fact is sometimes expressed by saying that each person has his own time and his own space, as forms of his physical perception. A large amount of attention has been devoted by Psychologists to the precise character of the individual's intuitions of time and space, and to the mode in which his powers of judgment involved in such intuitions are developed. Every actual presentation in what the percipient regards as the present involves both duration in time and spatial extension. Whatever part of his time does not belong to this duration of an act of attention appears to him as the past, in which he had earlier presentations, or to the future in which he anticipates further presentations. There is thus in his intuition of time a qualitative distinction between the past and the future. The present always has some actual duration, and can only be regarded as an instant of time without duration, by an abstraction representing the conceptual limit of a duration indefinitely diminished. Such a conceptual instant of time is accordingly not an element in his actual perceptions, and thus his time cannot, except by abstraction, be regarded as composed of instants. There has been some discussion of the question whether perceptual time must be regarded as continuous or discrete. Public time, that which we employ in social life, and in which from the point of view of Science physical processes are regarded as taking place, is a concept constructed by means of intersubjective intercourse. An event perceived by different percipients is, under certain conditions, regarded as one and the same event capable of being perceived by any normal percipient in suitable circumstances; it is then regarded as occurring in time, that is in public time, which may be correlated with the private time of any particular percipient. I leave here out of account the questions which have recently aroused great interest as to the simultaneity of an event for different observers. For all ordinary purposes, not only of every-day life, but also of Science, the older conceptions on this matter are sufficient. By consideration of sequences of such events, or of processes, regarded as independent of a particular percipient, a general correlation is set up between this public time and the private time of a percipient; the notions of past, present, and future being transferred to public time. The measurement of time is founded upon that of some standard measurable process, sometimes that of the rotation of the earth. Thus equal intervals of time may be defined as intervals in which the earth turns through equal angles; and the time of a complete rotation of the earth may be taken as a unit of time, the sidereal day. It is a result of many forms of empirical observation that there exist a number of different physical processes which, when employed in the manner I have indicated, give rise to one and the same measure of time, at least when a suitable averaging is resorted to. Thus for example, a nearly constant number of oscillations of a properly constructed pendulum take place in equal intervals of time measured by the earth as time-keeper. The ordinary units of time, the year, the mean solar day, the minute, and the second, are obtained by a somewhat more complicated construction dependent both on the average rotation of the earth round the sun, and on the rotation of the earth relative to the sun. We possess, in spring watches, pendulum clocks, sand-glasses, etc., the means of measuring time which, with greater or less accuracy, correspond with the standard time founded on the mean solar day.

For the purposes of an abstract conceptual scheme time, as empirically measured, must be idealized. For this purpose, Newton suggested the conception of absolute time, as that which flows uniformly. Such a definition however can only be taken as generally descriptive, and not as really defining a precise conception, because the expression "flows uniformly" implies already an underlying conception of time. Thus his definition is open to the fatal objection of circularity. In abstract Dynamics, and in fact in any purely abstract conceptual scheme, the rôle of time is played by an independent variable which, on account of the function it has in a scheme to be applied to the description of actual physical processes, may be spoken of as the time-variable. The field of this time-variable is taken to be the arithmetic number-continuum; the aggregate of all real numbers. Thus a particular time, in the abstract scheme of Dynamics, is simply a number; a particular value of the time-variable; and an interval of time is the difference of the numbers which represent the two ends of the interval. In this completely abstract conception of time the generic distinction between future and past which exists in perceptual time has disappeared; there exists only an ordered aggregate of a particular type, of which the elements are conceptual instants, the relation of elements of lower rank in the order to those of higher rank being all that corresponds to the original distinction of past from future. When the conceptual scheme is applied to describe actual physical processes, as for example, the motions of bodies, the time-variable is correlated with the public time, as measured by the standard physical process, in such a manner that equal intervals of the arithmetic continuum which is the field of the time-variable, correspond with equal measurable amounts of the standard process. This procedure adopted by Science, of representing abstract time by the arithmetic continuum, has been spoken of by Bergson as the spatialization of time, no doubt on account of the fact that the linear spatial continuum can also be correlated with the arithmetic continuum. It should however be observed that, in doing this, a process of abstraction is applied to the space of perception of a character similar to that which has been employed in the case of time. Thus a more accurate account of the procedure of Science would be to say that both time and (linear) space are represented conceptually by one and the same abstract scheme, that of the arithmetic continuum. Both have in fact been represented as ordered aggregates of the same type. It must be remembered that the arithmetic continuum is a construction not arrived at by idealizing the notion of linear space[1], but that it is logically independent of the notion of spatial magnitude. One of the chief applications of the arithmetic continuum is to provide an ideally exact theory of the measurement both of spatial and of temporal magnitudes.

As in the case of the temporal perception of an individual, his spatial perception is only separable by abstraction from his whole physical experience. His spatial perceptions are dependent on a variety of sensations, partly visual, and partly tactile and motor. The spatial perceptions which arise from these two sources considered separately differ notably from one another in character; the representative space of the individual is a synthesis of the two, in which habit, that is past spatial experience, plays a large part. The geometrical space which we employ when we reason about the spatial properties of ideal bodies is of a conceptual

[1] The logical, not the historical, order is here referred to.

character, and differs in important respects both from visual space and from tactile and motor space. Geometrical space is of three dimensions, infinite, or at least unbounded, homogeneous, and isotropic.

It is impossible for me to analyse in detail the characteristics of visual space, and of tactile and motor space. It must suffice to say that visual space is neither homogeneous nor isotropic, and that, in it, distance is only indirectly appreciated by convergence of the two eyes, with muscular sensations due to accommodation between them. This space has three dimensions, and the particular kind of Geometry adapted to the conceptual description of it is what is known as projective Geometry. In tactile and motor space we make direct experimental estimates of distances and measures. Each muscle gives rise, when it is contracted, to a special sensation, so that motor space may be said to have as many dimensions as we have muscles. The conceptual Geometry which corresponds to tactile and motor space is what is called Metric Geometry. The private, or representative, space of an individual percipient, like his private time, is finite, but with an indistinct boundary. Private space, like private time, is sensibly continuous, as any portion of it is conceived to contain lesser portions, without any definite limit of smallness. As in the case of time, it is not atomic, for it does not consist of points without extension: the notion of a point of space is, like an instant of time, a pure abstraction. In spatial, as in temporal, presentations, there are qualitative distinctions of direction. The frame of spatial reference of the individual is provided by his own body, and the qualitative distinctions of up and down, right and left, depend upon this frame of reference. The private space of the individual has a certain absoluteness, because the sense of effort which he has when he moves relatively to his material environment is lacking when he himself remains quiescent and the external objects are moved relatively to his body; the relative motions of his body and the objects being the same in the two cases.

The space employed in social intercourse, sometimes called physical space, is a construction of such a character that the private spaces of individuals may be systematically correlated with portions of it. The possibility of such correlation depends upon the fact that an identity is attributed to a physical object of such a kind that it is regarded by all percipients as one and the same object. Thus physical space is the complex of the spatial relations of physical objects regarded as a system of objects capable of being perceived by any percipient in suitable circumstances. It is in physical space that all the phenomena with which Natural Science is concerned are regarded as happening. The frame of reference of the single individual having been eliminated, the only meaning that can be ascribed to the rest or motion of a body in physical space is conservation, or change, of spatial relations with other bodies. For any measurement of position, or of the motion of a body, some standard frame of reference in some standard body, such as the earth, or the sun, or the walls and floor of the room, must be assigned. Thus all spatial relations in physical space are relations of extension of perceptual objects and between different objects; physical space cannot be regarded as empty space because, with the disappearance of perceptual objects, the whole scheme of relations which constitutes physical space would disappear.

The opinion has been prevalent that external objects are localized by us in a geometrical space of a unique character with definite properties; this geometrical space has been frequently regarded as a kind of ready-made framework in which we localize our physical perceptions. When I discuss the possible forms of geometrical space, it will, I think, appear that this is by no means the case. To give a systematic scheme descriptive of the relations in physical space is the first object of the Science of Geometry, although in some of its developments the Science has so extended itself as to transcend this primary object. In order to attain this object, the spatial relations are idealized, and transformed into a precise form, by means of a system of definitions and postulations. By this process of abstraction and idealization, conceptual space, the space of abstract geometry, has been created. It is in this conceptual space that all the ideal objects of Geometry are regarded as situated, and as subject to a scheme of relations specified by a system of postulates. In accordance with the general method of scientific procedure such a scheme must satisfy the condition of logical coherence; that is of freedom from contradiction. It must also satisfy the test of applicability to the description of the spatial relations which are observed to hold in physical space. As a result of the efforts of Mathematicians, prolonged through many centuries, and supplemented by the critical examination to which the foundations of the subject have been subjected during the last few decades, the Science of Geometry has attained to a degree of coherence which may be held to justify the designation of it, that I made in an earlier lecture, as a model to which other departments of Science may tend to conform. Geometry is better fitted than Arithmetic to be regarded as a model, in this sense, because, although Arithmetic, as we saw in the last lecture, has been completely conceptualized, it did not originate from the necessity of describing conceptually any one particular class of physical properties; whereas Geometry had its origin in the effort to describe one particular class of physical relations, namely the extensional properties of perceptual objects.

I propose to give some account, necessarily brief, of the process of growth by which Geometry, in its present highly developed state, has come into being. A remarkable illustration of the fact that a valid conceptual scheme, adequate for the purposes of description for which it was devised, is not necessarily unique, is afforded by the fact that different conceptual schemes of Geometry have been constructed. These systems are all logically coherent; they are inconsistent with one another as regards their several postulations; and yet several of them are adequate, at least when they are suitably restricted, for the purpose of description of percepts. The nature of our knowledge of spatial relations has been a problem for Epistemology and Psychology, which has been very widely discussed. In this connection, the history of Geometry is of great interest, especially since, as a result of modern investigations, a decisive refutation has been provided, from within the Science, of Kant's celebrated views as to our spatial intuition. These views are still maintained in some philosophical circles, often owing to an insufficient comprehension of the nature of the modern developments of abstract Geometry. I reserve a discussion of the latest theory of spatial and temporal relations, that connected with the names of Minkowski and Einstein, for a later lecture. It is sufficient here to remark that, if that theory be finally established, it does not affect the validity of the previously existing theories of Geometry, or their applicability as descriptive schemes sufficient for ordinary purposes. The older Geometry will accordingly never be completely superseded by the more comprehensive theory to which I have referred.

The earliest rigorous treatment of the Science of Geometry was that of the Greek Mathematicians; of this we possess a systematic account in that great text-book of the subject, Euclid's *Elements of Geometry*. This contains an account of the current Geometry in Euclid's time; it is the most ancient text-book of Science that we possess; and the fact that it has been used for many centuries is a testimonial to its excellence. To a large extent, the Greek Geometry exhibits, in dealing with physical extension, the true scientific method. The simplest regularities and uniformities observable in the shapes and spatial relations of actual bodies were singled out and then conceptualized. Thus points, straight lines, planes, rectilineal figures, circles, spheres, pyramids, and other objects, transformed and idealized from percepts into concepts, are the geometrical objects with which Euclid deals in his abstract Geometry. Later writers, especially Apollonius and Archimedes, treated of the Geometry of cones and conic-sections; an example of non-utilitarian scientific investigation which bore useful fruit in the hands of Kepler. The type of Geometry, developed by the Greeks, known as Euclidean Geometry, and for a long time regarded as the only possible type of Geometry, still remains the standard Geometry for practical purposes and for ordinary scientific purposes. It forms the conceptual basis for all our actual spatial measurements, and notwithstanding the increased generality of modern geometrical schemes, it will certainly continue to be employed for the more ordinary purposes of Science, even if it be superseded for the purposes of certain general theories relating to gravitation and Electrodynamics. In order to distinguish the Euclidean Geometry from other more modern Geometrical schemes of a divergent character, it is frequently spoken of as the Geometry of Euclidean space. In view of the fact that geometrical space cannot properly be regarded as an entity endowed with special properties, since it in reality represents a mere possibility of spatial determinations, it is more accurate to speak of Euclidean Geometry as a Euclidean system of spatial relations, or as Geometry with a Euclidean Metric. In the nominal definitions of the geometrical objects with which his scheme deals, Euclid gives a somewhat elaborate descriptive account of characteristics of those objects. In his deductive treatment, many of these characteristics are not in any way employed in the argument. In some of the definitions, postulates lie hidden which later criticism has brought to light and stated explicitly. An example of this is the fourth definition of the fifth Book, which implicitly contains the postulate known as the axiom of Archimedes. What we now regard as the postulated or hypothetical scheme of relations in Euclidean Geometry appears in part, in Euclid's elements, in his axioms, or common notions, which he regards as self-evident; and in part in his postulates which are taken to be facts that are unproved, but the assumption of the truth of which is necessary for the purposes of his theory. Other postulates, derived from intuition, are made implicitly in the course of his deductions. In Euclid's own form, although his Geometry has many of the characteristics of a valid conceptual scheme, the treatment is far from exhibiting that example of a flawless deductive scheme, for which it has frequently been accepted. From Euclid's own time onwards there has been much discussion and criticism relating to the true character of the definitions, axioms, and postulates of the scheme. This has resulted during recent decades in a restatement of the foundations of the subject, of such a kind that synthetic Geometry may now be regarded as conforming to all the requisites of a conceptual scientific scheme.

The objects with which Euclid deals, points, straight lines, planes, etc., are obtained as the result of idealization of actual percepts, in which some constituents of the percept are removed by abstraction. Thus a point is the ideal limit, postulated as existing, of an object from which its extension is completely abstracted. It retains spatial relations with other objects, that is, position. A line is an object in which we abstract from the thickness of perceptual lines; and a straight line is an idealized object which arises from our observation of empirical straightness. Euclid gives no complete conceptual representation of relations of magnitudes. Thus, equality of magnitude, and the relations of congruence of segments of straight lines, of angles, and of areas, are not made to depend upon a logically complete system of postulations; but recourse is had to intuition of magnitude in the region of percepts. This defect is exhibited prominently in Euclid's use of the method of superimposition employed in the theorems relating to congruent figures. The apparent reluctance with which Euclid employs this method, restricting it, as he does, to cases in which its use seemed unavoidable, would appear to indicate that he had misgivings as regards its logical validity. As it stands, no complete defence is possible against the charge of circularity which has been made against this reasoning. In the modern form of synthetic Geometry, a system of definitions and postulations relating to congruence is introduced into the foundations of the conceptual scheme itself, rendering the intuitional method of superimposition unnecessary.

Euclid's theory of parallels has formed the starting point, from his own day onwards, of discussions which have ultimately led to a generalization of the whole theory of Geometry. It is remarkable that the so-called axiom of parallels was given by Euclid himself not as an axiom or self-evident truth, but as a postulate, an assumption necessary to his scheme; and thus it may be interpreted as having a hypothetical character. Since every actual observation of relations in physical space is confined to some region which is necessarily finite, whereas abstract Geometry deals with relations extended into indefinitely great regions, it is clear that any postulation relating to parallels must be incapable of complete empirical verification of a direct kind. But the postulate may have consequences, such as that the sum of the angles of a triangle is two right angles, which can, subject to inevitable errors of measurement, and with certain physical assumptions, be regarded as capable of empirical verification. From an early time onwards, attempts have been made to dispense with the use of the postulate relating to parallels, as an independent assumption, by showing that it can be proved as a deduction from the rest of Euclid's scheme. Proclus (410–485 A.D.), in his commentary on Euclid's *Elements*, gave a detailed account of attempts, made by Ptolemy and by himself, to effect this deduction. These attempts, and many others made in modern times, were, as we now know, doomed to inevitable failure. The first Geometer who appears to have contemplated the possibility of an hypothesis, relating to parallels, inconsistent with that of Euclid, was Girolamo Saccheri (1667–1733), a Professor at Pavia. But his invincible prejudice in favour of the Euclidean hypothesis, as a necessary constituent of the only possible Geometry, prevented him from recognizing the true implications of his investigations. Lambert (1728–1777) went further in the same direction as Saccheri; he showed that the area of a triangle is proportional to the difference between the sum of its three angles and two right angles, in the two cases corresponding to what we now call hyperbolic and spherical Geometry. John Wallis (1616–1703) remarked that Euclid's postulate may be replaced by the equivalent assumption that similar triangles of different magnitudes exist. It is in fact sufficient to assume the existence of only two similar triangles, with different magnitudes; as was observed by Laplace and by Carnot. Legendre (1752–1833) carried out investigations, of permanent value, in connection with various long-continued attempts to prove the truth of Euclid's postulate. These investigations prepared the way for the great change in the whole position of Euclidean Geometry which resulted from the labours of Lobachewsky, Bolyai, and Gauss; the last of whom was the first Mathematician to express a definite conviction that the postulate is incapable of proof.

The first publication of a synthetic geometrical theory

in which a postulate was employed that differs essentially from Euclid's postulate relating to parallels was in the form of a treatise by Lobachewsky, which appeared in 1829; this was followed in 1832 by the publication of a similar theory discovered independently by Bolyai. In accordance with Euclid's postulate, through a point outside a straight line, one and only one coplanar straight line can be constructed which does not intersect the first straight line, but, in the theory of Lobachewsky and Bolyai, a whole sheaf of such straight lines can be constructed. This sheaf is bounded by two straight lines said to be parallel to the one considered. It was shown that, when this postulate is substituted in the Euclidean scheme, for that of Euclid, it is possible to build up a systematic Geometry in which the properties of figures will be in some respects very different from those in the Euclidean Geometry. For example, the sum of the angles of a triangle is less than two right angles, the amount of the difference depending upon the size of the triangle, so that, in this system, there can exist no relation of similarity between two rectilineal figures of different dimensions. This Geometry is a species of what is called non-Euclidean Geometry, and in view of later discoveries, this particular non-Euclidean Geometry, developed by Lobachewsky and Bolyai, is now called hyperbolic Geometry. That Geometry in which no postulate relating to parallels is employed, and which therefore includes what is common to both Euclidean and non-Euclidean Geometry, is frequently called absolute, or general, Geometry. A later discovery was made by Riemann, of which I shall presently have more to say, that a Geometry is possible, in accordance with which all coplanar straight lines intersect one another, so that no parallel straight lines exist. In this Geometry, of which two distinct forms are now recognized, called respectively spherical and elliptic, a straight line is always a closed figure of finite length. In place of the well-known Pythagorean theorem relating to the sides of a right-angled triangle, which is fundamental in Euclidean Geometry, more general metrical relations were developed by Lobachewsky and Bolyai for the hyperbolic case. In spherical and elliptic Geometry the corresponding metrical relations are identical with those expressed by the ordinary formulae of spherical Trigonometry. The two important questions which arise as regards the systems of non-Euclidean Geometry, sometimes called meta-Geometry, are first, that of the logical validity of the schemes, and secondly, that of their applicability to the description of actual spatial relations in physical space.

Before, however, I discuss these questions, it is convenient to refer to two developments of Geometry, different in kind, but both of the most far-reaching importance, and both of such a character that, as a result of them, the whole Science of Geometry can be viewed from standpoints much more general than that of the older traditional scheme of synthetic Geometry. The first of these new departures in the Science was the introduction of Analytical Geometry by Descartes, in his method of employing coordinates to represent the positions of points of space. An essential element in Descartes' coordinate Geometry is that sets of three numbers (or in plane Geometry, two numbers) are correlated uniquely with points of space. This method of correlation, taken in conjunction with the conception of current coordinates, in which variables are employed, each of which is capable of taking up the values of the elements of a number-continuum, has the effect of reducing the statement of all geometrical relations, and properties of figures, to purely algebraical statements. Thus all Geometry—in Descartes' original scheme this is restricted to Euclidean Geometry—is reduced to arithmetical Algebra. Every theorem of synthetic Geometry is correlated with a corresponding theorem of a purely arithmetic nature, that is, one in which the objects to which the theorem relates are sets of elements, each of which consists of a triplet of numbers. We possess a modern Mathematical theory known as the Theory of Aggregates, in which, when an aggregate of elements is taken as fundamental, the properties of selected sub-aggregates, or portions of the fundamental aggregate, are classified and developed. If the fundamental aggregate be taken to be the set of all triplets of numbers, each of which numbers may be any number whatever of the arithmetic continuum, this fundamental aggregate may be taken as corresponding to all the points of geometrical space. All geometrical constructs, such as straight lines, planes, circles, spheres, etc., correspond to particular sub-aggregates, or parts, of the fundamental aggregate; and their geometrical relations correspond to properties of such sub-aggregates. It thus appears that Geometry is capable of a further stage of abstraction, in which synthetic Geometry is replaced by a scheme of relations between sets of objects, each of which is merely a triplet of numbers. Geometry thus becomes, in fact, a purely arithmetic scheme. The purely technical advantage of the consequent reduction of the ascertainment of particular geometrical properties to calculation by means of equations is, for most purposes, enormous. But besides this simplification, other consequences follow from the substitution of relations of variables and numbers for the objects and relations of synthetic Geometry. In plane Geometry, a curve is represented by an equation involving as variables the coordinates of a point on the curve; this equation is a relation which is satisfied whenever the variables take the values of the coordinates of any particular point whatever on the curve. When however we consider the arithmetic number-system to be extended so as to become the system of complex numbers, we find, in general, that there exist pairs of complex numbers which satisfy the equation of a curve, and thus the equation of a curve determines relations between complex numbers besides those of real numbers which alone have reference to the curve as originally conceived. It is convenient so to extend the language of synthetic Geometry as to take account of the extended interpretation of which the equations employed in the analytical scheme are capable. We speak therefore of a pair of complex numbers as the coordinates of an imaginary point in the plane. Thus, besides the original curve that is correlated with its equation, there exists, in general, a set of imaginary points in the plane, the coordinates of which satisfy the equation, as do those of the real points. Another extension of the use of geometrical language to denote arithmetical or algebraical facts concerning the equations of analytical Geometry is that involving the employment of infinite numbers, pairs of which, in a certain sense, may satisfy the equation of a curve. We then say that the curve contains points at infinity, which may be real or imaginary. The great advantage of this last extension of the use of geometrical language to denote arithmetical facts is apparent in the greater generality of form which it enables us to give to the statement of geometrical properties. To take a very simple case, instead of the statement that two circles in a plane may either not intersect one another, or may touch at one point, or may intersect one another in two points, we may assert that any two circles, as also is the case for any two conics, intersect one another in four points. The correctness of this statement implies that two points may be coincident, and that any point may be real or imaginary, and either in the finite part of the plane or at infinity. The comprehensive statement is in reality an expression of a property of the equations of the circles or conics. Without this kind of generalization of the use of geometrical language it is impossible to make a general statement about the spatial relations of figures of certain classes without reference to a number of special cases which may arise, some of them exceptional in character. Thus the statement that two straight lines in a plane intersect one another holds good, for Euclidean Geometry, without mentioning the exceptional case of parallels, because, on the algebraical ground I have referred to, parallels intersect one another at a point at infinity.

The full import of these extensions of the elements with which Geometry deals appears however only in connection with the second great development to which I have referred, that of Projective Geometry. Any precise description of the scope of this kind of Geometry would necessarily be of so technical a character that I cannot attempt to give one here. It may however be observed that a property which is distinctive of Projective Geometry is that two coplanar straight lines always intersect one another. Euclidean Geometry is accordingly not projective, but when the new entities which I have spoken of as points at infinity are introduced into it, it can then be expressed in the projective form. In the general sense of the term Geometry that is now employed, the fundamental elements in each Geometry consist of points, considered as a class of primitive elements; particular sub-classes of this fundamental class defining straight lines, planes, etc. One Geometry differs from another one in accordance with the nature of the relations that are postulated to exist between the fundamental elements, such as those relating to the order of points on a straight line. A method of introducing coordinates, independent of the ordinary Euclidean notions of congruence, was introduced by von Staudt; and this forms the basis of the analytical treatment of Projective Geometry, which is essentially independent of metrical considerations. But metrical relations of distance and of angles have been introduced into Projective Geometry, in the form of purely descriptive relations, by Poncelet, and later, in a very general form, by Cayley, in connection with the theory of invariants. The method of Cayley has an important bearing on the question of the validity of the non-Euclidean Geometries which we call Lobachewskyan and Riemannian. It has in fact been shown that all the relations in either of these systems are capable of being represented as relations within the Euclidean scheme. It follows that, if there arises any logical contradiction from the postulations made in hyperbolic, or in elliptic, Geometry, there must be exhibited a corresponding contradiction in Euclidean Geometry. Thus, if Euclidean Geometry be assumed to be a scheme free from contradiction, it is demonstrable that this is also true of non-Euclidean Geometry, whether hyperbolic or elliptic. All three Geometries stand therefore on the same footing, as conceptual schemes free from internal contradiction.

Before I consider the question whether the non-Euclidean Geometries have the same applicability as the Euclidean, for the purpose of describing relations in physical space, it is necessary to refer to a mode of considering the matter which was devised by Riemann and Helmholtz, and which has produced an epoch-making effect upon the whole theory of the foundations of Geometry. This new development was explained in two memoirs, the one by Riemann, the Mathematician, bears the title *On the hypotheses which lie at the base of Geometry*, the other by Helmholtz, the Physicist, is entitled *On the facts which lie at the base of Geometry*. Thus, what Riemann regards as *hypotheses* in a scheme of conceptual space relations correspond to what Helmholtz regards as *facts* of observation in physical space. The fact of experience which is regarded by Helmholtz as of significance in relation to the theory of Geometry is that freely movable rigid bodies exist in physical space; their dimensions remaining unaltered during the motion.

This may be stated in the more precise form that, if a pair of particles $A$, $B$ of any one such body can be brought into coincidence with a pair $A'$, $B'$ of another such body, then the coincidence of congruency remains unaltered when the pair of bodies are moved in any manner. Of course it is assumed that certain conditions as regards temperature and absence of strains are satisfied, so that approximately rigid bodies, which are the only ones that exist, may be regarded as perfectly rigid. The statement is equivalent to the one that a rigid body is freely movable, and so that the measurable distance of any pair of points of the body remains unaltered, the measure of distance being estimated by means of some standard body. In order that a system of abstract Geometry may be applicable to describe actual spatial relations, in which these facts of experience are taken into account, the measure of distance between any two points of the geometrical space should be such as to be an invariant for a certain set of transformations which shall represent mobility. In any transformation of this set, points $P$, $Q$ are made to correspond to other points $P'$, $Q'$ respectively, and the metric system of the Geometry should be such that the measures of the distances $PQ$, $P'Q'$ should be identical, for every pair of points and for every transformation of the set. Any Geometry founded upon a metric system in which this condition is satisfied can be applied to the representation of spatial relations in physical space, in such wise that the numerical measure of the distance between any pair of particles of a rigid body remains unaltered as the body is freely moved, without strains or changes of temperature.

Riemann's theory is based upon an extension of Gauss' general theory of curved surfaces in Euclidean space, that is when the ordinary Euclidean metric system is employed. In Gauss' theory, the position of a point on a given surface is specified by two numbers, the values of two variables, which may be regarded as the coordinates of the point in a widely generalized sense of the word. There is a large degree of arbitrariness in the choice of these variables; when to either of these variables a constant value is assigned, the other one remaining variable, the point represented by the coordinates lies on a curve upon the surface. Thus a particular coordinate-system on a surface is defined by a mesh-system formed by two families of curves lying upon the surface; this mesh-system being arbitrary, but subject to certain conditions of continuity. The element of distance of a point from a neighbouring point of the surface can then be shown to be the square root of a quadratic function of the differentials of the coordinates of the point, the coefficients depending in general upon the particular point. Gauss established the existence, at each point, of a certain function of these coefficients, which is invariant for all systems of meshes, and depends only on the nature of the surface in the neighbourhood of the point; it is called the absolute curvature of the surface at the point. He further showed that, in order that a certain set of transformations might exist, which represent the motion of a portion of the surface into a new position, in which every element of length in a first position corresponds to an equal element in a second position, the necessary and sufficient condition is that the absolute curvature of the surface should at all points have the same value. Interpreted in physical language this means that a portion of a material surface can be freely shifted along the surface without stretching any part of the material, bending being however permissible. This condition is, for example, realized by the surface of a sphere, on which all figures are movable without alteration of dimensions; it would however not be satisfied, for example, by the surface of an ellipsoid, for which the property of free-mobility is accordingly lacking. In

Riemann's abstract theory, a manifold of elements is taken as fundamental, each of which is specified by a set of $n$ real numbers. Each of these $n$ numbers may be any number in the arithmetic continuum, or in some specified portion of it; thus the fundamental manifold is ordered, and has an $n$-fold order. The particular case which provides a Geometry applicable to physical space is that in which $n$ has the value 3. The manifold of elements is purely abstract, and free from any conception directly dependent upon spatial intuition, as is emphasized by the fact that $n$ may have any integral value. The question was considered by Riemann, as also, in the case $n = 3$, by Helmholtz, what system of metric relations must be introduced into the manifold, in order that the system which consists of the manifold subject to this metric system may be regarded as the space of a Geometry, capable, when $n = 3$, of affording a conceptual representation of the physical space in which measurements of physical bodies are made. Denoting an element of the manifold by the term point, and the numbers which specify that point by the term coordinates, both Riemann and Helmholtz consider the question what is the most general form of an element of distance between two points of the manifold, expressed as an integral, of which the integrand, or element of distance, involves the differentials of coordinates, in order that continuous transformations may exist, in which the distances between corresponding pairs of points remain unaltered in any of these transformations. The answer to this question is that, in the first place, the square of the element of length must be defined as a quadratic function of the differences of coordinates of its extremities, the coefficients in this quadratic function being continuous functions of the coordinates of the point from which the element of length is measured. Further, there exist a certain number of functions (in the case $n = 3$, this number is 6) of the coefficients in the quadratic function which must have one and the same value, invariant for continuous transformations of the coordinates. It can be proved that (in case $n > 2$) if this holds good at one point it holds at all points. Thus there must exist a certain constant which has one and the same value at all points of the manifold. This constant may be either positive, negative, or zero. In the last case the metric system introduced into the manifold is Euclidean, and the manifold is then, when $n = 3$, said to be the space of Euclidean Geometry. When the constant is negative, the space, for $n = 3$, is that of the hyperbolic geometry of Lobachewsky and Bolyai. When it is positive, the space is of the new kind discovered by Riemann, and may be either spherical or elliptic. It is unbounded, but in a certain sense finite, and Riemann has drawn special attention to the fact that a space being unbounded is quite consistent with its being, in the sense referred to, finite. This is, for example, the case with that two-dimensional space which consists of the surface of a sphere.

On account of the analogy with Gauss' case of the two-dimensional space which forms a surface in ordinary Euclidean Geometry, Riemann termed the constant which I have referred to, the curvature of the particular space. This term has proved a somewhat unfortunate one, as it has led to much misunderstanding. It has given rise to the idea that non-Euclidean three-dimensional space is itself curved, and this notion has been strengthened by the usual illustration of the case of a non-Euclidean two-dimensional space regarded as embedded in a three-dimensional space, for example in that of the spherical surface to which I have already referred. The fact is however that the so-called curvature is not curvature of the space, but represents only a property of the metric system introduced, with a considerable degree of arbitrariness, into the manifold. It is, for this reason, better to speak of this constant as the space-constant of the particular geometrical space. The manifold, as such, has no curvature, and no metric properties; these latter are introduced into it, when it becomes a space, for the purpose of representing conceptually our system of actual measurements in physical space which depend upon physical properties of objects in that space. Instead of speaking of Euclidean or non-Euclidean three-dimensional space, it is more accurate to speak of geometrical space with an imposed Euclidean, or with an imposed non-Euclidean, metric system with an assigned space-constant, positive or negative. Moreover it is unnecessary to regard such space as a section of a space of higher dimensions, since the metrical scheme introduced into it does not require the consideration of a manifold with more dimensions than that of the space considered. The element of length being assigned as the square root of a quadratic function of the differentials of coordinates, the distance between any two points of the space is determined as the stationary value of the integral of the element of length taken from one of the points to the other. In the case, taken for purposes of illustration only, of the two-dimensional Geometry of a surface in ordinary Euclidean space, the distance between two points is the length of a geodesic which passes through the two points. The term geodesic has been frequently used in connection with the distance between two points in space with a non-Euclidean metric; but again this use of the term, although convenient, is apt to be misleading, as it suggests that the space must necessarily be regarded as a "surface" in space of four dimensions.

From the point of view that the difference between so-called Euclidean, and so-called non-Euclidean, geometrical space does not refer to any distinction of property of the point-manifold itself—that manifold being regarded as a mere field of possible metrical relations to be imposed upon it—the question whether our physical space is Euclidean or non-Euclidean, in itself, would appear to have no immediate meaning. The real question can only be taken to be whether, or under what restrictions, if any, the actual observed relations in physical space can be described by means of an abstract geometrical scheme with a non-Euclidean metric.

Euclidean Geometry, and also non-Euclidean Geometry with either a positive or a negative space-constant, are all self-consistent conceptual schemes. The crucial question, of very considerable theoretical interest, which arises in connection with them is that of their applicability to the description of actual spatial relations in physical space. Are they all so applicable, or is there anything in physical phenomena which compels us to assign to Euclidean Geometry a position, in relation to such applicability, which does not attach to non-Euclidean Geometry? No doubt Euclidean Geometry is the simplest for the purpose, but is non-Euclidean Geometry a possible system for application to physical space, if we are prepared to sacrifice simplicity? It should be observed that the question as here discussed is considered from the pre-Einstein point of view, in which no attempt is made to include gravitational or electromagnetic phenomena in the geometrical scheme.

In the first place it may be observed that, in sufficiently small portions of physical space, the results of adopting one or other of these schemes will be, in view of the approximative character of all our measurements, indistinguishable from one another. It would thus appear that all our ordinary spatial measurements are consistent with a non-Euclidean geometrical scheme, provided the numerical measure of the space-constant, when correlated with our ordinary scales, is sufficiently small; no deviations being then observable from the results of applying a Euclidean metric. For example, in either hyperbolic or elliptic Geometry, the sum of the angles of a triangle of sufficiently small dimensions is indistinguishable from two right angles.

It has however been maintained by Poincaré and many other Geometers that the applicability of non-Euclidean Geometry is theoretically possible, independently of any such restriction on the value of the space-constant; that it is in fact fundamentally a matter of convention and convenience, not of absolute necessity, which system we may employ. It is held by these Geometers, and supported by cogent reasoning, that no crucial experiment, consisting of the measurement of lengths and angles, can be made which, whatever its results may be, is inconsistent with a non-Euclidean scheme of representation, provided a requisite readjustment of the statement of physical laws be made, especially of the laws of Optics. If points of physical space be suitably correlated with points of geometrical space whose coordinates are assigned in accordance with either of the systems in question, the measurement of lengths by means of measuring rods will be consistent with the assumed principle of the existence of rigid bodies freely movable with unaltered numerical dimensions. If angles are determined indirectly by means of formulae connecting them with measured lengths, these formulae will differ with the system adopted. When angles are measured directly by methods which involve the use of rays of light, the determinations will depend upon the assumptions made as to the paths of such rays. For example, as we have already seen, in either hyperbolic or elliptic Geometry, the sum of the angles of a triangle of sufficiently small dimensions, measured in physical space, will be indistinguishable from two right angles. In order to measure the angles of larger triangles, we have to make use of rays of light; and the comparative simplicity of the Euclidean metric system, for purposes of application, arises from the fact that, if we employ it, we can assert that the path of a ray of light is a straight line.

It has frequently been suggested that astronomical observation might be employed to decide the question which kind of Geometry is the true one as representative of our physical space; in particular by the measurement of the angles of a triangle with very long sides. We might suppose, for example, that by such measurement, a triangle was discovered for which the sum of the angles differed from two right angles by an amount which could not be accounted for by instrumental errors. We should then have a choice of two interpretations of the observed fact. We might either say that physical space is only describable by a non-Euclidean Geometry, or we might say that it is Euclidean, but that the paths of rays of light are not strictly straight lines, but curved paths of such a character that the triangle with curvilinear sides was such as to explain the observed amount of the deviation of the sum of the angles from two right angles. Again, if we found that the sum of the angles was two right angles, we might either affirm that physical space is Euclidean, or else that it is non-Euclidean, but that the path of a ray of light is not a straight line. As an illustration we may take the fact that, on a spherical surface, the sum of the angles of a triangle of which the sides are geodesics exceeds two right angles; but that there exist triangles of which the sides are not geodesics, for which the sum of the angles is equal to two right angles.

Whilst admitting the strength of the case in favour of the view that physical laws are capable of being so stated that our actual spatial measurements are capable of being described by means of a Geometry with a non-Euclidean metric, it would certainly be more satisfying, as a confirmation of this view, if we were in possession of a detailed statement of the precise modification of physical laws, of our habits in relation to spatial intentions, and of our practical modes of measurement, which would be rendered necessary by the adoption of an abstract non-Euclidean Geometry as the mode of description of actual spatial relations.

That the Euclidean metric is the simplest for all ordinary applications to physical space, because it admits of greater simplicity in the statement of physical and dynamical laws, is abundantly clear. For all ordinary purposes it will not be superseded, but the possibility of the employment of other schemes, as theoretically applicable, is also clear. The interest in the development of non-Euclidean Geometries, apart from their technical interest for Mathematicians, lies in the distinction it has laid bare between those elements in our Geometry which are introduced as definitions and conventions, admitting variety of detail, and those which are fixed by facts of observation.

It appears to have been held by Kant that our Euclidean system of Geometry is present in the mind *a priori* as a necessary presupposition of physical experience. The development of systems of Geometry, of logical validity equal to that of the Euclidean, and capable of being applied, although with great loss of simplicity, to describe our actual experiences of spatial relations, would appear to provide a definite refutation of the Kantian view of space as an *a priori* form, at least as regards so precise a form as that of the Euclidean scheme. It has frequently been suggested that the study of non-Euclidean Geometry, and still more general schemes of Geometry which have occupied the attention of Mathematicians during more than half a century, is of purely technical interest, and can lead to nothing which has any relation to physical phenomena. The rise of the Einstein theory of relativity, which essentially depends upon the ideas developed by Riemann, Helmholtz, Minkowski, and others, affords a sufficient proof of the hazardous character of all prophecies as to the non-applicability of abstract theories and their generalizations for the purpose of representing physical phenomena.

## VII

# CORPUSCULAR THEORIES OF MATTER

BOTH in ancient and in modern times two divergent conceptions of the constitution of matter have been employed in scientific theories. In the first of these, matter is regarded as consisting of groups of discrete entities separated from one another by empty spaces; these are the atomic, or corpuscular, theories of matter. In the rival view, matter is regarded as continuous, and indefinitely divisible. So long as either of these types of theory of the constitution of matter is regarded realistically, as a constituent part of an ultimate philosophical view of the world, they are of course mutually exclusive. However, in accordance with the view of the true character of all scientific theories which I am endeavouring to explain and illustrate in these lectures, any theory of the constitution of matter will be taken solely as a conceptual representation of some assigned domain of natural phenomena; thus an atomic theory may legitimately be employed for some purposes, and a continuous theory of matter for other purposes. Frequently a theory of mixed type, in which both atoms and continuous substances occur as conceptual elements, has been made the basis of an attempt to represent classes of phenomena. The employment of different types of theory for different purposes has been a common procedure in modern times. For example, an atomic theory has been long prevalent for the purpose of representing facts relating to chemical combinations; and the kinetic theory of gases affords another example of such a corpuscular theory. On the other hand a continuous theory of matter is employed in the theory of Elasticity, and in the theory of the motions of fluids; the treatment of these subjects being thus made capable of the application to them of continuous Mathematical Analysis. Various attempts were made, especially in the first half of the nineteenth century, to represent these latter phenomena by means of corpuscular theories; but all such efforts were based upon complicated hypotheses as to the nature of interactions between corpuscles, and they can only be regarded as having attained a moderate degree of success. In any such case the simpler continuous theory is to be preferred, so far as it can be shown to have as large a range of applicability for the purposes of representing the phenomena, as has a corpuscular theory. It is of course always possible to regard a continuous theory as containing an idealization, by a process of averaging, of the particles of substances that are taken to be only sensibly continuous; and it has in fact been maintained that a continuous theory must necessarily be regarded in this way if it is to be considered as valid. The advantage of such idealization then consists in the fact that the theory becomes capable of expression in the form in which differential equations are used. In theories of the mixed type, imponderable substances have played a large part; we have for instance an example of such a substance in the modern ether of the electromagnetic theory.

The parentage of all atomic theories is to be found in Leucippus and Democritus, or even earlier; Democritus erected a Cosmology on the basis of the idea that the only existing objects are atoms in empty space; that these atoms are indestructible and eternal; and that all change consists in the aggregation and separation of these atoms. It may be observed that Aristotle, as an opponent of the atomistic theory, refused to admit the validity either of the conception of empty space, or of that of an indivisible atom. The Cosmology of Democritus leaves no room for contingency, or for teleology; since all change, being due to the motion of the atoms, is subject to an unconditional necessity. The atoms have an endless variety of form, and are infinite in number; in their eternal fall through infinite space, the greater atoms strike against the smaller ones, and, since the former have a greater velocity, the impacts give rise to lateral movements and vortices. These form the commencement of the growth of worlds, an innumerable series of which come into existence and perish. The atoms act on each other only by collision or pressure, and the variety in gross bodies is due to the variety in the number, size, shape, and arrangement in space, of the atoms of which they are composed. Democritus regarded such sensations as those of sweetness, bitterness, warmth, and colour, only as deceptive opinions; nothing but atoms and empty space were regarded by him as real. The phenomena of life are produced by fine, smooth, round atoms, like those of fire, which permeate the whole body, and constitute the soul, which is thus recognized as distinct from the body. The Physics of Epicurus was founded upon the conceptions of Democritus, and was employed by him to remove that dualism of mind and matter which is involved in the Philosophy of Plato and of Aristotle. He regarded space as infinite, and containing an infinite number of indestructible and indivisible atoms in perpetual motion. These atoms differ from one another only in size, shape, and weight, and they move with equal velocities. As they move they give rise to new worlds which perpetually tend to dissolution, and then to the production of fresh series of worlds. The soul is a fine substance, like warm air, distributed through the whole body. From the surfaces of external objects there is a constant stream of fine particles; and thus actual material copies of external objects enter the living body, which conducts these images to the soul, thus giving rise to sense-impressions.

These materialistic views of the world, which were taken up at a later time by Lucretius, whatever may be thought of their general tendencies, had the effect of eradicating from the minds of their adherents those animistic and magical conceptions which, together with the habit of constantly appealing to final causes, had done much to hinder the development of scientific methods of investigation. In the middle ages, the current view of matter, in accordance with the Aristotelian tradition, was that it consisted of underlying substantial forms which possessed accidental properties; our perceptions relating only to these latter. For a long period the attention of investigators was devoted to the occult properties of substances, but no interest was taken in their quantitative aspect. Atomism, as taught by the ancients, had been completely submerged by Aristotelianism, and it was not until the fifteenth century that atomistic conceptions again arose in connection with the metaphysical speculations of Nicolas de Cusa (1401–1464) and other writers who took part in the growing criticism of the dominant Aristotelianism. The most important influence upon later developments of Atomism was exercised by the speculations of Giordano Bruno, although their metaphysical character was such that their immediate effect on Physics was negligible.

The Cosmology of Giordano Bruno (1548–1600), which amounted to a metaphysical Monadology, was a scheme in which Being and Thought were coincident, so that the structure of time, space, and matter can be discerned by the operations of Thought alone. In accordance with his Philosophy, it is necessary to conceive time, space, and matter as composed of indivisible minima, that is, instants, points, and atoms. That there must exist, in each domain, a fundamental indivisible whole, with which our conception begins, he regarded as a necessary postulate of Thought. This notion of the atom, Bruno regarded as having a certain relativity; the magnitude of this indivisible, or minimum, is fixed according to circumstances. In Astronomy, for example, the heavenly bodies are the irreducible minima, and are thus to be regarded as atoms.

It has been the mechanistic view of the world of physical phenomena conceived by Descartes that has exercised the greatest influence on later conceptions of matter. It is difficult to reconcile Descartes' view of mind and matter as the two fundamental substances with his idealism; his theory of the physical world is, however, practically of a purely materialistic type. In extension and motion he recognized the source of all physical percepts, and thus all the occult properties with which medieval thought endowed substance were removed, the external world forming a purely mechanical system. He was not an atomist in the same sense as Democritus or Lucretius, for he regarded all space as one fundamental substance, infinitely divisible; thus, for him, spatial extension and substance were identical. The only differentiation of substance is due to the motion of its parts; thus a body, or portion of matter, is what can be moved as a whole relatively to the surrounding substance. Since space is a plenum, and since, on account of the identity of extension and substance, there can take place no diminution of the volume of substance in a portion of space, it follows that a circular streaming must form the basis of all motion. Descartes assumed that originally the substance of which the physical world consists was broken up into particles which were in rapid motions of rotation, and also in circular translational motion. By means of this motion, particles which were originally of irregular shape have become rounded, splinters having been broken off from them in the course of the gradual grinding. This process he regards as having given rise to three kinds of corpuscles. The first of these form elementary fire, and consist of splinters of various and varying size and form, in motion with enormous velocities. The sun and the stars are composed of this kind of elementary matter. The matter of the second kind, elementary air, consists of imperceptibly small spherical atoms, which move with great velocities in vortices; they fill all interplanetary space. The matter of the third kind, the elementary earth, consists of larger corpuscles of various forms, and in less rapid motion; the ordinary material bodies, and the earth and planets, consist of atoms of this species. The conception of ordinary matter which is in accordance with this scheme is that it consists of atoms which are in fact, but not in thought, indivisible. The interstices between these atoms are filled with atoms of the other two kinds, so that although the amount of ordinary matter in a fixed space may vary, the total amount of matter of all three kinds in that space is invariable. Plants and animals, like inorganic bodies, are machines; their vital spirits consisting of fine material in motion, as with Democritus. Descartes described the separate stages of a mechanism involving pressure and collision, as forming an uninterrupted chain of effects produced by external objects through the senses upon the brain, and back from the brain through nerves and muscular filaments. This view of a materially conceived world, subject to a rigorously determined sequence of causes and effects, would appear to be irreconcilable with Descartes' idealistic Metaphysics.

The difficulties relating to the possibility of the motion of atoms and bodies in a plenum filled with impenetrable substance were so great that a return was inevitable to the simpler conception of the ancient atomists, that atoms are surrounded by empty space. The writings of Pierre Gassendi (1592–1655) had an important effect in hastening the disintegration of the Aristotelian conceptions of matter, the authority of which had been already to a considerable extent undermined. Gassendi regarded empty space and atoms as the only principles in nature. All atoms he regarded as consisting of one and the same substance; they are only distinguished by differences of magnitude, shape, and weight. A limited number of different kinds of atoms suffices to explain the variation of bodies by their different groupings. In this view he was in agreement with that of Epicurus, but not with that of Democritus, who had regarded the number of different forms of atoms as infinite. The weight of the atoms he regarded as due to an inherent capability of self-determined motion; it is in the motion of the atoms that the explanation of all physical properties of bodies is to be sought. The main importance of Gassendi's work depends upon the fact that he was the first definitely to return to the ancient atomism, and thus to complete the breach with the medieval views of matter.

The Philosopher Thomas Hobbes (1588–1679) occupies an important position in the history of Materialism. Under the influence of Galileo, in regard to the theory of motion, he regarded the motion of bodies in space as the original phenomenon upon which all others depend, if they are to be subjected to scientific treatment. He was not an orthodox atomist, because he recognized the existence, not only of corpuscles, but also of a continuous fluid which fills all the interspaces between corpuscles, and in which motion is propagated. This conception of vibratory motion, or Conatus, is an attempt to objectify pressure or stress, and may thus be regarded as a step in the direction of the introduction of forces acting at a distance.

The breach with the Aristotelian conception of substantial forms, involved in the schemes of Descartes and Gassendi, was consolidated by the great practical experimenter Robert Boyle (1626–1691). Although his chief interest was in the ascertainment of facts by experiment, Boyle did not fail to recognize the necessity for a theory which would bind together the results of his experimental investigations in Chemistry and in relation to the weight, pressure, and elasticity of air. In his general Philosophy, Boyle was not a Materialist, but held a theory, formed out of Descartes' Physics and Gassendi's Metaphysics, which was designed not only to do justice to his scientific views as a Chemist and Physicist, but also to be compatible with his orthodox religious opinions. He distinguished between two orders of corpuscles, those of the second order, which form the constituents of matter, being formed by the aggregation of corpuscles of the first order. Between the corpuscles there are pores containing various effluvia. The primary constituents of matter, which we should now call molecules, are not absolutely indivisible, but the primary corpuscles of which they are composed are so firmly fitted together that they are only with difficulty separated from one another. Boyle set up various hypotheses relating to the corpuscles, for the purposes of explaining the constitution of air and other substances, and also of chemical combinations. Of these hypotheses, varying considerably according to the purposes for which they were designed, he appears to have recognized the tentative character. He attempted to explain all chemical changes mechanically, and laid considerable stress upon the quantitative determination of weights. He prepared the way for the modern chemical theory of elements, recognizing specific weight and chemical reaction as the distinguishing marks of a particular substance. On the theoretical side, he recognized the identity in character of the molecules which form the mass of the substance. The physical phenomena of heat, electricity, magnetism, and the transformations into one another of solid, liquid, and gaseous conditions of matter, he regarded as all capable of mechanical explanations.

The great discovery by Newton of the law of universal gravitation led ultimately to the great change in the nature of corpuscular and atomic theories which was produced when what is called action at a distance became an essential element in dynamical theories of matter. *Prima facie* Newton's discovery involves the discarding of the ancient notion that all action must be due to contact; a prejudice of which the origin is closely connected with the conception of efficient causation. The idea that the gravitation between two bodies is the resultant effect of the attractions upon one another of their individual corpuscular parts is inconsistent with the older atomic theories in which all interaction was regarded as due to impacts. For Newton and his contemporaries, under the influence of the notion that impact, or contact of some kind, is the only species of admissible explanation of physical action, the law of gravitation as it stood was incomplete without an indication of some mechanism by which the gravitational attraction can be deemed to be produced. A cause of gravitation must be sought for, and discovered, before the law could be regarded as, in the true sense, embodying a physical theory of the phenomenon. Newton himself declared "The reason of these properties of gravity I have not, as yet, been able to deduce"; and again, in a letter to Bentley, he writes[1]:

> It is inconceivable that inanimate brute matter should, without the mediation of something else which is not material, operate upon and affect other matter, without mutual contact, as it must do if gravitation, in the sense of Epicurus, be essential and inherent in it. And this is the reason why I desired you would not ascribe innate gravity to me. That gravity should be innate, inherent and essential to matter, so that one body may act upon another at a distance, through a vacuum, without the mediation of anything else by and through which their action may be conveyed from one to another, is to me so great an absurdity that I believe no man, who has in philosophical matters a competent faculty of thinking, can ever fall into it. Gravity must be caused by an agent acting constantly according to certain laws; but, whether this agent be material or immaterial, I have left to the consideration of my readers.

[1] *Opera*, Horsley's edition, Vol. IV, p. 438.

Newton's contemporaries for the most part shared his view, fearing the reintroduction of occult causes into Physics. Thus Huygens declared that "Newton's principle of attraction appeared to him absurd." John Bernoulli, who attempted to explain the motions of the planets by means of a modified form of the Cartesian theory of vortices, proclaimed "the two suppositions of an attractive faculty and a perfect void" to be "revolting to minds accustomed to receiving no principle in Physics save those which are incontestable and evident." Euler insisted on the necessity of supposing that gravitation is due to some subtle material medium; and D'Alembert regarded the real cause of gravitation as unknown, in contradistinction to action by impact, of which we have a clear mechanical conception. Until far into the nineteenth century this rejection of the notion of action at a distance held sway. Thus for example[1] E. Du Bois-Reymond writes:

> Forces acting through void space are in themselves inconceivable, nay absurd, and have become familiar concepts amongst physicists since Newton's time from a misapprehension of his doctrine and against his express warning.

Again, Balfour Stewart and Tait[2] write:

> Of course, the assumption of action at a distance may be made to account for anything; but it is impossible (as Newton long ago pointed out in his celebrated letter to Bentley) for anyone who has in philosophical matters a competent faculty of thinking for a moment to admit the possibility of such action.

Since Newton's time, very numerous attempts have been made to account for the phenomena of gravitation by propagation through a fluid or an elastic medium, or by means of impacts. All these theories must be pronounced to have failed in the purpose for which they were designed; in some cases, such as in the impact theory of Le Sage, on account of the nature of the

[1] *Ueber die Grenzen des Naturerkennens*, p. 11.
[2] *The unseen Universe*, 3rd ed., p. 100.

assumptions made in them, and in other cases on account of their inability to represent the known facts relating to gravitation. In the first place, gravitation is propagated instantaneously, or at least with a velocity which has been estimated by Laplace to be at least fifty million times that of light. Moreover all bodies appear to be absolutely transparent to gravitational action; and it is not subject to any kind of reflection or refraction. It appears also to be independent of the structure, or physical and chemical conditions, of the bodies between which it acts; its energy is unchangeable and inexhaustible.

Of all the attempts to account for gravitation on a corpuscular theory, that of Le Sage is the most ingenious, and is perhaps the only one which has been so far developed that its inherent weakness can be fully exposed. This theory is that the gravitation of bodies towards each other is due to the impact upon them of corpuscles or atoms moving in all directions through space. Each of these so-called ultramundane corpuscles is so small that collisions between pairs of them are of rare occurrence. If a body is not in any way screened from the bombardment of these corpuscles it would not acquire any motion, since the effects of the bombardment on all parts of its surface would neutralize one another. If there are two bodies in space, each acts, to a certain extent, as a screen against the bombardment of the other; thus, for each body, a smaller number of corpuscles will strike it on the side which is towards the other body than on the further side. Each body will appear to be attracted towards the other body, owing to the effect of the excess of the impacts it receives on the side furthest from the other body. Leaving out of account those corpuscles that have already struck some mundane body, and taking account only of those that come from infinite space, it can be shown that the force of attraction between two bodies whose dimensions are small compared with the distance between them will vary as the product of the sections of the bodies taken normal to the distance, and inversely as the square of that distance. In order that this may coincide with the attraction, as given by the law of gravitation, it is necessary that the effective areas of the bodies be proportional to their masses. Le Sage shows that, in order that this may be the case, whether the body be large or small, it must be assumed that the size of the solid atoms of the body is very small compared with the distances between them, so that a very small proportion of the corpuscles are stopped by even very dense and large bodies. Maxwell has shown that the energy of the corpuscles that is spent in maintaining the gravitation of a single pound of matter towards the earth must be millions of millions of foot-pounds per second. It can be shown that, on the assumption of perfect elasticity of the corpuscles, so that they rebound from the body with the same velocity with which they struck it, there will be no excess of the impacts on any other body on one side over the other side. On the other hand, if the velocity after impact is less than that of approach, although the attraction between the bodies will be accounted for, the excess of the energy which the corpuscles brought with them over that which they carry away remains to be accounted for. If any appreciable part of this excess appears in the form of heat in the body, it will, as is stated by Maxwell, in a few seconds raise it, and in like manner the whole universe, to a white heat.

I have emphasized the extreme reluctance which, from Newton's time onwards, men of Science have shown to regard the law of gravitation as anything but a mere stepping-stone on the way to the construction of a genuine scientific theory of gravitation, because, for us, Newton's law has all the characteristics of a genuine theory to represent a certain class of phenomena. In the hands of Newton himself, the law of gravitation, together with his scheme of Dynamics, proved a sufficient basis for the mathematical deduction of Kepler's laws of planetary motion, and for a deductive treatment of the principal inequalities in the motion of the moon. By the labours of a large number of Mathematical Astronomers, lasting into our own time, it has been shown that Dynamical Astronomy, with the Newtonian law as its base, is sufficient to afford an accurate representation of very nearly all the observed motions of the bodies of the solar system, and that the condition of predictability of such motions for a considerable time is satisfied. The law itself could hardly be surpassed in point of simplicity and of comprehensiveness of statement. It will not lose its importance as the foundation of a valid conceptual scheme if, as seems at present not unlikely, the new law of gravitation given by Einstein supersedes Newton's law as a more accurate law of the phenomenon. The detailed consequences of the two laws are indistinguishable from one another in the motions they assign to the bodies of the solar system, except apparently in the single instance of the motion of the Perihelion of Mercury's orbit. The idea that matter cannot act where it is not has no relevant meaning for us; since the notion of efficient action is no longer part of the stock of notions which Natural Science employs. The acceptance of Newton's law as a genuine scientific law depends upon the ascertained fact that it resumes in a formula, with sufficient degree of approximation for nearly all purposes, a whole class of observed facts relating to the motions of actual bodies.

In the first half of the eighteenth century, owing to the influence of Newton's theory, the conception that atoms exert an attractive force on one another rapidly modified the older atomic theories. The attractions between atoms made it unnecessary to conceive of them as having rough surfaces provided with hooks, since impacts no longer played the all-important part which they did in the older theories. There also arose the conception of imponderable atoms exercising repulsive forces on one another, filling the interstices between the gravitating atoms. These imponderable atoms were regarded as very much smaller than the ponderable atoms. Gradually these imponderable atoms came to be replaced by a continuous atmosphere of light and heat surrounding the ponderable atoms.

The views which were accepted in England, late in the eighteenth century, when Dalton was developing his conceptions of chemical actions, were stated[1] by him as follows:

> These observations have tacitly led to the conclusion, which seems universally adopted, that all bodies of sensible magnitude, whether liquid or solid, are constituted of a vast number of extremely small particles, or atoms of matter bound together by a force of attraction, which is more or less powerful according to circumstances, and which, as it endeavours to prevent their separation, is very properly called in that view *attraction of cohesion*; but as it collects them from a dispersed state (as from steam into water), it is called *attraction of aggregation*, or more simply, *affinity*. Whatever names it may go by, they still signify one and the same power....Besides the force of attraction, which, in one character or another, belongs universally to ponderable bodies, we find another force that is likewise universal, or acts upon all matter which comes under our cognisance, namely, a force of repulsion. This is now generally, and I think properly, ascribed to the agency of heat. An atmosphere of this subtile fluid constantly surrounds the atoms of all bodies, and prevents them from being drawn into actual contact.

Dalton was led to the fundamentally important conception of atomic weight by an interpretation, in terms of the atomic theory, of the observed fact that chemical combinations take place in definite and simple numerical proportions. He regarded all the atoms of a single substance as identical in size and weight; and thus the weight of a body is the product of the weight of an

[1] *New System of Chemical Philosophy*, Pt I, pp. 141–143.

atom by the number of atoms in the body. If one atom of the one substance always unites with one, two, etc. atoms of the other, then the regularity in the combining weights is made intelligible.

An important modification of Dalton's atomic theory was made early in the nineteenth century by the molecular theory initiated by Avogadro. This theory did not at first attain to general acceptance, on account of the rise of the electrochemical theory, in accordance with which the cause of affinities is to be found in the electrical relations of the atoms. Gay-Lussac had discovered that the various gases, under equal pressures and temperatures, combine in simple volumetric proportions. Avogadro explained this by the assumption that the numbers of smallest particles in equal volumes of different gases are the same, when they are under equal pressures and temperatures. He supposed that, in compound gases, and at least partially in simple gases, there exist combinations of two or more atoms; thus the smallest particle of a chemical body is not the atom but the molecule, a group of atoms. Chemical change, by combination or separation, he regarded as due to a change of place of atoms which grouped themselves into molecules of altered atomic composition.

These molecular and atomic theories of chemical combination were inspired by a spirit of physical realism in which sensuous images accessible to the imagination played a preponderating part. Although atoms and molecules are not perceptual objects, it is undoubtedly the case that this attitude of mind has been advantageous in facilitating the formation of theories which we may now regard as having a purely conceptual character. This use of the sensuous imagination attained a luxuriant growth after the discoveries of dimorphism and of isomerism, when it was found that substances of like chemical constitution appear in very different forms. Elaborate schemes for the localization in the molecule, and the transposition and various groupings of atoms arose in this connection. But there was soon a reaction against the idea that anything more than a convenient symbolism was implied in these formulations. Thus, for example, Liebig[1] declared in 1838 that:

> we know nothing as to the condition in which the elements of two compound bodies are, so soon as they have united in a chemical combination, and the way in which we conceive these elements as grouped in the combination rests merely upon a convention which has been consecrated by habit under the prevailing theory.

Again Kekulé, in his manual of Organic Chemistry, 1861, speaking of the proportional numbers of combining weights, as representing *fact*, says[2]:

> If to the symbols in these formulas another meaning is assigned, if they are regarded as denoting the atoms and the atomic weights of the elements, as is now most common, the question arises: what are the sizes or (relative) weights of the atoms? Since the atoms can be neither measured nor weighed, it is obvious that we can only be led by reflection and speculation to the hypothetical assumption of determinate atomic weights.

After the time of Dalton, who may be regarded both as a Physicist and a Chemist, the researches of Chemists and of Physicists led them along very divergent paths. Physics was treated in a manner which involved Mathematical Analysis, whereas Chemistry remained for a long time inaccessible to such methods. In our own day the barrier between these two branches of Science is being broken down.

Returning to the purely physical side of the atomic theory, we observe that the plausibility which attached to the notion that matter consists of solid, hard, impenetrable atoms which come into collision with one another loses its cogency when we inquire closely what happens as the result of such collisions. It seems impossible to attribute to such atoms the property which

[1] See Kopp's *Entwickelung der Chemie*, p. 597.

[2] Vol. 1, p. 56.

in sensible bodies is called elasticity, because that involves relative motions of the parts, which cannot take place in a perfectly rigid atom. The atom must be therefore regarded as inelastic. But in that case every collision between a pair of atoms would entail a loss of energy of motion, and the kinetic energy of an aggregation of such atoms would in consequence gradually disappear. Efforts of various kinds have been made to surmount this difficulty of the apparent impossibility of attributing elasticity to the atom, in the sense in which the term is employed in connection with gross bodies. One method is to regard the smallest particle of matter not as the atom, but as the molecule consisting of a group of atoms; in fact to follow the procedure adopted on other grounds by the Chemists. The suggestion is then that, in a shock between two molecules, the atomic constitution of the molecules may be such as to admit of their behaving like elastic bodies, although the constituent atoms are inelastic. The necessity of attributing perfect elasticity to the molecules has been recognized by those physicists who have developed the modern kinetic theory of gases. Both Clausius and Maxwell have emphasized this view. Lord Kelvin[1] has asserted that:

> we are forbidden by the modern physical theory of the conservation of energy to assume inelasticity, or anything short of perfect elasticity, in the ultimate molecules, whether of ultramundane or of mundane matter.

That the introduction of the molecular scheme, for the purpose of getting over this difficulty relating to elasticity, is a simple shifting of the difficulty further back to the atoms constituting the molecules, and not a solution of it, was the opinion that led Secchi to consider a different mode of dealing with the matter. He suggested that the apparent repulsion of the atoms and their reciprocal collisions can be simply referred to an

[1] *Phil. Mag.* (4), Vol. XLV, p. 329.

appropriate motion; it being sufficient for this purpose to suppose them to be in rotation. Relying on a theorem of Poinsot relating to the reflection of a rotating body from a resisting obstacle, Secchi attempted, without success, to show that, taking into account both the energy of rotational and of translatory motion, the total energy of two atoms is unaltered by an impact. Secchi also made unsuccessful attempts to explain the aggregation of atoms so as to form molecules, and the phenomenon of gravitation.

One of the most important attempts to evade the difficulty of defining precisely the character of the interaction between atoms which impinge upon one another, or come into contact, consisted of the radical step of depriving the atoms of all extension, and supposing them to be mere centres of attractive or repulsive force. In the middle of the eighteenth century an atomic theory of this kind was propounded by Boscovich. According to his theory matter consists of a swarm of atoms, each of which occupies a geometrical point of space, is capable of motion, and possesses a certain mass, so that a certain force is required in order to give such an atom a given acceleration. Two atoms at a distance from one another exceeding a certain small length attract one another with a force varying as the inverse square of the distance. For smaller distances the force is attractive for some distances and repulsive for others. In order to obviate the possibility of two atoms ever being in the same position, Boscovich supposes that, for all distances below a certain minimum, the force is repulsive, and increases indefinitely as the distance is diminished. The system of atoms which constitutes a material body occupies a certain region of space, by reason of the forces between the component atoms of the system and any other atoms which may be brought near them. No second body can come to occupy the same region of space because, before it could do so, the mutual actions of the atoms of the two systems would produce a repulsion between the two bodies, too great to be overcome by any force which we can apply. In this scheme, all action between bodies is action at a distance, and there is no such thing as actual contact between two bodies, although they may be so close to one another that the atoms of the two which are nearest exercise a great force of repulsion. When atoms were no longer regarded, in accordance with the view of the earlier atomists, as acting immediately on one another by contact, but by forces of attraction and repulsion acting at a distance through empty space, there seemed no longer any sufficient reason to attribute the property of extension to the atom. Even if it was retained, it was merely in deference to a desire to satisfy the sensuous imagination by making the atom resemble the bodies perceptible by our senses. Ampère, Cauchy, and Faraday all regarded the atoms as unextended, or as simple centres of force.

Another interesting and remarkable theory of the nature of the atom, of quite a different character from those to which I have referred, is Lord Kelvin's theory of vortex atoms. He imagines all space to be filled by an absolutely homogeneous, incompressible, frictionless fluid. It had been shown by Helmholtz that, in such a fluid, vortex tubes or filaments can exist in which the fluid is in permanent rotational motion, and that such a filament can form a closed ring, which may be called a vortex ring. Such a ring at all times consists of the same portions of the fluid, and is consequently of invariable volume. It is indestructible, and could not be formed in any portion of the fluid that is in irrotational motion. Two such rings could never amalgamate, or come into contact with one another. As such rings might be knotted, and two or more of them might be linked together by the passage of one ring through another without contact of their cores, the possibility occurred to Lord Kelvin that a new atomic theory might be founded on the existence of such rings; their convolutions and linkings admitting of an endless variety of forms. If such a convoluted, or linked, ring be regarded as an atom, such an atom would have permanence in magnitude and strength, capability of internal vibrations, and indestructibility. Thus, not the original fluid, although it possesses inertia, but only the vortex rings in it are regarded as having the character of matter. Such a vortex ring would seem to have more of the properties requisite to the atom than any of the earlier kinds of atoms. The difficulty of explaining the inertia of what is only a mode of motion of a substance, and not a substance itself, was pointed out by Maxwell. No substantial progress has been made in the direction of showing that the phenomenon of gravitation or the thermal and optical properties of matter are explained by this theory of vortex atoms.

The only theory of the constitution of matter which really comes to close quarters with the thermal and mechanical properties of the substance, and that can be regarded as an atomic theory, although the elementary parts of matter with which it deals consist of molecules, assumed to behave, on impact with one another, like perfectly elastic spheres, is the dynamical theory dealing with gases, known as the kinetic theory of gases. It is impossible here to enter into the complicated details of this theory, any adequate description of which would involve mathematical formulae. It has had very considerable success, although it would seem not complete success, in coordinating a large number of known facts relating to the mechanical and thermal properties of gases, and it may consequently be regarded as being the most comprehensive theory, of the atomic type, that has been proposed, for coordinating the physical properties of matter, in at least one of its forms. The theory has even been pushed so far as to assign rough estimates for the size and mass of a molecule of a gas such as hydrogen. As early as 1738, Daniel Bernoulli conceived the idea that a gaseous mass consists of a large number of perfectly elastic molecules in rapid motion of translation. On this hypothesis the pressure on the walls of the containing vessel is due to the impact against them of the molecules of the gas. This leads very simply to an explanation of the law of Boyle and Mariotte, that the pressure in the gas varies inversely as its volume, the temperature being unchanged. The further working out of the theory entails the use of calculation by probabilities, or in other words the statistical method. This method was introduced by Maxwell, and further developed by Willard Gibbs and Boltzmann. The molecules are divided into groups, in each of which the molecules are in the same state of motion. It is shown that, in every gas, every molecule has on the average, for a given temperature, the same kinetic energy, and that this is a fixed multiple of the absolute temperature of the gas. The theory has been applied to explain the diffusion of gases, and to the theory of their spectra.

If we take a general survey of the ideas which have inspired atomists both in ancient and in modern times, we see that the leading motive has been to reduce the complexities of the various forms of matter, and of the phenomena connected with them, by assuming that all matter is made up of parts, each of which has only the simplest of the perceptual properties of matter in bulk, those of indestructibility and of motion through space. The assumption that the world consists of atoms and of empty space has in appearance the supreme merit of simplicity, and it satisfies the instinctive craving for unification of heterogeneous and complex elements in our actual perceptions. The idea that all interaction requires contact has the merit of apparent accordance with our own experience, and thus by habit it has acquired the great advantage of picturability to the imagination.

The roughnesses and hooks which atoms were often supposed to possess served to increase the feeling of atomists that they possessed a picture which appealed to what was familiar in actual experience. In fact all the assumptions as to the nature and properties of atoms were inspired by the desire to explain the complex forms and properties in the material world by reducing everything in the smallest parts of matter to simple processes of the most familiar type. The extreme reluctance that was exhibited to assign to the atoms any properties which were not familiar features of matter in bulk is characteristic of the thoroughgoing realistic spirit which has dominated the minds of nearly all atomists through the centuries. The supposed necessity that atoms, or corpuscles, should satisfy this condition of picturability has naturally exercised a considerable restrictive influence upon the possibility of developing atomic theories which should really be adequate for describing the more complex phenomena connected with matter; and this may account for many failures. Those of us who do not feel bound by such requirements are entitled to regard a purely conceptual atomic theory as admissible, whatever properties are assigned to the atoms, provided the requirements of a self-consistent scheme are satisfied, whether these properties are directly copied, or not, from perceptual properties of gross bodies. Subject to this condition of applicability, the success of the scheme in representing actual properties of matter, as the result of a synthetic process of combining the effects of the conceptual atoms or molecules in an aggregation, is the only criterion which an atomic scheme need satisfy; a reasonable degree of simplicity being presupposed. This view of atomic theories, or one closely approaching it, is certainly held by those Chemists who regard the number and arrangement of atoms in molecules as having a symbolic meaning only.

When Chemists found it necessary to regard the

molecule, and not the atom, as the smallest part of a particular substance, the question whether the atom could be regarded as the ultimate constituent of all matter became insistent. It seemed that it would be necessary to assume the existence of as many different kinds of atoms as there are chemical elements. If we stop there, the notion that all matter is reducible to atoms consisting of one primitive substance has to be given up. The knowledge obtained of the relations between the atomic weights of different elements made it difficult to stop short at the recognition of the existence of some seventy elements. The idea then suggested itself that the atom must be regarded as consisting of a group of smaller atoms, of the second order. It seems difficult to stop at any particular stage of the indefinite regress into which we are launched, when an atom of any order is regarded as composed of smaller atoms of the next higher order. If, at any stage, we substitute extensionless centres of force, the whole scheme loses that character of picturability which was one of its chief recommendations.

As soon as the conception of forces acting at a distance came to be accepted, in consequence of the Newtonian doctrine of gravitation, Atomism, in the primitive meaning of the term, underwent a profound modification. The theory of matter became a dynamical scheme in which the notion of central forces is introduced as a necessary conception, on a parity with the older conceptions of atoms or corpuscles, and empty space. Of this Dynamical scheme of relations I shall speak in a later lecture. I must also postpone any consideration of the modern electron theory of matter, as also of the still more revolutionary theory associated with the name of Einstein, in accordance with which matter and its relations are represented by a purely geometrical conceptual scheme.

The reluctance with which the notion of force, as an

independent conception, was accepted, is accounted for by the inveterate materialistic prejudice, in accordance with which matter is regarded as the only ultimately real object. That this view of the unique reality of matter is no longer tenable has been concisely expressed by Helmholtz, who wrote[1]:

> It is just as inaccurate to try and explain matter as something real, and force as a mere notion to which nothing real corresponds; both are rather abstractions from the real, formed in exactly the same way. We can perceive matter only through its forces, never in itself.

[1] *Ueber die Erhaltung der Kraft*, Wissensch. Abh., Vol. 1, p. 14.

# VIII

## DYNAMICS

THE mechanical theory of natural phenomena consists of a formulation of the idea that all such phenomena can be viewed as essentially consisting in changes in the motions of parts of material systems. A scheme in which, under given conditions, all these motions can be numerically calculated, in accordance with a set of fundamental laws, is known as a dynamical scheme. The first postulate of such a scheme is that matter can be regarded as being of such a character that certain aggregations of it, or certain parts of such aggregations, remain unchanged through all changes of distribution and configuration in space; thus retaining a certain identity. The shape which this idea of conservation of matter through all changes has taken in modern times is formulated in the principle of the conservation of mass. Of the gradual emergence of this principle from less definite attempts to fix the character of the unchanging elements in the mechanical theory, I shall give an account in the next lecture. In the present lecture I propose to give an account, to some extent critical, of the Classical Mechanics which is associated with the names of Galileo and Newton, in which the fundamental concepts employed are those of Force, Mass, Time, and Space. The Classical Mechanics, as originally conceived, is the Mechanics of gross bodies, or Molar Mechanics; it is non-atomic in that it does not assume matter to have an atomic, or a molecular constitution, although, as we shall see, it requires the assumption that matter is indefinitely divisible into parts. The notions of force and mass which influenced the building up of the Classical Mechanics were for the most part realistically and concretely conceived; and the mode in which they were employed was largely directed by the notion of efficient causation. Consequently, a considerable amount of alteration of what have been the traditional modes of presenting the theory is requisite, in order that it may be stated in the form of a conceptual scheme in which it is made clear what parts of that scheme consist of definitions, and how far the postulations made in it have been derived from direct observations of the behaviour of actual bodies in motion under certain conditions.

The ideas as to the motion of heavy bodies which prevailed before the time of Galileo were so confused and conflicting that we may regard the foundation of the Science of Dynamics as due to Galileo (1564–1642). It should however be observed that a scientific treatment of Statics, which is concerned with the conditions under which bodies remain at rest, had been initiated by Archimedes, and possessed some considerable body of doctrine at the time of Galileo. The first problem which Galileo set himself to solve was that of describing the mode in which heavy bodies actually fall; and this independently of any attempt to answer the question why they fall. The first attempt which he made to describe the motion of a falling body consisted of a guess that the velocity attained by a body falling freely from a state of rest is proportional to the height through which it has fallen. This hypothesis Galileo abandoned, not because of a failure to verify it experimentally, but because he convinced himself that it led to contradiction. The idea that there is anything self-contradictory in such a motion, is however erroneous; such motion is conceivable, although it is not that of a falling body. Galileo then made the hypothesis that the velocity acquired is proportional to the time of the fall, and he correctly deduced as a consequence of this hypothesis that the height fallen would be proportional to the square of the time. In order to verify that this is the actual law of the motion of falling bodies, he first experimented on a smooth ball rolling down an inclined plane. As no pendulum clock then existed he measured the time of the motion by weighing the water which flowed through a small orifice at the bottom of a very large vessel full of water, during the motion of the ball. In this manner he confirmed his surmise that the height moved by the rolling sphere would be proportional to the square of the time of the motion from rest.

He then made the assumption, confirmed by an experiment in which he employed a simple filar pendulum with a heavy ball attached to it, that the velocity acquired depends only on the vertical height through which a body has moved from rest. He was then able to connect the acceleration of a freely falling body with that of a body moving along an inclined plane. In this manner Galileo ascertained the law of the fall of a body independently of any theory; thus obtaining a genuine scientific law by observation. A most important new conception introduced into Dynamics by Galileo is that of acceleration, the gradient of velocity with respect to time. He perceived that, when a body is in such circumstances that it is set in motion, or has its motion altered, it is the acceleration that characterizes the immediate manner in which these circumstances exhibit themselves. Before Galileo, the immediate condition of the production of motion was recognized in pressure on the body due to contact with another body, but it was quite unknown that it is acceleration, and neither velocity nor position, that is determined by the pressure. Thus Galileo's discovery paved the way for reaching the modern conception of force, as determining acceleration. Of great importance are Galileo's investigations of the motion of projectiles, which led him to the conception that the projectile has two independent motions, a horizontal uniform motion, and a vertical uniformly accelerated motion. Thus he introduced in this case the principle of compounding motions in accordance with the parallelogram law.

The next great contributor to Dynamical Science after Galileo was Christian Huygens (1629–1695), who invented the pendulum clock. One of his greatest discoveries is that of the existence and magnitude of the acceleration of a point which describes a circle uniformly directed towards the centre of the circle; its so-called centripetal acceleration. This is a case of acceleration in which the magnitude of the velocity remains unchanged, the acceleration representing the effect of the change in its direction. Huygens was the first to ascertain the magnitude of the acceleration due to gravity by means of pendulum observations. In connection with his determination of the centre of oscillation of a compound pendulum Huygens was led to a particular case of the principle that *work* is what determines velocity, in fact to a particular case of the modern principle that the change of kinetic energy in a system is equal to the work done upon the system. What was afterwards called the moment of inertia was also introduced by Huygens in this connection.

Apart from Newton's supremely important discovery of the universal law of gravitation, but in close connection with that discovery and his deductions from it, he laid down in a complete form the essential principles of Dynamics as they have been accepted by succeeding generations, although, as I have already stated, the form in which abstract Dynamics, as a conceptual scheme, is now presented differs, in some important respects, from Newton's own formulation in his laws of motion and the accompanying scholia. Although Newton had by no means emancipated himself from the idea that a complete scientific theory must provide an explanation of phenomena in accordance with the law of efficient causation, as is shown by his refusal to accept the notion of so-called action at a distance, his method of procedure is that of the ascertainment of actual facts which he then employs for the formulation of scientific laws. His view of scientific method is formulated in a set of rules[1] for the conduct of natural inquiry (the Regulae Philosophandi).

*Rule I.* No more causes of natural things are to be admitted than such as truly exist and are sufficient to explain the phenomena of these things.

*Rule II.* Therefore, to natural effects of the same kind we must, as far as possible, assign the same causes; *e.g.* to respiration in man and animals; to the descent of stones in Europe and America; to the light of our kitchen fire and of the sun; to the reflection of light on the earth and on the planets.

*Rule III.* Those qualities of bodies that can be neither increased nor diminished, and which are found to belong to all bodies within the reach of our experiments, are to be regarded as the universal qualities of all bodies.

If it universally appear, by experiments and astronomical observations, that all bodies in the vicinity of the earth are heavy with respect to the earth, and this in proportion to the quantity of matter which they severally contain, that the moon is heavy with respect to the earth in the proportion of its mass, and our seas with respect to the moon; and all the planets with respect to one another, and the comets also with respect to the sun; we must, in conformity with this rule, declare, that *all* bodies are heavy with respect to one another.

*Rule IV.* In experimental physics propositions collected by induction from phenomena are to be regarded either as accurately true or very nearly true, notwithstanding any contrary hypotheses, till other phenomena occur, by which they are made more accurate, or are

[1] *Opera*, Horsley's edition, Vol. III, pp. 2–4.

rendered subject to exceptions. This rule must be adhered to, that the results of induction may not be annulled by hypotheses.

I have already in an earlier lecture pointed out the defect in Newton's definition of absolute time, and urged the view that, in a purely conceptual scheme, an independent variable which takes as its values the numbers of the arithmetic continuum must be employed; in the application of the scheme to percepts, an interval of this continuum must be taken to correspond to the duration of some standard process, an interval of public time.

Newton's views concerning space and motion he has stated substantially in the following form[1]:

> Absolute space, in its own nature and without regard to anything external, always remains similar and unmovable.
>
> Relative space is some movable dimension or measure of absolute space, which our senses determine by its position with respect to other bodies and which is commonly taken for immovable space.
>
> Absolute motion is the translation of a body from one absolute place to another absolute place: and relative motion, the translation from one relative place to another relative place. And thus we use, in common affairs, instead of absolute places and motions, *relative* ones; and that without any inconvenience. But in physical disquisitions, we should abstract from the senses. For it may be that there is no body really at rest, to which the places and motions of others can be referred.

In these statements we can recognize, in a somewhat involved form, the distinction which we now make between conceptual space, that of abstract geometry, and physical space, the space of perceptual bodies. The conceptual space of ordinary Dynamics, that which Newton calls absolute space, is a three-fold ordered aggregate in which metrical relations of the Euclidean type are employed. As basis of the system of measurement in this space we postulate the existence of a

[1] *Opera*, Horsley's edition, Vol. II, pp. 6–12.

definite frame, usually conceived as a set of coordinate axes. In the conceptual scheme, a particle is regarded as being at rest or in motion according as its coordinates remain unaltered, or not, when the independent time-variable takes up varying values. This is then the conceptual definition of rest and motion.

The frame of reference being once for all fixed, conceptual motion may then be regarded as absolute. As to what we are to understand by a particle or a body in this absolute conceptual space, I shall speak later. Abstract Dynamics consists of a scheme of rules by which, having given certain specifications, it is possible to calculate the positions of all the conceptual bodies of a system of such bodies in this purely conceptual space. The whole set of such rules is so devised that these calculated motions may be employed for the description and approximate determination of the actual motions of a system of perceptual bodies in physical space. In order that the conceptual scheme may be applicable in this manner, it is necessary to define a mode in which the positions of conceptual bodies in conceptual space are to be made to correspond to positions of actual bodies in physical space.

The only meaning which we can attach to the motion or rest of a body in physical space is that it is in motion or at rest relatively to some other body, taken as a standard; in other words, motion and rest in physical space are purely relative. In order then to employ the conceptual scheme for the description of actual motions, it is necessary to fix some frame in physical space which shall be taken to correspond to the conceptual frame, or coordinate axes, the existence of which has been postulated. This can only be done by taking some particular body, regarded as unchangeable in its shape and dimensions, to correspond to a conceptual body at rest, and defining the frame as fixed in this body. Or some more complicated process may be employed for fixing upon a frame of reference, the basis of which however always rests upon a choice of actual material bodies. The success of the conceptual scheme of abstract Dynamics consists in the fact that it is possible to determine such a frame of reference in physical space that, when the positions of the bodies whose motions are under investigation are measured with reference to the frame, their actual positions at different instants of public time are found to correspond, with a sufficient degree of approximation, to the calculated positions of the corresponding conceptual bodies in the conceptual scheme. What frame of reference is taken in any particular case depends upon the particular motions to be investigated, and upon the degree of precision that is requisite in the determination of those motions. For the purposes of representing motions of bodies in this room it will often be sufficient to take a frame fixed relatively to the walls of the room, say a vertical line and two perpendicular lines fixed on the floor. For more delicate observations this will give an insufficient determination of motions; when, for example, the fact of the rotation of the earth must be taken into account, we take a frame determined by the directions of the so-called fixed stars. For the determination of the motions of the planets we take a frame of reference in the sun, fixed relatively to the stars, to correspond to the conceptual frame at rest in conceptual space. The fact that such actual motions are sufficiently described in this manner by the corresponding motions in the conceptual scheme of Newtonian abstract Dynamics is the only ground upon which that scheme can be accepted as adequate for its purpose. There is no *a priori* reason why that scheme may not have to be superseded, in whole or in part, by some different scheme, in case it fails to describe with sufficient degree of approximation any actually observed set of motions. In particular there is no *a priori* reason for assuming that Newtonian

Dynamics is sufficient for the purpose of describing the motions of the sub-molecular parts of bodies, as all the direct verifications of the adequacy of the scheme have reference to the motions of molar bodies. Any extension of the scheme to the representation of motions in the microcosmic region, in which the motions cannot be directly observed, is a hypothesis, the value of which must be judged by those deductions from it which are capable of direct verification by actual measurements.

In the Newtonian Dynamics there is embodied a principle, that of inertia, which is closely related to the Newtonian conception of Force. Newton has stated this principle in one of the definitions which precede his laws of motion, and has further stated it in the first of his laws of motion. His definition takes the form that: "The resident force of matter is a power of resisting, by which every body, so far as in it lies, perseveres in its state of rest or of uniform motion in a straight line." This definition may be regarded as rendered superfluous by his later definitions of force, as they include the notion that all accelerations are dependent upon impressed forces.

Before we examine the precise meaning that can be attached to the principle of inertia, and its position in the scheme of abstract Newtonian Dynamics, it is advisable to glance briefly at the history of the principle; bound up as it is with the conception of force. The conception of a body persisting indefinitely in uniform rectilineal motion was quite unknown to the ancients. For example, Aristotle employs the impossibility of such motion in an argument of *reductio ad absurdum*. He believed that a body can only be moved by the action of another body which is continually in contact with it. Under ordinary conditions this continuous contact is veiled from our eyes, because when we project a body, we at the same time impart a certain motion to the air, and this continues to act upon the projectile; in a vacuum this would not take place. In accordance with another theory, advanced by Hipparchus, a body that is set in motion has received from another body an impulse which continues to reside in the projected body after the contact with the other body has ceased. This impulse keeps it in motion, in a straight line, although that motion is not uniform, but gradually diminishes and finally ceases. There was no idea, at that time, that the velocity of the projectile could maintain itself without action from without. The same conception was expressed by Themistius, a commentator on Aristotle, who compared the impulse received by the body to the communication of heat to a body, which remains in the body for some time whilst the body gradually cools down to its original temperature. Another notion which was commonly held by ancient physicists is that uniform circular motion is a natural kind of movement which persists unchanged when not interfered with.

The idea of the relativity of motion, which by later Physicists was connected with the principle of inertia, was distinctly conceived by Cardinal Nicolas de Cusa (1401–1464) in the first half of the fifteenth century. De Cusa, who tried to demonstrate that the earth can move without our perceiving it, uses as an illustration the fact that a boat in rapid motion may appear to be at rest to persons in it who do not see the banks; but he failed to connect the idea of relative motion with the principle of inertia. Although he affirmed the possibility of indefinite motion in a straight line, he still accepted the Aristotelian doctrine of natural circular motion. He explained the fact that a smooth ball started in motion on a smooth floor continues in motion, by the persistence of the tendency of the ball to rotate in its rolling motion. It is the perfection of rotundity that causes the perpetuity of the motion. The persistence of the motion of the ball is, with him, persistence of the rotation, and not that of the translation.

Copernicus, like Aristotle, attributed to the celestial bodies a natural circular motion, and denied that this gives rise to the appearance of a centrifugal force. On the other hand, Kepler believed that the movements of the planets were due to a material emanation from the sun, and that each of them would stop dead in its orbit if the sun ceased to act. Benedetti, a precursor of Galileo, still believed that the impulse communicated to a projectile decreases continually with the time, but he regarded the motion of a projectile as compounded of the motion due to the original impulse which started it and of the natural motion due to its weight, although he did not understand the cumulative effect of the weight. Even Galileo in his earlier utterances seems not to have rid himself of the idea of the gradual diminution of the impressed impulse. He never ceased to regard the circular motion of the heavenly bodies as a natural motion, just as Copernicus and the Greek Philosophers had done, but he distinctly affirmed the perpetuity of rectilineal motion in a horizontal direction, and appealed for confirmation to the example of a ball rolling on a plane. The principle of inertia, although implied in his works, was not stated by him in its general form.

The first such statement in a complete form of the conservation of velocity in a straight line is due to Descartes, who emphasized the fundamental importance of the principle in the general theory of motion. It is not unlikely that he was influenced by the writings of Galileo which had already been published when he formulated the principle for the first time. He divided the principle into two statements, the first referring to a body at rest, and the second to a body in motion. Thus he says:

> A body when it is at rest has the power of remaining at rest and of resisting everything which could make it change. Similarly when it is in motion, it has the power of continuing in motion, with the same velocity and in the same direction.

A similar statement of the principle in two parts was

given by D'Alembert, who attempted to demonstrate it by means of the principle of sufficient reason, without any appeal to experience except as regards the mere existence of motion.

Other attempts to give an *à priori* proof of the principle have been made by Kant and by Maxwell. The latter advanced what has at least the appearance of being a proof[1] of the principle by a *reductio ad absurdum*. He supposes that the movement of a body left to itself might gradually cease, in which case it would have a negative acceleration. This would be changed into a positive acceleration if we considered the motion relative to some body to which an appropriate motion was assigned. Maxwell infers that the law has no meaning unless the possibility of defining absolute rest and velocity be admitted, and argues that the denial of the law contradicts the only doctrine of time and space which we can form. The same argument might however be applied to prove the impossibility of uniform motion. The fact is that, in physical space, there is no meaning in the assertion that a body moves uniformly in a straight line, unless some material frame is specified with respect to which the motion is measured. However, a definite meaning exists when the body is taken to be a conceptual body moving in conceptual space, in which all positions are assigned absolutely by numerical specifications. The reluctance which Natural Philosophers have shown sharply to disentangle the conceptual statement of scientific laws and theories from statements relating to percepts has not only obscured the real nature of Science but has introduced much confusion into the formulation of results. It is essential for clarity of meaning in scientific theories to distinguish quite clearly between statements of facts of observation and postulations of the conceptual scheme which is designed to summarize and describe those facts.

An example of the inconvenience I have referred to

[1] *Matter and Motion*, Larmor's ed., pp. 28, 29.

is exhibited in proposals which have been made by C. Neumann, Streintz, and others, to do away with the difficulties which arise in making statements about the motion of actual bodies in physical space by assuming the existence of some standard body. C. Neumann assumes[1] the existence, somewhere in physical space, of a body, called the body Alpha, which is completely immovable; and all motions of other bodies are referred to this body Alpha. Such a body is a mere figment of the imagination, and can certainly serve no purpose in actual measurements of motion. This notion of a body Alpha occurred to both Newton and Euler, but was rejected by them as unsatisfactory. Streintz gives[2] the name "fundamental body" to some body which can be regarded as independent of the bodies that surround it, and is ascertained by the aid of pendulum experiments not to be in rotation. He then regards the principle of inertia as affirming that, when any body is not subject to external influence, it describes uniformly a straight line when referred to "fundamental coordinates" fixed in the standard body. It seems preferable in every way to regard the "fundamental body," or "the body Alpha," merely as the postulated ideal frame with reference to which rest and motion in conceptual space are reckoned, instead of regarding it as an actual body in physical space. If luminiferous ether could be regarded as an actual substance filling all physical space, it would be natural to regard it as affording the means of describing the motion of actual bodies; a body at rest would then be one which was at rest relatively to the ether. But in accordance with the view which I have maintained in these lectures, the ether is, if indeed it continues to retain any place in the Science of the future, a concept introduced for the purposes of scientific theory, and not

[1] *Ueber die Prinzipien der Galilei-Newton'schen Theorie*, Leipzig, 1870.

[2] *Die physikalischen Grundlagen der Mechanik*, Leipzig, 1883.

a percept. Moreover, as I shall explain in a later lecture, all attempts to detect the supposed velocity of a body relatively to a material ether have proved fruitless.

The question arises as to what is the true nature of the principle of inertia. The view that it is an *à priori* truth is untenable unless the principle of efficient causation, or that of sufficient reason, is employed; and the employment of such principles is not necessary for the purposes of Natural Science. It is hardly possible to maintain that the principle of inertia is a direct description of observed facts. We are not acquainted with any actual bodies which satisfy the conditions of validity of the principle. No actual body is, or can be, so isolated from other bodies, as to be removed from conditions dependent on those other bodies; and as we have already seen, when an actual body is said to be at rest, that means relatively to some other material system. Uniform motion in a straight line relatively to, say, the earth, would neither be uniform nor rectilinear relatively, say, to the sun. The observation of a ball moving on a smooth floor, which is usually appealed to as an experimental proof of the principle, does not really establish it as an empirical law, or even as an approximate fact of observation, partly on the ground I have just mentioned, but also because the ball, having weight, and being in contact with the floor and the surrounding air, does not satisfy the condition of being under no forces. That, apart from friction, the weight and the pressure of the ground, taken together, do not affect the horizontal motion, can only be inferred when other parts of the dynamical scheme are taken into account. Thus the law of inertia is not a direct result of facts of observation; and this probably accounts, to some extent at least, for the length of the period during which it was not discovered, or was not generally accepted. It follows then that the law of inertia cannot be taken in isolation, but must necessarily be regarded as forming a part of a complete conceptual scheme of Dynamics. The exact position of the principle can only be determined when the whole of Newtonian Dynamics is taken into account, including the doctrine of the relation of motion with force. We shall see that the principle appears in this conceptual scheme in the form of a definition. On the impossibility of regarding the principle of inertia either as an *a priori* truth, or as embodying a direct result of observation, Poincaré has remarked[1]:

> If it be said that the velocity of a body cannot change, if there is no reason for it to change, may we not just as legitimately maintain that the position of a body cannot change, or that the curvature of its path cannot change, without the agency of an external cause? Is, then, the principle of inertia which is not an *a priori* truth, an experimental fact? Have there ever been experiments on bodies acted on by no forces? and, if so, how did we know that no forces were acting?

One of the main points in which Newton's theory of Dynamics contains an advance beyond the conceptions of Galileo or Huygens is in the generalization of the conception of force. For the origin of the notion of force we must go back to the feeling of exertion which we have when our muscles are employed in changing the position of actual bodies. The notion was extended to cover the case of the interaction of two bodies in contact with one another; the force then acting on one of the bodies being regarded as the efficient cause of the motion produced in it. The explanation of motion as due to contact action appeared to explain the phenomenon in the sense to which I have referred in an earlier lecture, that it reduced it to the familiar case of contact of the human body with another object. That a precise examination of what occurs in case of contact between two bodies must be undertaken in order to explain the effects of pressure was not recognized until modern times. We now know that such an examination

[1] *Science and Hypothesis*, 1905, p. 91.

involves conceptions as to the relations between the smallest parts of the bodies in contact, and that this reduction to the corpuscular domain gives rise to problems as difficult as the original one which led to the whole investigation. The futility of the attempt to indicate the point at which efficient causation is to be found is exhibited by the endless regress to which we are led when we attempt to account for the interactions of the smallest parts of the bodies. If we refuse to regard a body as consisting of corpuscles of any kind, but take it to be absolutely homogeneous, we are unable to begin to understand the nature of contact action.

Involved in the discovery by Newton of the law of gravitation is the extension of the notion of force to such cases as the so-called forces of attraction between the sun and the planets. Newton himself, under the influence of the ancient idea that all forces must be due to contact, was, as I have already observed, never satisfied with the bare fact of the existence of these forces, but thought that they were inconceivable without some mechanism, such as a homogeneous ether, which should condition the forces by ultimate reduction to contact action. However, the existence of forces, regarded as due to the action of one body on another body at a distance, came to be accepted as part of the apparatus of Dynamics, and was accepted by Newton himself with the reservation just indicated.

Newton's mode of measuring a force depends upon the notion of the mass of a body as a measurable quantity. The formulation by Newton of the definition of mass as the quantity of matter in a body measured by the product of its volume and density has the defect that it is circular, since the conception of density is meaningless unless that of mass has been previously defined. Newton ascertained by means of experiments that a body has a property, the mass, distinct from its weight, whereby the quantity of motion of the body is determined. He also showed that the mass of a body may, in one and the same place, be measured by its weight, and that the ratio of the weight to the mass is independent of the chemical composition of bodies. There is a lack of clarity in the deductions from these experiments, owing to the want of a satisfactory definition of the mass, or quantity of matter which a body contains.

Newton's first two laws of motion contain a statement of the relation of an impressed force to the change of motion of the body, which is really equivalent to a definition that he had previously given; and this we now express in the form that the force is proportional to the acceleration it produces in its own direction. As, according to this definition, there is no acceleration without force to produce it, the first law is really contained in the second. Newton's third law of motion, that the forces between two bodies which influence one another are equal and in opposite directions, only attains a precise meaning when the notion of mass as having numerical measure has been made precise by means of a definition. If we consider the action and reaction between two bodies as equal, the ratio of their relative masses may be taken to be the inverse ratio of their accelerations in opposite directions. This only amounts to the definition of the mass-ratios of the two bodies. If now we consider a third body, we can then define the mass-ratio of any pair of them. It was pointed out by Mach that observation is required to establish the fact that to each one of the bodies a number may be assigned, such that, for each pair of the bodies, their relative masses are given by the ratio of the two numbers assigned to the bodies of the pair. Thus a regular and consistent mode of assignment of mass-numbers can be set up, such that the law of action and reaction between any pair of bodies will hold good. That the mass-number assigned to a body is independent of change in its form, or chemical and thermal conditions, is a fact which must be verified by observation.

The rule that the force acting on a body is equal to the product of its mass and its acceleration, depends upon the possibility of measuring three magnitudes, the force, the mass, and the acceleration. I have pointed out that mass is not capable of measurement independently of the notion of the equality of two forces. The impossibility of an independent measurement of force has been trenchantly signalized[1] by Poincaré as follows:

> When we say that force is the cause of motion, we are talking metaphysics; and this definition, if we had to be content with it, would be absolutely fruitless, would lead to absolutely nothing. For a definition to be of any use it must tell us how to measure force; and that is quite sufficient, for it is by no means necessary to tell what force is in itself, nor whether it is the cause or the effect of motion. We must therefore define what is meant by the inequality of two forces. When are two forces equal? We are told that it is when they give the same acceleration to the same mass, or when acting in opposite directions they are in equilibrium. This definition is a sham. A force applied to a body cannot be uncoupled and applied to another body as an engine is uncoupled from one train and coupled to another. It is therefore impossible to say what acceleration such a force, applied to such a body, would give to another body if it were applied to it. It is impossible to tell how two forces which are not acting in exactly opposite directions would behave if they were acting in opposite directions. It is this definition which we try to materialise, as it were, when we measure a force with a dynamometer or with a balance.

The impossibility of measuring independently of one another the three magnitudes, force, mass, and acceleration, as they are employed in Newton's scheme, and of then verifying experimentally that force is proportional to the product of mass and acceleration, leads to the conclusion that force must be defined as the product of mass into acceleration, and thus does not appear in the scheme as an independent conception. Thus the first

[1] *Science and Hypothesis*, 1905, p. 98.

two laws of Newton, regarded as part of a conceptual scheme, consist in reality of a definition of force as the product of the mass-number of a body into its acceleration in a particular direction, which direction is taken to define the direction of the force. The fact of observation which leads to the formulation of Newton's third law of motion is that a consistent set of mass-numbers can be assigned to a set of bodies so that any two of the bodies have towards each other accelerations inversely proportional to the mass-numbers of the two bodies.

Newton gave a deduction of the principle known as the parallelogram of forces from his second law of motion. But in order that this deduction may be valid, the assumption is required that the accelerations produced in a body by two other bodies are independent of each other; so that the acceleration which $B$ produces in $A$ is independent of the fact that $C$ is also producing an acceleration in $A$. That this is the case in actual bodies is not self evident, and can only be established by experience; in the conceptual scheme the assumption must appear as a postulate. If it be assumed that all the forces between bodies are central forces, depending in magnitude only on the distances between the bodies, Newton's deduction will be justified; but if we do not make this assumption we have no means of resolving the acceleration of $A$ into components, one of which we regard as due to $B$, and the other as due to $C$; and in that case we have no means of assigning definite mass-numbers to the three bodies, and therefore the principle of equality of action and reaction would no longer be valid.

It would thus appear that the validity of the Newtonian scheme is dependent on the assumption that the forces between any two bodies are central. But this assumption has really been implied in the statement of the laws of motion. For the acceleration of a body has no definite meaning unless the body be either of negligible dimensions, or unless the acceleration be taken to mean that of some one particular point in the body; for in general a body can move not only translationally but also rotationally, and thus different parts of a body may have different accelerations. In order that Newton's laws of motion may have a precise meaning, it must be assumed either that the bodies referred to in them are considered as masses concentrated at points, or as bodies of non-negligible size which are equivalent to such concentrated masses at their centroids; the forces between any pair of such bodies being along the straight lines joining their centroids. This was realized with sufficient degree of approximation in the cases of motion of bodies of the solar system to which Newton applied his Dynamics.

The extension of Newton's Dynamics to the general case of bodies of various shapes, acting and reacting on one another, can only be made by a process of integration, in which the bodies are divided up into a large number of very small parts, so that, very approximately, any two of such parts can be regarded as two points at which the masses of the respective parts are concentrated; and it is assumed that the forces which act between two such parts, whether they belong or not to one and the same body, are central, that is along the line joining these parts. By means of the procedure of passing to a limit, that is of integration, the dynamical scheme can be built up in a form applicable to a system of bodies. This scheme takes the form of a set of differential equations called the equations of motion of the system. In the conceptual scheme, the bodies consist of defined portions of geometrical space, which are movable in that space, retaining in the case of a rigid body their dimensions unaltered; the mass of each such body appears in the equations of motion only as a coefficient, to which in any special case a numerical value is assigned. The position for each value of the time-variable, of any ideal rigid body, is denoted by a set of numbers which represent coordinates relatively to the frame of reference with respect to which all positions are referred. The motion of a point is denoted by a functional relation between its coordinates and the time-variable. As we have seen, the principle of inertia and the conception of force appear in the scheme as definitions, and the masses of bodies or of parts of bodies appear as coefficients to which numerical values may be assigned. The hypothesis of central forces between elementary parts of bodies, of magnitudes dependent only on the distance, is also an essential part of the scheme. The equations of motion of the conceptual bodies suffice to determine their motions, and the results obtained suffice for the approximate determination of the motions of the perceptual bodies in an approximately isolated system, provided a correlation between measurements in physical space and in geometrical space can be set up.

To do this it is requisite to assign, for each special problem, a frame of measurement in physical space which may be taken to correspond with the absolute conceptual frame by means of which all positions in the geometrical space are measured. This frame of reference must be fixed relatively to some actual body, and when it consists of rectangular axes it may be spoken of as a Newtonian set of axes. The utility of the conceptual dynamical scheme for its purpose of describing the motions, with sufficient accuracy, of an approximately isolated material system, depends upon the verifiable fact that it is possible to determine, with sufficient degree of approximation for the purpose on hand, a Newtonian set of axes[1] suitable to the particular system. If such a Newtonian set of axes has been determined in any particular case, any other set of axes parallel to the

[1] By some writers the terms "Galilean axes" or "inertial frames" are employed.

former will also be a Newtonian set, provided its origin moves with uniform velocity relatively to the origin of the first set. This principle is subject to the test of experience, but it is an essential principle of Newtonian Dynamics. But a set of axes in rotation with regard to a Newtonian set will not be itself a Newtonian set, unless additional forces, including the so-called centrifugal forces, are introduced into the system.

In this connection, it has frequently been maintained that, in physical space, the rotation of a body is absolute, whilst its motion of translation is relative; and thus that absolute directions, unlike absolute positions, can be defined. Newton's celebrated experiment with the water in a bucket is frequently appealed to in support of this contention. As long as the bucket is at rest the water in it has a flat surface, but if the bucket be made to rotate about its vertical axis, a motion of rotation is gradually set up in the water, when its surface is no longer plane, but has the form of a concave surface of revolution. It is then suggested that there is an absolute difference between axes fixed relatively to the bucket when it was at rest, and axes fixed relatively to it when it is in motion. The only relevant fact would however appear to be that it is possible to determine with sufficient approximation, in any concrete case, the directions of a Newtonian set of axes; and other axes in rotation relatively to these do not form a Newtonian set. It can however be shown that the directions of the axes of a Newtonian set have no valid claim to be regarded as absolute.

In the first place a Newtonian set of axes, determined for a particular isolated system, need no longer be a Newtonian set for an enlarged system. For example, in this room, a vertical line and two horizontal lines, fixed relatively to the walls, would form a Newtonian set of axes, for the purpose of describing the motions of ordinary bodies in this room; and in particular the rotation of Newton's bucket would mean rotation relatively to these axes. But for the purpose of describing the motion of Foucault's pendulum, these axes would no longer form a Newtonian set. For that purpose we should have to take the axis of the earth and two other axes fixed in direction relatively to the stars. Again, if we had to take account of the slow precessional motion of the earth's axis, these last would no longer form a set of Newtonian axes. However far we may proceed in this manner to fix Newtonian axes which shall be sufficient for the motions of ever more extensive, or complicated, systems of bodies, we are unable to assert that the axes obtained will necessarily be a Newtonian set for the purpose of measuring the motions of every system of bodies whatever, which we may at any time have to consider. Moreover the approximate determination of a Newtonian set of axes is insufficient to determine, even approximately, absolute directions, because the smallest error in the determination will be of a cumulative character, involving the existence of a large, and no longer negligible, error after a sufficiently long time. The rotation of a set of axes, however slow that rotation may be, will, in a sufficiently long time, produce a large deviation of the axes from their original position. There is thus no warrant for the assertion that it is possible to assign directions in physical space which may, for every possible purpose, be regarded as Newtonian axes. On this ground, and also because the Newtonian system of Dynamics, although the simplest system for the descriptive purposes for which it was devised, is not the only possible system that might be employed, and may not even suffice for all purposes, it is not correct to assert that absolute directions in physical space can be determined by any means at our disposal; nor are we compelled to conceive absolute directions in physical space as existent.

To regard the earth as rotating round its axis is

convenient for the purposes of Descriptive Astronomy, and it is also convenient for the purposes of the dynamical description of the motions of bodies in the solar system. But however convenient is the assertion that the earth rotates, it still remains a convention, although the opposite assertion would vastly complicate both the astronomical and the dynamical schemes which would then have to be employed. In connection with the new general theory of relativity, both spatial relations and dynamical theory have to be considered from a point of view which necessitates a rediscussion of such matters as the introduction of centrifugal, and other, forces when non-Newtonian sets of axes are employed in connection with the motions of a material system.

# IX

# THE CONSERVATION OF MATTER AND ENERGY

THE notion that, in all the multifarious changes which we perceive to take place in the material world, there must be permanent elements which persist unaltered through all these changes, has been, throughout the history of Science, one of the guiding ideas which have ultimately given rise to such formulations as those contained in the expressions Conservation of Matter, Conservation of Weight, Conservation of Mass, and Conservation of Energy. The principle of permanence expressed in such a formula as that "nothing is created and nothing destroyed" has usually been regarded as an *a priori* principle closely related to the principle of causation. The very generality of this *a priori* principle has prevented it from functioning as an efficient guide to the determination of the precise elements in the perceptual world to which the characteristic of persistence through all transformations appertains. Thus the actual successes of scientific investigation in this order of ideas have consisted in the ascertainment, by experiment and observation, of empirical laws, the law of conservation of mass, and the law of conservation of energy. Such empirical laws are ascertained to express definite facts relating to a considerable, but limited, range of observed phenomena, the conservation which they express being of such a character that it is expressed quantitatively, in a numerical form. The laws are then adopted as hypothetical principles in conceptual theories which relate to ranges of phenomena wider than those to which the empirical verification has been in the first instance applied.

The value of the laws in their general form must depend upon their success in performing their functions of description and prediction in relation to new classes of phenomena to which they are tentatively applied. The *a priori* principle in its general form, to which I have referred, as a metaphysical principle expressing a supposed necessity of thought, need not be accepted as a part of the foundations of Natural Science, whatever its actual influence may have been upon the minds of scientific investigators in the past. The main difficulty as regards the principle of the conservation of matter, the principle that matter is neither created nor destroyed, is that of forming a clear conception of what it exactly is that is conserved. If we regard matter as a construct including a complex of physical properties, extension, colour, hardness, conductibility of heat and electricity, etc., we have ample and obvious evidence that these properties do not persist unchanged, but are subject to the largest changes in what we regard as one and the same material system. What then is to be understood by the statement that matter can be neither created nor destroyed; that is, by the principle of the conservation of matter? If we assert that what persists unchanged is a sub-stratum, substance itself, not identified with any or all of these physical properties, but regarded as their bearer, not only do we reduce the principle to one dependent upon a metaphysical theory, but we remove from it all possibility of verification. It then becomes a bare philosophical assertion which has no direct relation with the world of percepts, and is thus outside the domain of Natural Science. A real scientific law of conservation must contain an indication of some measurable quality or property of matter, which can be ascertained to remain unaltered in magnitude during the actual chemical and motional transformations occurring in the physical world. Not even what is regarded as a primary quality of matter, that of extension, is conserved as a measurable quantity unaltered through all transformations.

There is however one other property which we have come to associate with all matter, that of weight; this is estimated by the balance, the systematic employment of which by Lavoisier brought about what has been described as a revolution in chemical Science. If however we understand conservation of matter to mean conservation of weight, we are at once met with the difficulty that the weight of what we regard as one and the same piece of matter, when estimated by the spring balance, varies with the latitude of the place at which it is measured. Moreover, in accordance with the theory of gravitation, the weight would be very greatly altered if the matter were transported to another planet. However, in the chemical transformations which take place in any one locality, a verification of the principle of conservation of matter consists in a verification of the principle of conservation of weight. For different localities the differences of weight of one and the same object are eliminated, in accordance with the Newtonian and Galilean Dynamics, by dividing the weight by the acceleration due to gravity; this division yielding the measure of the mass of the body. Thus the conservation of matter now comes to be taken to mean the conservation of mass.

With the concept of mass I have already dealt more fully in connection with Dynamics. The actual mass of a body can be regarded as a quality which can be measured as derivative from the two measurements of weight and of acceleration. That the mass of a body is the amount of matter in it is a tautological statement which can only be taken to denote that the meaning assigned to the term quantity of matter is that it is the mass regarded as a measurable quality of the body. The principle of the conservation of matter, regarded as mass, has however a much wider meaning than that it is unalterable for one and the same body in whatever position it be placed, or however it be moving. It includes the assertion that mass, as a measurable quantity, is unchanged in amount throughout all the chemical and thermal changes that may take place in an isolated material system.

Thus the principle implies that matter, however it be sub-divided actually, or conceptually, may be regarded as having a quality, the mass or quantity of matter, which is measurable, and remains unchanged in total amount, during all motions, and all chemical, thermal, or other transformations. The only means we have in an abstract conceptual scheme of representing this assumed quality is by the employment of numbers in relation to conceptual bodies in geometrical space. In its abstract form, the principle asserts an invariant property of the sum of such numbers for the conceptual elements of a limited system. Only to a limited extent has this general law been verified experimentally; for the difficulties of measurement, and of securing complete isolation of the substances which undergo chemical transformation, are very great, being subject to errors difficult to take completely into account. Moreover, the conservation of mass through all motions is capable only of indirect verification, in connection with the verification of the adequacy of a particular dynamical scheme.

Accordingly the principle must be regarded as an hypothesis, which has been verified approximately in a large number of cases and is assumed by the Chemist and the Physicist to hold good as descriptive of relations in a large range of actual phenomena, but as subject to possible refutation in cases in which more refined methods of measurement are employed in connection with electromagnetic or other phenomena.

In the new electron theory of matter, of which I shall speak later, mass occupies a wholly different position from that which is assigned to it in the mechanical theory which has been here discussed. In accordance with the electron theory, the mechanical masses of bodies are no longer constant, but have a sensible variation when the bodies are set in motion with velocities comparable with the velocity of light.

A sketch of the history of the doctrine of the conservation of matter is sufficient to show that the establishment of the principle, in the modern form of that of the conservation of mass, was the result of a gradual evolution. The ancient atomists, not being in possession of the principle of inertia, made no distinction between mass and weight. In this they are in accord with the popular view held even to-day. For example, Lucretius, following Democritus in this order of ideas, appears to have regarded weight as an unalterable characteristic of all matter; in his poem the atoms are in motion on account of their weight; and this involves an identification of their masses with their weights. On the other hand, the school of Aristotle, radically opposed as it was in almost all respects to the views of the atomists, did not believe that all matter has weight. With the Aristotelians, the notions of matter and weight are kept quite distinct from one another; weight being regarded as an accidental quality of matter, like colour or temperature. In their view, weight is the resultant of two opposed qualities, heaviness and lightness. Fire has no heaviness, and earth no lightness. Water and air have both, one being preponderant in water, and the other in air. Plato had observed that the four elements are constantly transformed into one another; thus air and fire are concerned with the transformation of matter, as when water boils or when wood burns.

In the middle ages the Aristotelian views about matter were prevalent, although traces are to be found of the influence of the ancient atomists. Some of the Alchemists used reasoning founded on the consideration of weights; but they did not attach to weight the primary importance which it came later to possess, and they did not believe in its unchangeableness during transformations. Indeed some of them expressly mention change of weight as occurring in the transmutation of matter. Thus, for example, Geber writes (? eighth century A.D.): "By our artifice we easily form silver out of lead; in the transformation the latter does not preserve its own weight but changes into a new weight." This view of the Alchemists cannot be explained away as involving merely a reference to change of specific weight, or to the change of weight produced by absorption of matter from air or fire; indeed specific weight and absolute weight were constantly confused with one another, even as late as the seventeenth century. As long as the weight of a body was regarded as a merely accidental quality, like its colour, it was quite natural to suppose that it might be changed without the addition or subtraction of matter. Even Bacon held opinions not very different from those of the Alchemists. In some statements he affirmed the existence of absolutely light bodies, and that change of weight may accompany a change of state; but in other statements, probably following the ideas of the atomists, he affirmed the constancy of weight. It must always be remembered that, with the Alchemists, and generally with those under the influence of the Aristotelian conception of substantial forms, the question whether the weight remained quantitatively constant or not, during a transformation, seemed a matter of subordinate importance; the smallest change of quality was in their eyes of much greater interest. A knowledge of the relations of quantity being of little importance to the adherents of the philosophical doctrine of substantial forms, it is hardly possible to regard the notion of the permanence of mass, considered quantitatively, as a part of the stock of ideas of those who were under the influence of Aristotelian conceptions. For the notion of matter as underlying substance, differentiated from the accidental quality of weight, did not admit of quantitative measurement, although the substance was regarded as in some sense persisting through all changes. As the gradual emancipation from the Aristotelian conceptions took place, the conception of mass became clarified. It was formulated with tolerable clearness by Kepler, and also by Descartes, although with the latter, as with the Aristotelians, weight remained an accidental property, not possessed by all matter. That air has weight was generally recognized in the time of Descartes, but fire was still regarded as devoid of weight. With Descartes, matter being a plenum, the quantity of matter was indicated by its volume; thus he makes the assertion that "when a jar is full of gold or of lead, it does not contain more matter than when we think it empty." But he regarded the terrestrial matter as the only kind to be taken into account in mechanical action, and thus in reality he made a distinction between mass and volume, the mass appertaining to terrestrial matter only. He did not however recognize that mass and weight are in a fixed ratio; in fact the title of one of the chapters of his *Principles* is "*That their weight has not always the same proportion to their matter*," and here matter is to be understood as terrestrial matter. For a long period after the time of Descartes, the notion that weight is an accidental quality of matter prevented the general acceptance of the conservation of weight, although opinion on the point was by no means unanimous even in Descartes' own time. Jean Rey, in his essays, which were published before the actual appearance of Descartes' *Principles*, attempted to give an *a priori* demonstration that weight is conserved in every transformation. Moreover he gave an experimental proof that air is heavy, and that in the formation of lime the increase of weight is due to material taken from the air. The persistence of disbelief in the invariability of weight is exhibited in the utterances of many writers, even until the end of the eighteenth century. Thus Hobbes declares that "all accidents other than greatness or extension can be engendered or destroyed," thereby leaving no room for the conservation of weight or of mass. Leibniz, who had a clear conception of mechanical mass, states that "water contains in equal volume as much matter as mercury, only to the matter belonging to the water there is added a foreign non-heavy matter which is between its pores" for "it is a strange fiction to make all matter heavy."

Newton, who did not admit the existence of imponderable matter, showed experimentally that the weight is proportional to the mass of a body; and Huygens stated definitely that quantity of matter is measured by its weight. The separate lines of work of Physicists and Chemists make it difficult to ascertain the views of the Physicists of the seventeenth and eighteenth centuries on the nature of chemical phenomena. Almost the only exception to this separation was the work of Robert Boyle, both Physicist and Chemist, who appears to admit the principle of conservation of weight, without however explicitly formulating it. In the seventeenth century, although it was generally admitted that air has weight, it was not generally believed that this is the case for fire. In the eighteenth century, however, we find that Berkeley regarded the increase of weight of some metals when heated, for example in the case of antimony, as due to the fire in the sun's rays; he remarked that we do not know the weight of a solar ray. Diderot stated that "the fire of our furnaces considerably augments the weight of some matter, such as calcinated lead."

The special form which the notion of imponderable substance took in the minds of the Chemists of this period was the theory of phlogiston, a substance invented to account for thermal phenomena. Phlogiston was endowed with negative weight, and since it intervened in all chemical reactions there was no difficulty in conceiving that its admixture with matter prevented the conservation of weight in a chemical transformation. How little attention was paid by Chemists to all questions relating to quantities is illustrated by the fact that one of the chief French Chemists of the day, Macquer, on hearing that Lavoisier was preparing an attack upon the theory of phlogiston, stated that he had been disquieted for a moment, but was reassured when he learned that Lavoisier's objections were based solely on quantitative considerations. The definite establishment of the principle of conservation of matter, by the systematic use of the balance, is mainly due to Lavoisier, and may be dated from his memoir on *The change of water into earth*, published by the French Academy in 1773. This was followed in 1774 by a work in which, by the employment of the balance, he decided between the rival theories of Black and Meyer as to what happens in chemical transformations. First applying the principle without explicitly formulating it, he afterwards gave a precise statement of it. In his elementary treatise on Chemistry he remarks that "any matter can furnish nothing in an experiment beyond the totality of its weight," and further "the determination of the weight of materials and their products before and after experiments is the basis of everything useful and exact in Chemistry." "In every operation there is an equal amount of matter before and after the operation."

In 1774, Lavoisier described the commencement of his fundamental discoveries relating to combustion. He verified that various metals, when heated in a closed vessel, receive an increase of weight, and that the amount of air in the vessel is diminished; he showed that the loss of weight of the air is nearly equivalent to the increase of weight of the metal. A slight increase in the weight of the whole vessel he properly attributes to an exterior deposit due to the fire. In this way he provides a refutation of the idea of the intervention of the element fire, and shows that the increase of weight can only come from the air. Even after the composition of water became known, and the phenomenon we call oxidization, the new conceptions of Lavoisier only triumphed slowly. They do not appear to have been completely accepted by Priestley or Cavendish. Scheele regarded heat as a compound of phlogiston and oxygen; both of them he thought of as heavy, but supposed that together they give rise to an imponderable substance. Heat united with very little phlogiston is transformed into light, but united with a great quantity it becomes inflammable air, that is hydrogen. Even Lavoisier shows traces of analogous conceptions. He regarded oxygen as resulting from a combustion of ponderable matter and an imponderable fluid, caloric. Heat he regarded as a material element contained in a gas, and the conception which he had of gases was related, by means of intermediate hypotheses, with that of imponderable fluids.

After the vicissitudes which I have sketched, the principle of the conservation of matter, regarded as measured by dynamical mass, has come to be accepted as an empirical law which is applicable within a large range of phenomena, although some Chemists have maintained that it is possible to detect deviations from the law, which cannot be assigned to the effect of instrumental errors or to disturbing factors not easily taken fully into account. Moreover the flood of light which has lately been thrown upon the properties of radioactive substances, has suggested views in which dynamical mass no longer holds its former position as fundamental and irreducible. It has indeed been suggested that, in accordance with the electrical theory of matter, there would be nothing surprising in a change of weight owing to chemical reactions.

The origin of the principle of the Conservation of Energy is much more modern than that of the principle of the Conservation of Matter. In its general form, the principle of the Conservation of Energy dates back only to the middle of the nineteenth century, but in its restricted form, as a principle of Mechanics in the narrower sense of the term, it is in the ideas of Descartes, Leibniz, and especially of Huygens, that we find its origin. The notion of matter is one formed by common sense, but the conception of energy has been created by Science for its own special purposes. It seems therefore quite natural that the doctrine of the conservation of energy should have arisen at a much later stage in the history of Science than that of the conservation of matter, at least in a crude form. The notion of *work*, as measured by the product of a force into the displacement, in the direction of the force, of the body on which it acts, is due to Galileo, who showed that in simple mechanical machines the work of the resistance in a displacement is equal to that of the power. He concluded that, by the aid of such machines, it is impossible to create work, but he did not show that work cannot be destroyed. For the case of a falling body, he gave the formula which expresses the principle of energy.

The next step in the direction of setting up a general principle relating to the movement of bodies was taken by Descartes, who attempted to set up a principle of the conservation of motion through all changes in the physical world. In this attempt he made the mistake of taking the sum of the products of the masses into their velocities, instead of the squares of the velocities, as representing the quantity which is conserved. This error was pointed out by Leibniz in a treatise bearing the title "*A short demonstration of a Remarkable Error of Descartes and others, concerning the Natural Law by which they think that the Creator always preserves the same Quantity of Motion; by which, however, the Science of Mechanics is totally perverted.*" Leibniz distinguished between simple pressure (*vis mortua*) and the force of a moving body (*vis viva*), but he confused the question of the right measure of force with that of the constancy of momentum and of the kinetic energy of a system. Neither the Cartesian nor the Leibnizian measure of the effectiveness of a body in motion is, as Leibniz observed, to be identified with the Newtonian measure of force. Leibniz, like Descartes, regarded the principle formulated by him as embracing all the phenomena of the Universe. He justified the principle of the conservation of *vis viva* by an appeal to the principle of causation, in the form that the effect is equal to the cause.

To attempt to demonstrate this law (he writes) would obscure it. Indeed, everyone regards it as an incontestable axiom that every efficient cause cannot perish, either totally, or in part, without producing an effect equal to the loss. The idea that we have of the *vis viva*, as it exists in a body in motion, is something absolute, independent, and positive; that it remains in the body even if the rest of the Universe were annihilated. It is then clear that, if the *vis viva* of a body diminishes or increases on impact with another body, the *vis viva* of this other body must change, increase, or diminish by the same quantity.

His view of the scope of the principle appears clearly in the following passage:

I had maintained that active forces are conserved in the world. It has been objected that two soft or inelastic bodies, when they collide, lose part of their force. I answer that this is not so. It is true that the "wholes" lose it in respect of their total motion, but the parts receive it, being agitated internally by the force of the collision. Thus the loss ensues only in appearance. The forces are not destroyed, but dissipated amongst the minute parts. That is not as if they were lost, but it is like the changing of large coins into small ones.

The truth of the principle had previously been demonstrated by Huygens, who had however formulated it without indicating the great generality of its scope. He distinguished between the conservation of *vis viva* and that of momentum in his statement:

> The quantity of motion possessed by two bodies may be augmented or diminished by their encounter; but there remains always the same quantity on the same side, if we subtract the quantity of opposite motion. The sum of the products of every hard body multiplied by the square of its velocity is always the same before and after the encounter.

It must be observed that, both for Descartes and for Leibniz, the world consists only of matter in motion, and there exists no action at a distance. Consequently they did not admit the existence of what we call potential energy, so that, for them, the principle of energy consisted in the constancy of the total kinetic energy. It should also be observed that Leibniz, in the passage I have quoted, in speaking of the dissipation of the molar energy amongst the smallest particles of a body, does not seem to have considered this transformation as equivalent to the production of heat.

In the eighteenth century, the conception of heat as a substance gradually gained upon the Cartesian idea of heat-motion. The amount of this substance was supposed to be conserved in its passage from one body to another. When it ceased to manifest itself by means of the thermometer, heat was regarded by Black as still present, but as latent heat, capable of manifesting itself in certain conditions, and thus his conception of latent heat was analogous to our conception of potential energy. Even the invention of the steam engine produced no immediate change in this conception of the substantiality of heat. Watt and his successors failed to attain to the view that thermal changes indicate any relation between heat and mechanical motion. However, towards the end of the century, Lavoisier and Laplace tentatively related the production of heat by friction with the conception of heat-motion, and defined the amount of heat as the sum of the products of the masses of the molecules into the squares of their velocities.

Early in the nineteenth century, direct experimental demonstrations were obtained, by Rumford and by Humphry Davy, of the transformation of motion into heat. The concept of latent *vis viva*, called by Poncelet *work*, was formulated (1803) by Lazare Carnot; this concept is that now known as potential energy. In 1839, the engineer Séguin, in a work on the construction of railways, remarked that:

> As the theory at present adopted would lead however to this result (perpetual motion) it appears to me more natural to suppose that a certain number of calories disappear in the act of producing mechanical force or power, and conversely; and that the two phenomena are bound together by conditions which assign to them invariable relations.

It appears however that, before Séguin, the principle of the equivalence of heat and mechanical energy had been conceived in its generality by Sadi Carnot, who obtained by calculation an estimate of the mechanical equivalent of heat. In his earlier work, Carnot had employed the material theory of heat, and his later formulation of the modern theory was preserved only in manuscript notes which remained unpublished until 1871. But it was the experimental researches of Joule, published in 1843, that brought prominently before the scientific world the theory of the equivalence of heat and mechanical energy. The earlier estimates obtained by Joule of the number of foot-pounds of work equivalent to the heat required to raise the temperature of a pound of water one degree Fahrenheit were widely discordant, varying between 742 and 1040; but as the result of a later series of experiments he obtained 770 foot-pounds as the equivalent; and this is not very different from the value now accepted. Joule did not however doubt that the value of the equivalent exists as a definite number, notwithstanding the considerable variation in his experimental determinations of it. His certitude on the matter was derived from his conviction of its *a priori* necessity.

We might reason *a priori*, (he writes)[1] that such an absolute destruction of living force cannot possibly take place, because it is manifestly absurd to suppose that the powers with which God has endowed matter can be destroyed any more than that they can be created by man's agency; but we are not left with this argument alone, decisive as it must be to every unprejudiced mind.

The work of J. R. Mayer on the conservation of energy, which appeared in the year before the first publication of Joule's experiments, was the earliest publication on the subject in its modern form. In 1843 there also appeared a work on the same subject by a Danish savant, A. Colding. The work of Mayer was mainly guided by his philosophical ideas, and did not include any experimental verifications of the principle, such as those of Joule. He appeals to the old idea that forces are causes, and that the cause is equal to the effect. His determination of the equivalent of heat by a calculation presupposes the existence of the constant relation of equivalence, as was the case in the similar calculation made by Sadi Carnot. Colding makes the assumption that energy persists as a kind of indestructible, non-material substance. Thus he writes[2]:

Since forces are spiritual and immaterial beings, since they are entities which are known to us only by their empire over nature, these entities must doubtless be very superior to every existing material thing; and as it is evident that it is by forces alone that the wisdom that we perceive is expressed and that we admire in nature, these powers must be in relation with the spiritual, immaterial, and intellectual power itself which guides the course of nature; but if this be so it is absolutely impossible to conceive that these forces should be anything mortal or perishable. Without doubt consequently they must be regarded as absolutely imperishable.

[1] *Scientific Papers*, Vol. 1, p. 269.
[2] *Annales de Chimie et de Physique*, (4), Vol. 1 (1864), p. 467.

So far, the scope of the principle of the Conservation of Energy has been confined to the mechanical domain and to the equivalence of mechanical energy and heat, but in the well-known treatise published by Helmholtz in 1847, there is consistently developed the doctrine that the conservation of energy is applicable to all departments of Physics. This work gives ample evidence that the author, like Joule, Mayer, and Colding, originally regarded the principle as one which follows from the principle of causation.

In a very interesting passage in the introduction to his treatise Helmholtz writes:

It is the object of these sciences (the physical sciences) to seek for the laws by which the different processes in nature are reduced to general rules, and from these rules can be re-determined. These rules, for example the laws of the refraction or reflection of light, and that of Mariotte and Gay-Lussac for the volume of gases, are clearly nothing but general notions by which all the phenomena concerned are embraced. The search for them is the business of the experimental part of our sciences. Their theoretical part on the other hand attempts to find the unknown causes of the processes from the visible effects; it attempts to bring them under the law of causality. We are compelled and authorized to do this by the principle that every change in nature must have a sufficient cause. The immediate causes to which we attribute phenomena may be invariable or variable; in the latter case the same principle compels us to seek for other causes of this variability, and so on, until we have arrived at ultimate causes which work according to an invariable law, which consequently at every time with the same external conditions produce the same effect. The final goal of the theoretical sciences is thus to search for the final invariable causes of the processes in nature. Whether all processes can be reduced to such (causes), that is, whether nature is completely comprehensible, or whether there are changes in her, which do not obey the law of necessary causation, and which therefore fall into the domain of spontaneity, or freedom, cannot here be decided. It is certainly clear that science, whose aim it is to comprehend nature, must start from the hypothesis that she is comprehensible, and must examine and form conclusions in accordance with this hypothesis until it is perhaps compelled by irrefutable facts to acknowledge the limits of the hypothesis.

It will be observed that Helmholtz's opinions as to the search for efficient causation in nature being the function of the theoretical parts of Natural Science are in divergence with the view that I have maintained in these lectures. It is however very interesting to notice that, at a later time, Helmholtz declared that he had modified the opinions expressed in the passage I have quoted. In fact, in a note appended to a later edition of his work he says:

> The philosophical discussions in the introduction were more strongly influenced by Kant's epistemological views than I at the present time would recognize as correct. I have later made clear to myself that the Principle of Causality is in fact nothing else than the hypothesis that all natural phenomena are subject to law.

It would appear from this statement that Helmholtz emancipated himself from the idea that efficient causation is to be found in nature by the aid of science, and that he finally identified the term causality with the recognition of invariability in sequences of phenomena.

Leaving aside the supposed demonstration of the principle of energy by means of the *a priori* principle of causation, it is possible to deduce the principle from the classical system of Mechanics, if the assumption be made that all the phenomena of motion are governed by central forces; in which case they form what is called a conservative system. In accordance with this assumption the forces acting between every pair of corpuscles of a system are along the line joining them, are equal in magnitude and opposite in direction; and the magnitude of such a force depends solely on the distance between the particles. The whole energy of such a system consists then of two parts, the kinetic energy, or energy of the motion of the system, and the potential energy, or energy of position, which represents the capacity of the forces of the system to do mechanical work. That the principle, in the form that the sum of the kinetic energy and the potential energy of the system is constant for such a system during its motion, follows as a mathematical consequence of the dynamical scheme was demonstrated by Helmholtz in his treatise. But as regards the application of the principle in general physics, it should be remarked that it is exceedingly doubtful whether, in the molecular or sub-molecular domain, it is possible to restrict the forces to those of the central type. For example, it does not seem possible to regard the phenomena of permanent deformation and of crystallization as involving such forces only.

We have already seen that the desire to form picturable images of mechanical phenomena resulted in a reluctance to accept the notion of forces acting at a distance as part of a mechanical scheme. The same tendency has led to attempts to explain what is apparently potential, or latent, energy as really reducible to the kinetic energy of small parts of bodies, or of a medium, and thus ultimately to abolish the distinction between kinetic and potential energy; the latter being regarded as less concrete or picturable than the former. But in accordance with the view of the character of conceptual schemes which has been adopted in these lectures, the concept of potential energy as a measurable quantity is really on the same footing as that of kinetic energy; and consequently even if it still appears to be a desirable simplification to reduce the two concepts to one, such reduction has no longer the same urgency as with those who feel bound to a more realistic interpretation of the concepts of dynamical science.

The principle of energy has frequently been regarded as a consequence of the principle of the impossibility of perpetual motion. Very numerous attempts, many of them very ingenious, have been made to construct machines which should actualize perpetual motion. The conviction of the impossibility of perpetual motion however became gradually very strong among men of science. It was affirmed by Leonardo da Vinci, Galileo, Stevinus, and by Leibniz, the last of whom employed it to establish the principle of *vis viva*. In 1775, the French Academy of Sciences directed that solutions submitted to the Academy of the problems of the duplication of the cube, the trisection of an angle, the quadrature of the circle, and of the construction of machines involving perpetual motion, should no longer be examined. In the case of the last problem the Academy based its decision upon *a priori* grounds stated as follows:

> The construction of perpetual motion is absolutely impossible: even if friction and the resistance of the medium did not ultimately destroy the resistance of the moving force, that force can only produce an effect equal to its cause; if then it is desired that the effect of a finite force shall continue indefinitely, it is necessary that the force should be infinitely small in a finite time. Making abstraction of the friction and the resistance, a body on which motion has been once impressed will conserve it always; but it would be by not acting on other bodies, and the only perpetual motion possible on this hypothesis (which moreover cannot be realized in nature) would be absolutely useless for the object at which the constructors of perpetual motion aim.

Helmholtz showed that the principle of energy is deducible from that of the impossibility of perpetual motion. That impossibility he regarded however as a fact of experience established by the numerous vain attempts to construct a *perpetuum mobile*. It has however been pointed out by Poincaré that it is only in the case of reversible phenomena that the conservation of energy follows from the impossibility of perpetual motion. The general principle of the conservation of the energy of an isolated system, in all its various forms, through all the physical and chemical changes which the system may undergo, can only be regarded as an hypothetical principle to be used tentatively as a guide in our attempts to describe conceptually the various processes in the transformations. The fact that we have no assurance that all the possible forms of energy which may occur in physical phenomena are known to us makes it impossible to conceive that the principle should admit of anything like complete empirical verification.

The history of science exhibits the discovery of various forms of energy previously unrecognized. In particular, our knowledge of the phenomena of electricity, except in merely trivial manifestations, dates only from the investigations of Gilbert, three centuries ago. The recent discoveries in the last few decades; those of Hertzian waves, Roentgen rays, and of radioactive substances, have made us acquainted with forms of energy whose existence had been previously unsuspected. Are we even now certain that we are acquainted with all the forms of energy that may be discovered in solar radiation, and which may be beyond the known limits of the luminous, thermal, and actinic rays? The discovery of the Roentgen rays is one illustration of the fact that a form of energy may long remain undiscovered when it is, so to speak, before our eyes; for Crookes' tubes had been employed for a quarter of a century before Roentgen's discovery of the rays to which they give rise. We have seen that, in the case of a system consisting of particles between which central forces act, dependent only on their relative distances, the principle of energy takes the simple form that the sum of the kinetic energy and the potential energy is constant; the former depending only on the velocities of the particles, and the latter upon their positions, but not on their velocities; so that the total energy can be resolved only in one manner into the sum of the two components. But if, as in the case of Weber's law of mutual action of two electric molecules, the mutual action depends not only on their distance but also on their velocities and on their accelerations, the second part of the energy would depend on the velocities, and it might contain terms depending on the squares of the velocities. In such a case we have no means of distinguishing between terms which belong to the two parts respectively of the total energy. Poincaré has pointed out[1] that we then have no means of defining the energy of the system, because, if the total energy of the system is constant, so also is any function of that total energy; and such a function might be substituted for the energy itself, and made the basis of an amended definition of the energy of the system. There exists in such a case no means of fixing upon a precise definition of the energy, as such that it may be divided into two parts, each of a specified form. Moreover, if the principle of energy is to be of any use, it is necessary to take account of the distinctions between the mechanical energy of molar bodies and the other forms of energy, such as heat, chemical, and electrical energy. This can only be done if it is possible to divide the whole energy of the system into parts which are absolutely distinct in form; a part involving only the squares of the velocities of the bodies, another part independent of these velocities and of the thermal and electric states of the system, and a third part independent of the velocities and the positions of the bodies, and dependent only on their internal states. But the case of electric energy, due to the mutual electric action of the bodies, suffices to show the impossibility of this division into such separate parts. For the electrostatic energy depends, not only on the electric charges of the bodies, but also upon their positions. If the bodies are in motion, their electrodynamic energy depends, not only on their states and positions, but also on their velocities. We have therefore no obvious means of

[1] *Science and Hypothesis*, 1905, p. 126.

selecting and separating out the different parts of the total energy in the desired manner. The conclusion drawn by Poincaré from these considerations is that, when an attempt is made to extend the principle of the conservation of energy so as to embrace all the phenomena with which Physics has to deal, we are faced with the difficulty of defining the energy of the system in a unique manner, so that different parts of it may be identified as referring to the different phenomena which occur in the system. He remarks that, when this extreme generality is aimed at, there appears to be nothing left of the principle, except an enunciation: "there is something constant," and that, in this form, the principle lies outside the bounds of experiment and is reduced to a kind of tautology.

This criticism is pertinent in relation to the attempt made by Ostwald and others to set up a science of Energetics, based upon the Principle of Energy, and that of Least Action (or some other similar principle), with the view of avoiding the difficulties connected with the hypothesis of the existence of atoms. The fundamental conception of Energetics is that every change in an isolated system is regulated by two laws. The first is that the sum of the kinetic and potential energies is constant through all the transformations of the system. The second is that, if the system passes from one configuration at one time to another configuration at another time, the passage always takes place in such a manner that the mean value of the difference of the two kinds of energy in the interval of time between the two specified times is a minimum.

The lessons to be drawn from the history of the varying conceptions that have arisen at various times, in connection with the sustained efforts that have been made to attain clear conceptions of what it is that is conserved in matter and its various transformations, are mainly those of the inadequate character of *a priori* conceptions such as the principle of causation, and of the partial character of the empirical verification of the principles. It would appear that, when the utmost has been attained as regards clearness of statement of these principles, as conceptual laws, there remains an element of doubt and uncertainty, and of tentativeness as regards the range of applicability, both in practice and in theory, of these laws, in their function of describing the actual changes and transformations in the perceptual world.

# X

# MECHANICAL THEORIES AND THERMODYNAMICS

THE aim of a Mechanical theory of a special class of natural phenomena has been to represent the changes, of which the phenomena consist, in terms of mass and motion. Other concepts, those of force, work, and energy, are also employed in most such theories; and each one of these may be taken to be an independent concept, or else as derivative, according to the special form in which an abstract theory is stated, and also according to the special class of phenomena with which the theory deals. In the case of the changes of position and motion of molar bodies, the Mechanical theory employed is founded upon the Classical Dynamics of Galileo and Newton; and, as I have previously shown, this theory may be stated in a form in which force, work, and kinetic energy are *derivative* conceptions defined in terms of the fundamental concepts of mass, space, and time. In the more developed forms of this theory, certain general principles have been deduced as consequences of the fundamental assumptions of the theory, and one or other of these principles has been tentatively assumed as the basis of more extended mechanical theories, independently of the original assumption that all the forces of a system of bodies are central forces. I propose in the present lecture to give some account of more general mechanical theories which have been set up in this manner. Mechanical theories have been devised to describe other physical phenomena, such as the elastic deformations of solid bodies, the phenomena of sound and vibrations, optical, thermal, and electromagnetic phenomena. For some purposes, not ponderable matter, but an imponderable ether of some special type, has been assumed as the field of masses in motion. In such cases the character of the relation between ordinary matter and the assumed ether forces itself upon the attention as urgently requiring elucidation, if ether and matter are to be considered together in one mechanical scheme. But, as we shall see, it is possible to set up an abstract mechanical theory without making a complete set of detailed assumptions as to the precise nature of all the connections of the system. The relations between ether and matter have been speculatively conceived in a variety of ways. For example, in Kelvin's vortex atom theory, the basis of ordinary matter consists of vortex rings in an ether conceived as having the properties of a perfect fluid. A view which was propounded by Riemann makes the smallest element of matter a singular point in the ether at which a continuous annihilation of ether takes place. In Larmor's rotational ether, an electron is a point of the ether at which a special kind of singularity exists. In some theories in which ether and matter are combined, the Newtonian law of the equality of action and reaction is not satisfied when the matter alone is taken into account, so that the interaction of matter and ether forms an essential part of the dynamical scheme.

In all mechanical or quasi-mechanical theories, the masses, whether of ponderable matter, or of imponderable ethers, have usually been conceived in accordance with the realism of common sense, not as pure concepts, but as independently existent entities. As I have before urged, however, the validity and the usefulness of these theories are in no way invalidated if we refrain from all realistic assumptions, and regard the elements employed in the theories as concepts only. A strong underlying motive which has usually dominated investigators who have built up these theories has been the desire to give explanations which should make the stages of physical processes accessible to the sensuous imagination. Any elements in a theory which fail to satisfy this requisite have been accepted with reluctance. I have already referred to a striking instance of this in the unwillingness which has been shown to regard the notion of action at a distance as anything more than a provisional makeshift to be reduced, if possible, to the apparently more plausible notion of action by contact, by means of some articulated mechanism involving the propagation of action through some medium connecting the bodies whose interaction is to be explained. That action by contact is itself in need of elucidation has, however, gradually forced itself upon the minds of men of Science, though apparently not in all cases upon the minds of scholars for whom Greek Philosophy contains the quintessence of all wisdom. The failure to attain to anything except an indefinite regress, of attempts to reduce contact action to a form which would satisfy the craving for efficient causation, has been one of the factors which have led to the removal of the category of efficient causation from Natural Science. A great advantage of the modern view, that a scientific theory is a purely conceptual scheme, is that such a theory is emancipated from the somewhat narrow limitations imposed by the necessity that its form should be such that a sensuous representation of phenomena is provided. Of this freedom, some modern theories, especially those of Geometry, and the theory known as the Einstein theory of relativity, have availed themselves in a strikingly large measure. The latter theory is not a mechanical theory, in the sense in which I here employ the term, and I accordingly postpone any discussion of it. In the Newtonian Dynamics, as originally conceived, mass and force were taken as fundamental and independent concepts; force being regarded as a cause which produced, as its effect, change in the amount of motion, defined as product of mass into velocity. I have, however, in an earlier lecture, pointed out that this causal relation must be removed from its position in the theory, in order that the theory may be stated as a conceptual scheme which becomes a deductive one when the requisite definitions and postulations have been fixed in precise form. The notion of force is then not independent of the concepts of mass, time, and space, since force is used as a synonym for the product of mass into acceleration, if indeed the term continues to be employed. In fact, if the scheme originally stated by Newton in his laws of motion, his definitions and deductions, be taken in a revised form, as the basis of a kinetic theory of the motions of molar bodies, the existence of forces, and the existence of accelerations, are not two facts of observation, but one only. That the earth and the sun move towards one another with accelerations in a definite ratio is the one fact, as regards their relation to one another, which is relevant to the dynamical theory. The supposed cause of this, their so-called mutual attraction, is not an independent fact, but merely an assumption made in accordance with the supposed necessity of accounting for the first fact as due to causation. Thus, in Newtonian Dynamics, as a theory of the motion of such bodies as those of the solar system, force is not an independent concept; the theory operates with the three independent concepts of time, space, and mass; motion being regarded as a transformation involving elements of space and time.

There exists, however, the department of Statics which is much more ancient than the Dynamics of Galileo and Newton. In accordance with the Newtonian Dynamics the weight of a body is the product of its mass into the acceleration with which it falls to the ground. But if the body be supported by a spring balance, or in other manner, in accordance with the principles of Statics the weight is regarded as a force still in existence, but balanced by another force due to the support. If we refrain from formulating any theory as to the state of the support, such as a kinetic theory of its corpuscles, the notion of force, as a stress or pressure, is requisite as an independent concept. Moreover, when the small-scale phenomena which occur in the smallest parts of bodies are taken into consideration, the notion of force, as an independent concept, is indispensable, in default of a complete kinetic theory of the corpuscles or smallest parts of such bodies. Especially in the theory of the elastic deformations of solid bodies, the bodies are frequently treated as continuous distributions of mass; and the conception of force, in the form of systems of stresses between adjacent parts of a body, is employed as a necessary element in the construction of a theory of the strains or elastic displacements within the body.

In order to comprehend the character of the mechanical schemes that have been tentatively applied to the small-scale phenomena of Physics, it is necessary to consider the later developments of Newtonian Dynamics, and the mode in which some of these have been extended by generalization into schemes of Dynamics that are free from some of the special restrictions involved in the original treatment of the subject by Newton and Galileo. In view of what I have said, the notions of force and mass as independent conceptions, or else the equivalent conceptions of work and mass, are in general required in the basis of mechanical schemes. A conception which is usually regarded as requisite in the more general formulation of Dynamics, as for example in the system developed by Hertz, is that of inexorable constraints in a system of particles or of gross bodies; these constraints have the effect of diminishing the number of possible independent modes of motion of which a system is capable. They are usually represented, on the perceptual side, by rigid connections, inextensible cords or rods, connecting different parts of a system. A method of formulating Dynamics so that it may be applicable to systems which include such inexorable constraints was introduced by d'Alembert. He regarded the reversed mass-accelerations of all the elementary parts of a system as in equilibrium with the forces acting on the system, whether from without, or between parts of the system. All the conditions which determine the motions of the system are then included in one formulation, of a statical character; the condition, namely, that no work is done by the equilibrating forces in any displacements consistent with the preservation of the postulated set of connections of the parts of the system. The deduction of d'Alembert's principle from Newton's theory in its conceptual form is subject to considerable logical difficulties. A method has been given by Boltzmann, in his lectures on Dynamics, by which these difficulties may be overcome. He takes as the basis of his treatment a finite set of masses concentrated at points; between each pair of such points he considers forces to act, of equal magnitude and in opposite directions along the line joining them, so that each of the points has an acceleration towards the other, inversely proportional to the assigned masses of the points, in accordance with Newton's third law of motion. Boltzmann then builds up the equations of motion of a system of bodies, in which there may be inexorable constraints, by treating each body as consisting of a very large, but finite, number of mass-points. The constraints are represented in the same manner by sets of mass-points between which forces act that are functions of the distances between pairs of these mass-points, of such a character that these functions have very great magnitude whenever the standard distance between a pair of the points is changed, either by excess or by defect. External forces acting on a system he takes as also due to external mass-points. In this manner Boltzmann deduces d'Alembert's principle, and in fact the whole dynamical scheme for a system of bodies, in a manner which avoids the logical difficulties of the way in which that extension of Newtonian Dynamics has usually been made. But the greatest advance in the direction of setting up a unified scheme of dynamics, in a form so general that it is applicable to a system in which many of the details concerning the connections of the system may be unknown, was made by Lagrange, and published by him in 1788, in his great work, the *Mécanique Analytique*.

In the history of Science it is possible to find many cases in which the tendency of Mathematics to express itself in the most abstract forms has proved to be of ultimate service in connection with physical theories. A striking example of this is to be found in Lagrange's abstract formulation of Dynamics, as given in the *Mécanique Analytique*. In order to characterize the spirit in which this great work is conceived I cannot do better than quote the words of Lagrange himself from the Preface. He writes:

> We have already various treatises on Mechanics, but the plan of this one is entirely new. I intend to reduce this Science, and the art of solving problems relating to it, to general formulae, the simple development of which provides all the equations necessary for the solution of each problem. I hope that the manner in which I have tried to attain this object will leave nothing to be desired. No diagrams will be found in this work. The methods that I explain require neither geometrical, nor mechanical, constructions or reasoning, but only algebraical operations in accordance with regular and uniform procedure. Those who love Analysis will see with pleasure that Mechanics has become a branch of it, and will be grateful to me for having thus extended its domain.

Lagrange's procedure was to express d'Alembert's variational equation in a form in which a certain number of variables of the most general kind are employed. The number of the variables is the number of independent motions which are allowed to the system by the constraints or connections contained in it. This number is what we now call the number of degrees of freedom of the system, and the independent variables which specify the configuration of the system are called the generalized coordinates; the generalized force-component which corresponds to each of the generalized coordinates is defined as the coefficient of the variation of that generalized coordinate in the final expression for the virtual work of the forces which act on the system or between its parts. In the case in which the forces of the system form what is known as a conservative system, that is when they are the gradients of that single function of the generalized coordinates which we call the potential energy, Lagrange's equations of motion of the system are such that only a knowledge of the forms of two functions is required to make a determination of the positions of the system possible when its configuration and motion at one specified time are given. These two functions are the kinetic energy of the system expressed as a quadratic function of the generalized velocities, that is of the gradients of the generalized coordinates with respect to the time, and the potential energy, which is a function of the generalized coordinates only. It will be observed that a conservative system includes as a particular case that in which all the forces are functions of the distances only between pairs of particles between which they act, as in the original Newtonian scheme.

The Lagrangian equations of motion are equivalent to the statement that the possible paths of a conservative system are the extremals of the time-integral of the single function which is expressed as the difference between the kinetic energy and the potential energy of the system. An important modification of the Lagrangian dynamical scheme was made by Helmholtz, by E. J. Routh, and by Thomson and Tait, in order to make a formulation suitable to the case in which the system contains parts that are independently in rotation. By elimination of the coordinates and velocities corresponding to these freely rotating parts of the system, their effect is taken account of by a modification of the Lagrangian function. In the modified form the velocities of the rotating parts of the system give a contribution to the potential energy of the system. This so-called method of "*ignoration of coordinates*"—a technical term which has given rise to some misunderstanding—by indicating that part of the potential energy of a system may be regarded as really dependent upon the kinetic energy of motions that are concealed within the system, has been sometimes regarded as a step in the direction of reducing all potential energy to kinetic energy. To those who attach paramount importance to the direct correlation of all the conceptual elements of a scientific scheme with sensuous intuition the ultimate reduction of potential energy to kinetic energy has usually been regarded as an ideal to be striven after. This aim is, however, of subordinate importance for those who are willing to accept as valid and satisfactory a scientific conceptual scheme in which some of the concepts employed do not correspond directly with anything that is accessible to sensuous intuition.

The analytical Mechanics founded by Lagrange was extended and generalized in the two fundamentally important memoirs on the subject published by Sir William Hamilton. In accordance with the Hamiltonian scheme the whole of Dynamics is subsumed under what are known as the principles of least, and of varying action. In two alternative forms, a single function, either the action, or the characteristic function, according to the alternative adopted, has the property that the whole of the possible motions of a system are disclosed from a complete knowledge of the form of the function, by means of the variation of an integral in which the function is the integrand. Like most such general principles, the principle of least action has a previous history: it was originally formulated for simple cases of motion by Maupertuis, without any adequate foundation. On Maupertuis' discovery, Whewell[1] writes:

> Maupertuis conceived that he could establish *a priori* by theological arguments that all mechanical changes must take place in the world so as to occasion the least possible quantity of *action*. In asserting this it was proposed to measure the action by the product of velocity and space; and this measure being adopted, the mathematicians, though they did not generally assent to Maupertuis' reasonings, found that his principle expressed a remarkable and useful truth, which might be established on known mechanical grounds.

The Hamiltonian principle can only be deduced from the principles of Dynamics, as formulated by Newton and Galileo, by the employment of certain restrictions on the nature of the forces and the connections in the system. Although these restrictions are of a very general character, they imply various restrictions upon the nature of the motions which can be deduced from the Hamiltonian principle. When the principles of Mechanics, as formulated by Lagrange, Hamilton, and Jacobi, are taken as the basic principles of the Science, it is unnecessary to assume *a priori* that their applications are restricted in the manner to which I have referred. The principles of energy, and of least and varying action, may be accepted hypothetically for the purposes of conceptual description of actual motions; the test of the descriptive value of the principles, as in all such cases, can only be that of experience. The fundamental conceptions with which the scheme operates are those of space, time, energy, and mass; the last of these appears on the abstract side only in the form of coefficients in the energy-function. In this scheme the concept of force does not appear as an independent notion, but

[1] *History of the Inductive Sciences*, Vol. II, p. 94.

only as a derivative conception, that of a gradient of potential energy. Although this has the advantage of being free from the various difficulties connected with the conception of force in the Dynamics of Newton and Galileo, the identification and formal representation of the various forms of energy that are required in connection with various physical phenomena constitute the main difficulty in the employment of the general dynamical scheme in which the conception of energy is fundamental. The Hamiltonian principle, in either of its equivalent forms, gives a complete account of the transformations of the energy of a system between its various forms, by means of the employment of a single principle, whenever we are in possession of the formal expressions for the kinetic and potential energies of a particular dynamical system. One great advantage of this general dynamical scheme is that it affords the means of discovering the main features of the various motions that occur in a system, without requiring the possession of a complete knowledge of the details of the mechanism of the system. In fact an indefinite variety of actual mechanistic systems can always be imagined, for all of which the forms of the kinetic and potential energy-functions are identical; and the consequences of the fundamental principle are applicable to the description of the changes in all such systems.

Until about the middle of the last century most of the theories which were set up for the description of the various physical phenomena consisted of attempts to reduce them to cases of forces acting at a distance between material atoms. In fact the Newtonian system of gravitating forces between the bodies of the solar system formed the model and the inspiration of such attempts. In the second half of the nineteenth century the concentration of the attention of men of Science upon the principle of the Conservation of Energy, owing to the brilliant verification, by Joule, of that principle, in the case of transformation of mechanical work into heat, led to the adoption of the transformations of energy as fundamental in the newer physical theories. In these newer theories, especially in Thermodynamics, and in the hands of Maxwell, in Electromagnetics, the generalized scheme of Dynamics associated with the names of Lagrange and Hamilton found a wide field of application. As regards the Hamiltonian principle, in either of its forms, considered as an hypothetical scheme for the description of physical processes, the chief question which arises is as to its scope; that is, whether it is capable of representing all the motions which take place in an isolated physical system. There is one important restriction of the principle which appears to limit its applicability even in the case of the motion of gross bodies. The connections of a system are expressed by means of equations connecting the coordinates which represent the positions of the bodies of the systems; but in some cases these equations essentially involve the gradients of these coordinates with respect to the time, that is the velocities. In this latter case the Hamiltonian principle, at least in its original form, is not applicable, and if it be employed it may lead to results which are not in accordance with the actual motions which take place in such a system. Attention has been drawn by Hertz, in the introduction to his attempt to formulate anew the principles of Dynamics, to a comparatively simple case in which the Hamiltonian principle is in default. This is the case of a spherical body rolling freely on a horizontal plane sufficiently rough to prevent all sliding motion. If the initial and final positions of the body are arbitrarily assigned, there is always one mode of motion such that the Hamiltonian integral is a minimum. But in point of fact there are initial and final positions, such that, even with initial velocities at our free choice, the body will not move into its final position unless external forces are applied to it to compel it to do so. Even if the initial and final positions are so chosen that a natural motion from one into the other is possible, this motion is not the one for which the time of the motion is a minimum, as it should be in accordance with the Hamiltonian principle. We may attempt to explain this discrepancy by denying the possibility of a motion of rolling, absolutely without sliding, but there certainly exist natural motions in which this condition is very approximately satisfied, and we should consequently expect that the Hamiltonian condition would give in such cases an approximation to the actual motion, which appears not to be the case. The general result is that, in some natural motions, the connections of the system are of such a character that the Hamiltonian principle is not applicable to the description of these motions.

There is, however, another restriction on the scope of the Hamiltonian Dynamics, at least in its original form, which must be taken into account in attempting to form an estimate of its range of applicability. The kinetic energy of a system, expressed in terms of the generalized velocities, is a homogeneous quadratic function of those velocities, which accordingly is unchanged in value if the signs of all the velocities be reversed without changing their magnitudes. It follows that the motions in the system are all capable of being described in the reverse order, without being otherwise changed; in other words the motions described are all reversible motions. There is, however, evidence of overwhelming strength, which emerged originally in connection with the theory of Heat, that some small-scale motions which occur in nature must be regarded as irreversible, at all events that we are unable by any means at our disposal to realize such motions in the reverse direction. This difficulty Helmholtz endeavoured to combat in his investigations on cyclical systems. He showed that, by elimination of the coordinates which represent certain concealed stationary motions in a system, the kinetic energy is expressed in a form which involves not only quadratic, but also linear, terms in the remaining velocities of the system; and in that case the Lagrangian equation would lead to motions which are irreversible, in default of means for reversing the concealed motions of small parts of the system. In this connection Helmholtz has pointed out the necessity of considering more general forms, than the original one, of the expression for what he terms the kinetic potential of a system; by which is denoted in the original scheme the difference between the potential and the kinetic energy. The attempts made by Helmholtz and others to construct a mechanics based on the conception of energy, and on the hypotheses of the conservation of energy and the principle of least action only, without having recourse to atomic assumptions, which should be applicable to thermal, electrical, and chemical phenomena, have been only partially successful, especially as the difficulties connected with the interpretation of the irreversibility of many physical processes have not been overcome.

The most completely developed physical theory dependent upon the principle of the Conservation of Energy, supplemented by another principle not immediately obtainable from Classical Mechanics, is the theory of Heat, known as Thermodynamics. Before the time of Rumford and Davy, heat was regarded, for the most part, as an indestructible substance, caloric. It was supposed that when caloric entered a body, the effect of combination was in general an expansion of the body; even when contraction was the result of the combination, the analogy of certain chemical combinations, such as that of potassium and oxygen, could be appealed to. The phenomenon of conduction of heat, as transference of caloric, presented no difficulty. The difference of specific heat of various substances was explained by assuming that they required different amounts of caloric to be mixed with them in order to produce equal changes of temperature. Black's theory of the latent heat of water assumed that water differs from ice at the same temperature in containing an admixture of a definite equivalent of caloric which was represented by a molecular state of the body which does not exhibit itself in the form of a rise of temperature. Thus the theory of caloric provided a plausible explanation of the most prominent thermal phenomena, with, however, the important exception of the production of heat by friction or concussion, frequently recognized by the adherents of the theory as not capable of satisfactory explanation.

The theory that heat is representable as motion, instead of as a substance, was established by the experiments on friction of which Count Rumford published an account in the *Phil. Trans.* for 1798. He pointed out that friction led to an inexhaustible supply of heat, and that this is inconsistent with the theory of heat as a substance, but consistent with the idea that heat consists of motion. The theory of heat as motion was still more lucidly developed by Davy, in a tract published in 1799, in which he gives an account of his experiment in which two pieces of ice were rubbed together until both were almost entirely melted. The general law of the communication of heat was laid down by Davy in his *Chemical Philosophy*, published in 1812, in the proposition that "The immediate cause of the phenomenon of heat, then, is motion, and the laws of its communication are precisely the same as the laws of the communication of motion."

The foundations of modern Thermodynamics were first laid down by Sadi Carnot in his essay, *Réflexions sur la puissance motrice du feu*, published in 1824. He recognized the important fact that, in order to produce work by heat, it is necessary to have two bodies at different temperatures, and he pointed out the analogy of work done when there is a fall of temperature with the case of work done by a fall of water from a higher to a lower level. He introduced the notion of a cycle of operations in which the initial and final states of a body are identical as regards temperature, density, and molecular condition. Carnot's theory of the heat engine is injuriously affected by the fact that he still held the theory of heat as caloric, but he stated the important result that, to obtain the maximum of work in a cycle, that cycle must be reversible; which means that heat must only pass from a body to another body at very nearly the same temperature. He showed that the ratio of the work done by a reversible engine to the heat taken from the source is a function of the temperatures of the source and condenser only; when this difference is very small the ratio is that known as Carnot's function, which depends only on the temperature of the source. In 1848, this conception was made by Lord Kelvin the basis of his absolute thermometric scale, independent of the properties of any particular substance.

The development of Thermodynamics, based on the rejection of the theory of caloric, was carried out by Rankine, Clausius, and Kelvin, almost simultaneously. Rankine based his investigations on the hypothesis that the motion which represents the heat in a body consists of molecular vortices or circulating streams. He supposes that the whirling matter is diffused in the form of atmospheres round nuclei, and that radiation, whether of light or heat, consists in the transmission of a vibratory motion of the nuclei, by means of forces which they exert on one another. By means of this special hypothesis of molecular vortices he deduced from general dynamical principles what is termed the general equation of the mechanical action of heat. In later works he introduced the function known as the Thermodynamic function, and applied his principles of

Thermodynamics to various practical questions relating to the steam engine and other heat engines.

The theories of Clausius and Kelvin have the advantage of being independent of any special theory of the character of the motion which exhibits itself as heat, but instead of being deducible from the general principles of Dynamics they make use of a fact of observation known as the second law of Thermodynamics, or the principle of Clausius. The first law of Thermodynamics is taken to be the principle of Conservation of Energy, as applied to the equivalence of heat and mechanical work, the amount of which was determined by Joule and later experimenters. The principle of Clausius asserts that, in a series of transformations in which the final is identical with the initial state, it is impossible for heat to pass from a colder to a warmer body unless some other accessory phenomenon occurs at the same time. A more precise statement of the principle, called by Clausius "the law of the equivalence of transformations," is to the effect that:

> in all cases in which a quantity of heat is transformed into work, and the bodies by means of which that transformation is effected return at the end of the operation to their original condition; another quantity of heat must at the same time pass from a hotter to a colder body; and the proportion which this latter quantity of heat bears to the former depends solely upon the temperatures of the bodies between which it passes, and not upon the nature of the intervening bodies.

A function called "entropy" was introduced by Clausius, whose value is found by dividing the quantity of heat expended in producing a given change in a given substance by the absolute temperature as measured by a perfect gas thermometer. The conception of entropy is a case of a fundamental concept, essential to the scheme of Thermodynamics, which does not directly represent anything accessible to sensuous perception. The second law of Thermodynamics involves the postulation that the entropy of a thermally isolated system always tends to increase. The change in entropy is quite distinct from change in temperature, and from the change which consists in loss or gain of heat. For example, in chemical reactions, the entropy increases without any heat being supplied to the substances. When a perfect gas expands in a vacuum its entropy increases, and yet the temperature does not change, and the gas has neither yielded nor received heat. A difficulty in the conception of entropy is that it is not possible to define the equality of entropy of two bodies chemically different, although it is possible to compare the variations of entropy to which the bodies are separately subject.

The principle that entropy tends continually to increase has been stated by Perrin in the form that "an isolated system never passes twice through the same state." It involves the postulation that the course of physical phenomena is, so to speak, in a definite direction which is never reversed. This principle has been generalized by Clausius and Kelvin, in a form in which it is made to apply to the whole universe. It is said that the entropy of the universe is continually increasing. Thus, although the whole energy of the universe is regarded as remaining constant through all transformations, it becomes increasingly unavailable, since the energy is transformed gradually into heat uniformly distributed at an everywhere identical temperature. The final state of the universe would then be one in which nothing would happen, because no energy would anywhere be available for the purposes of the chemical and thermal transformations by which all change is conditioned.

This theory of the dissipation of energy is open to the very serious criticism applicable to all statements made about the physical universe as a whole. The extension of a principle, asserted in the first instance to apply to a finite isolated system, is made hypothetically to one which we are not warranted in regarding as finite. If we consider ever larger portions of the universe, it may be the case that the energy and the entropy of a portion increase indefinitely as the portion is continually increased, and in that case the assertion ceases to have a definite meaning. Besides, the range of validity of the principle, even as applied to an isolated finite system, has not really been ascertained. Kelvin himself expressly excluded living organisms in his statement that "it is impossible, by means of inanimate material agency, to derive mechanical effect from any portion of matter by cooling it below the temperature of the coldest of the surrounding objects." Moreover, the view is held by those who have considered the matter from the point of view of molecular or atomic theories that the principle of increase of entropy is only a statistical principle, based upon the laws of probability. The tendency of these statistical methods, developed by Willard Gibbs and by Boltzmann, is to regard the principle as pointing out that a given system tends towards the configuration presented by the maximum probability; and the entropy of the system is expressible in terms of the numerical value of this probability. This maximum probability increases with the number of molecules concerned, but does not reach absolute certainty.

Thermodynamics, in the hands of Willard Gibbs and later investigators, has attained to very great success in its application to the ascertainment of laws regulating changes of chemical constitution or of physical state. After Gibbs, Helmholtz introduced into the domain of Chemistry the conception that energy can be divided into two parts; the first, free energy, capable of undergoing all transformations and of producing external action; the second, bound energy, only manifesting itself by giving out heat. It is the variation of the free energy, not that of the total energy, which is efficient in determining chemical reactions. The utilization of the theory of Gibbs, expressed chiefly in what is known as the Phase Law, was due to Van der Waals, Van t'Hoff, and Roozeboom, in the discussion of complicated chemical reactions.

On the whole it may be asserted that the principles of Thermodynamics have proved a most valuable tool for the coordination of a great number of physical and chemical properties of matter; but the laws must still be considered as hypotheses the precise range of applicability of which is not known and has not yet been definitely delimited.

Both in those theories which depend upon corpuscular hypotheses and in those in which no molecular or atomic hypothesis is made, the conception of Energy as a measurable quantity capable of continuous transformation has been employed as a fundamental element. In any theory based upon the Classical Dynamics, even when extended in accordance with the conceptions of Hamilton and others, the transference of energy from one form to another is regarded as essentially continuous in amount. This is also the case when the fact of the irreversibility, or apparent irreversibility, of certain motions is taken into account and formulated abstractly in a theory of entropy. Certain facts have, however, recently emerged which throw very serious doubts upon the adequacy of any theory in which continuous transformations of energy are admitted for the purpose of representing certain classes of phenomena. A theory of Quanta, in which portions of energy are transformed by jumps, that is discontinuously, has arisen as the result of an attempt to represent these facts. This theory, associated with the name of Planck, arose in the first instance from an investigation of the spectrum of black-body radiation; and it also has bearings upon the theories of the line-spectra of the elements and of the specific heats of solid bodies. The result, up to the present time, of the discussion of this matter has been to show that there are exceedingly strong grounds for the assertion that certain kinds of phenomena involving the transformation of energy are incapable of being described by any conceptual scheme of the kind which we call Newtonian Dynamics, even in a generalized form; but the subject is still in the region of controversy. The theory of radiation starts from the assumption of the (conceptual) existence of an ether which must be regarded as continuous, or at least as very much more finely grained than ordinary matter. On this assumption it can be proved that, in accordance with the dynamical theory, a state of thermal equilibrium between, say, a piece of iron and the surrounding ether, can only be attained when all, or nearly all, the energy of the motion of the parts of the iron has been drawn from the body into the surrounding ether. There is in fact a tendency for the whole energy of moving systems immersed in a medium of any type like that which must be assigned to the ether to be transferred to the medium, and ultimately to be found in the shortest vibrations which that medium is capable of executing. It is now a matter for observation to determine whether this is what actually happens; and the observed facts relating to thermal equilibrium between a black-body and the surrounding medium appear to give decisive evidence that what should happen in accordance with the dynamical theory does not actually occur. A law of thermal equilibrium between a black-body and the surrounding medium was obtained by Planck from thermodynamical considerations; and this law is inconsistent with the total, or almost total, absorption of the energy by the ether. It has been shown that Planck's law is in close agreement with the results of observation. A demonstration has been given by Poincaré, which has been widely accepted as valid, that Planck's law of partition of energy between the black-body and the surrounding medium is not consistent with any scheme of continuous transference of energy, but necessarily involves the assumption that the energy is transferred discontinuously by jumps. This would also be the case if Planck's law be taken to express only an approximation to the actual law of partition. It would thus appear that the motion of the medium must be governed by laws which involve the quantum-theory; and this negatives the possibility of describing such motion in accordance with the Dynamics of Newton and Galileo, or with any extension of it in which the treatment of the transference of energy by continuous amounts is fundamental.

The general result of these recent investigations is to indicate that limits of the applicability of Newtonian and Post-Newtonian Dynamics exist. The dynamical scheme has been found sufficient, with certain reservations indicated by the Einstein theory of relativity, for the description of the large-scale phenomena of Physics; but there never existed any cogent reason for assuming that the scheme need necessarily prove adequate for the description of all the small-scale phenomena.

# XI

# ELECTRICITY, MAGNETISM, AND LIGHT

THE department of Physical Science with which the terms Electricity and Magnetism are associated has a history of which the interest is unsurpassed by that of any other branch of Science. Until the sixteenth century the knowledge both of electrical and of magnetic phenomena was exceedingly scanty, and it was not until the nineteenth century, the birth-time of Electromagnetism, that the two sets of phenomena were brought into relation with one another. The relative triviality of those facts of observation in connection with which the terms Electricity and Magnetism were employed which were known before modern times gave no ground for any expectation of the fundamental position which Electrical Science now holds in relation to our theories of matter, or for any anticipation of the fact that light itself would one day be regarded as an electromagnetic phenomenon. One of the greatest triumphs of the Science of the nineteenth century, directly on the theoretical side, and indirectly in relation to practical applications, consisted of the breaking down of the barriers which previously appeared to exist between the three sets of phenomena, those of Electricity, of Magnetism, and of Light. This is a remarkable instance of the general law of development of Natural Science; the growing together into one complex structure of what had previously been developed as separate edifices; the process of gradual unification. I propose to give an outline of the stages by which the studies of Electricity and of Magnetism were developed, at first separately, and later as one department, up to the time when the great step was taken by Clerk Maxwell of subsuming the phenomena of Light under an electromagnetic scheme. Of some of the later aspects of Electrical Science, in relation to theories of the constitution of matter, I shall speak in the next lecture.

The history of these subjects contains an account of the rise and fall of a multiplicity of theories; of these the earlier consisted mainly of crude conceptions in which effluvia, fluids, and corpuscles played their part. These theories were designed to appeal to familiar images of the sensuous imagination. Later on, the same rivalries are to be found, as we have already observed in other branches of Physical Science, between the notions of action at a distance, action by contact, and propagation through media. Such theories, imperfect, contradicting one another, and sometimes even self-contradictory, served in various degrees the all-important functions of fixing the direction of future observation of facts, and of predicting results by which the theories could be verified or refuted. This subjection of rival theories to the law of the survival of the fittest is an essential element in the life of Natural Science.

In ancient times the sole knowledge of Electricity and Magnetism was confined to the facts that amber (ἤλεκτρον), when rubbed, has the power of attracting light substances, and that a certain iron ore (λίθος μαγνῆτις) has the power of attracting small pieces of iron. The use of the magnetic needle for indicating directions at sea was known at the time of the crusades, and it appears also to have been known in China at an early period. One discovery of importance was made in the thirteenth century by Petrus Peregrinus; that of the polarity of the natural magnet, or loadstone. Experimenting with a loadstone of globular shape, Peregrinus showed, by placing needles in contact with it, in various positions all over the surface, that the positions of the needles all pointed towards two points at opposite ends of the stone; and these points he proposed to call the poles. He then observed that the way in which magnets set themselves, and attract each other, depends only on the position of the poles, as if the magnetic power were concentrated in these points. No corresponding increase of knowledge of electric phenomena was obtained until the sixteenth century; these were known only as a few bare facts connected with amber and one or two other substances. The first considerable increase of knowledge, both of electric and magnetic phenomena, was due to the researches of William Gilbert (1540–1603), of Colchester. He made the highly important discovery that the orientation of magnets may be accounted for by regarding the earth itself as a great magnet, with its poles in high northern and southern latitudes, in accordance with the general principle that the north-seeking pole of every magnet attracts the south-seeking pole of every other magnet, and repels its north-seeking pole. In Electricity, Gilbert discovered that a whole class of bodies, such as glass, sulphur, sealing-wax, etc., have the same property as amber, that of being electrified by friction. The contrast in various respects between magnetic and electric forces which Gilbert observed, especially in respect of screening and polarity, led him to set up the theory that electrical phenomena are due to an atmosphere of effluvia round an electrified body; the effluvia consisting of material liberated from the body by the process of friction. As he regarded action at a distance as inconceivable, the effluvia provided for the necessary contact with the attracted bodies, in analogy with his view of the phenomena of falling bodies as due to the atmosphere acting as an effluvium by means of which the earth draws all bodies towards itself. This theory was generally accepted by those Natural Philosophers of the period who were interested in the subject, but there was some difference of opinion amongst them as to the manner in which the effluvia acted on the neighbouring small bodies. Gilbert himself imagined that the effluvia had an inherent tendency to return to the body from which they emanated. Moreover the fact that electrified bodies exercise repulsion as well as attraction did not remain unobserved. Notwithstanding the rise of the Newtonian theory of gravitation, which shattered that emanation theory of falling bodies which had suggested the emanation theory of electricity, the latter theory remained unaffected into the eighteenth century.

The first attempt to set up a theory of Magnetism was due to Descartes, who connected it with his general theory of vortices. He postulated the existence of a vortex of fluid matter round the magnet; this fluid entering the magnet at one pole and leaving it at the other. The fluid matter was regarded as acting upon iron on account of resistance offered by the particles of iron to the motion of the fluid. Even in the eighteenth century, Euler and two of the brothers Bernoulli worked out theories of magnetism based upon the hypothesis of vortices.

A very important discovery, that of conduction of Electricity, was published in 1729 by Stephen Gray, a pensioner of the Charterhouse, who showed, as he describes in his own words:

> that the Electrick Vertue of a Glass Tube may be conveyed to other Bodies so as to give them the same Property of attracting and repelling Light Bodies as the Tube does, when excited by rubbing; and that this attractive Vertue might be carried to Bodies that were many Feet distant from the Tube.

No other mode of electrification than that by friction had been previously known. It was found that this property of conduction belongs only to a certain class of bodies, especially to metals. To these bodies Desaguliers, in 1736, gave the name of non-electrics or conductors. The view that electric effluvia are inseparably connected with the body that has been rubbed became untenable in consequence of the discovery of conduction; and thus the notion of an electric fluid of an imponderable character was substituted for the effluvium; and this fluid came to be regarded as one of the chemical elements, although it was thought by some Physicists to be closely connected with caloric, the substance of which heat was then supposed to consist. This last idea was refuted by an experiment of Stephen Gray, who showed that the electrification of two similar bodies, one of which was solid and the other hollow, produced exactly similar effects. From this it was concluded that only the surface of a body has to do with its electrification, whereas caloric is diffused through the whole substance of the body. The next important discovery, made by Charles François du Fay (1698–1739), was that there are two species of electrification. He showed that an electrified body repels another one that has been electrified in the same way as itself, but that two bodies whose electricities are of different species attract one another. To the two kinds of electricity the existence of which his experiments proved, he gave the names "vitreous" and "resinous," by which they are still known. In 1745, Pieter van Musschenbroek (1692–1761), a Professor at Leyden, as the result of an attempt to prevent the electric charge of a body from undergoing the decay which had been observed to take place when the charged body is surrounded by air, was led to the discovery of the Leyden phial, or jar, as a means of accumulating the effect of electrification; at the same time the physiological effect of the discharge of the phial through the human body was discovered. Soon after this discovery, a London apothecary, William Watson, in the course of experiments with the Leyden jar and its discharge, was led to propound, in 1746, the theory that the phenomena are due to the presence of an "electrical aether" which is transferred, but not created or destroyed, during the process of charging and discharging the Leyden jar. In accordance with his theory the electrification of a body is due to its receiving, at the expense of some other body, an excess of electric fluid over the normal amount that belongs to the body; the fluid in the other body being correspondingly depleted. The two species of electrification correspond, on this theory, respectively to an excess or a defect of the electric fluid belonging to the body that is electrified.

The same theory was independently proposed a few months later by Benjamin Franklin (1706–1790), of Philadelphia, as the result of various experiments. The principle of Watson and Franklin may be stated as that of the conservation of electric charge, indicating that, in any isolated system, the total quantity of electric fluid is invariable. In view of this principle, Franklin was led to regard electrification as being either positive or negative, according as it denoted an excess or defect of electric fluid in the electrified body; he attributed the positive sign to vitreous, and the negative sign to resinous, electricity. In considering the theory of the Leyden jar, Franklin was led to assume that the glass in the jar is impenetrable to the electric fluid, although the attractive effect between one electrified body and another body is not destroyed by interposing a glass plate between them. He was thus led to the idea that the surface of the glass nearest the electrified body is able to influence its other surface through the glass, and that this effect then accounts for the influence on the other body. In the case of the jar the excess of fluid on the inner face exercises through the glass a repulsion, which causes a defect of fluid on the outer face. This interpretation of fact was in accordance with a theory, the one-fluid theory of electricity, which, unlike the older conception of effluvia, involved the notion of action at a distance. The view that glass is impermeable to the electric fluid was extended by Aepinus (1724–1802) so as to apply to all non-conductors. In order to account for the repulsion between two resinously charged bodies, Aepinus, who also held the one-fluid theory, set up the hypothesis that the particles of ordinary matter repel each other, but that, when bodies are unelectrified, this repulsive force is balanced by the attraction which he, like Franklin, assumed to exist between matter and electric fluid. He suggested that gravitation might be due to a slight inequality between these attractive and repulsive forces. He applied his theory to the explanation of *the induction* of electric charges, a phenomenon which had been previously observed by others, and had been studied a little earlier by John Canton (1718–1772) and by Wilcke.

I have spoken of the facts known at an early period relating to electricity and magnetism as trivial. There is, however, one phenomenon that cannot be so described, although it fortunately occurs only occasionally. I mean the phenomenon of lightning in thunderstorms. The important discovery that lightning is an electric phenomenon, akin to the spark observed when a Leyden jar is discharged, is due to Benjamin Franklin. Experimenting during thunderstorms with kites, he was able to establish the electric character of the occurrence by charging a Leyden jar by means of electricity conducted from the kite.

The discovery of the law of force between two electric charges was made by Joseph Priestley (1733–1804), the discoverer of Oxygen. He showed experimentally that, when a hollow metal vessel is electrified, there is no charge on the inner surface, and no electric force in the interior; and he inferred from this fact that the law of force is that of the inverse square of the distance, the same as that of gravitation. This discovery of Priestley, taken in conjunction with Franklin's law of the conservation of electric charge, brought the phenomena of Electrostatics for the first time to a completeness of description sufficient to render them accessible to Mathematical calculation. The first to take advantage of the possibility of applying calculation to electrical phenomena was the Hon. Henry Cavendish (1731–1810). In a memoir presented in 1771 to the Royal Society he adopted the one-fluid theory of Aepinus and Franklin, which he had however discovered independently. In this memoir he assumed the law of electric force between charges to be inversely as some less power of the distance than the cube, and virtually introduced, under the term intensity of electrification, the notion of the potential, which later became fundamental in electrical theory, although its use was hindered by the fact that he did not definitely assume the law of force to be that of the inverse square. Unfortunately, Cavendish's researches remained for the most part unknown until 1879, when they were published at the instance of Lord Kelvin who had examined the manuscripts. It then appeared that Cavendish had not only rediscovered the law of the inverse square but had even determined the correct value for the ratio of the electric charges carried by a circular disc and a sphere of the same radius in metallic connection with one another; thus he introduced the conception of electrostatic capacity. Moreover, he anticipated the later discovery by Faraday of specific inductive capacity, and investigated experimentally the conducting powers of different materials for electrostatic discharges.

Whilst the progress I have described of knowledge of electric phenomena was made, the subject of Magnetism had not been neglected. The law of force between magnetic poles was discovered by John Michell (1724–1793), a Fellow of Queens' College, Cambridge, who published his researches in 1750. It had been previously believed that the attraction between opposite poles followed a different law from that of the repulsion between like poles. A theory of magnetic fluid similar to that of the one-fluid theory of electricity was propounded in 1759 by Aepinus, who regarded the two poles of a magnet as places where the magnetic fluid existed in excess and in defect of the normal amount respectively. He supposed that the particles of the fluid repel each other, but attract particles of iron and steel; moreover he saw that it was necessary to assume that the material particles of the magnet repel each other. By later investigators a two-fluid theory of magnetism was employed; these fluids were taken to have properties of attraction and repulsion similar to those of vitreous and resinous electricity.

Exact measurements, both electric and magnetic, were made by Charles Augustin Coulomb (1736–1806), who employed for this purpose the torsion balance. By this means he verified decisively the law that the force between two small globes charged electrically is inversely as the square of the distance between their centres, and is repulsive or attractive, according as the electricity is of the same or of the opposite kinds. Instead of the one-fluid theory which had been accepted as the basis of the explanation of electrostatic phenomena by Franklin, Aepinus, and Cavendish, Coulomb postulated the existence of two fluids, corresponding respectively to the two kinds of electrification. He supposed that, in an uncharged body, these fluids were both present in equal amounts, and thus neutralized each other; but when in an electric field, a decomposition of the neutral fluid takes place into equal amounts of the separated fluids, which can then be separately located. The controversy which took place between the upholders of the rival one-fluid and two-fluid theories was in reality one between the merits of two conceptual descriptions of the phenomena, and no experimental evidence was forthcoming which was capable of deciding between them in respect of their power of representing the facts of observation. Coulomb also verified the law of force of magnetic poles on one another by means of the torsion balance. He endeavoured to explain the fact that the two magnetic fluids, unlike the two electric fluids, cannot be divided between different substances, so as to obtain a magnetic pole in isolation. He propounded the view that the magnetic fluids cannot move from one molecule of the magnetic body to another, so that each molecule always contains equal amounts of the two fluids, which are separated within the molecule when the substance is magnetized, giving rise to two poles in each molecule.

At the end of the eighteenth century the theory of effluvia in Electricity and that of vortices in Magnetism had been eliminated, and had in each case been replaced by a theory which involved the postulation of the existence of either one or two fluids, and in which the notion of attractive and repulsive forces acting at a distance was a fundamental element. The Sciences had now arrived at a point which made them accessible to Mathematical Analysis. A complete mathematical development of Electrostatics on the basis of the two-fluid theory was published in 1812 by Simeon Denis Poisson (1781–1840). He showed that, on the basis of the theory of attractions and repulsions between particles of the fluids which were supposed capable of moving freely in conductors, there is no electric force in the interior of the conductor; and that a charge of such a body, which consists of an excess of one kind of fluid over the other, distributes itself over the surface of the conductor as a layer, of which the thickness at every point depends upon the shape of the surface. He showed that, in simple cases, it is possible to determine the distribution of electricity over the surface of the conductor; and for these purposes he transferred to electric theory various results which had been obtained in the theory of gravitational attraction, in which the same law of force is involved as in the case of electricity. Of special importance was the introduction into electrostatics of the potential function which had been previously introduced into gravitational theory by Lagrange, who showed that an attractive force at a point can be expressed as the gradient of this function. In all later developments of electricity, and in their practical applications, this notion of potential, or of potential difference, has proved to be of fundamental importance; upon it depends the whole theory of the distribution of electrical charges upon conductors, and it was seen later that this is but a small part of the function which this conception fulfils in electrical theory. In 1824, Poisson published a corresponding complete mathematical theory of Magnetism, of which the notion of the potential is the basis. He showed that the effect of a magnetized body can be represented by fictitious surface- and volume-distributions of magnetic matter. At the same time, he gave a theory of the temporary magnetization induced in a body made of soft iron by the approach of a permanent magnet. Very important developments and extensions of the mathematical theory of electricity and of magnetism were published in 1828 by George Green (1793–1841), to whom the term *potential* is due, which has ever since been employed to designate the function introduced into the theory by Lagrange and Poisson. The celebrated theorem known as Green's theorem was established in this memoir; of this theorem Poisson's resolution of the effect of magnetization of a body into the sum of parts due to surface- and volume-distributions is simply a particular case.

The theories of Electrostatics and of Magnetism had now been so far developed that they had been subsumed under schemes of conceptual description which represented the ascertained facts, and which were sufficient for mathematical calculation of the details of the phenomena, in cases which were sufficiently simple, and for the ascertainment in the form of mathematical theorems of a variety of the more general aspects of those phenomena. The stage of development of Electrical Science in general which comes next for consideration arose from the discovery, in the latter part of the eighteenth century, of a set of phenomena belonging to a quite new class.

Luigi Galvani, Professor of Anatomy at Bologna, by an accidental observation made in 1780 whilst dissecting a frog, was led to the discovery that, if the nerves and the muscles of the frog are connected by a metallic arc formed of more than one kind of metal, the limbs of the frog became violently convulsed. He was led to the conclusion that this was due to a flow of electric fluid, and he considered the phenomenon as essentially similar to what happens when a Leyden jar is discharged. This view did not receive universal acceptance; some physicists thinking that this so-called galvanism, or animal electricity, was a fluid different from the electric fluid which was regarded as functioning in electrostatic phenomena. In 1792, the opinion was maintained by Alessandro Volta (1745–1827), Professor at Pavia, that the essential element in Galvani's experiment was the connection of two different metals by a moist body, and that the supposed animal electricity, due to the nervous system of the frog, had nothing to do with the phenomenon. Shortly afterwards, Fabroni, of Florence, placed two plates of different metals in water and observed that, when they were put in contact, one of them became partially oxidized; from this he concluded that some chemical action is connected with the galvanic phenomenon. In 1800, Volta showed that the galvanic effect could be made much greater by constructing a pile in which a number of pairs of zinc and copper discs were used, each pair being separated from the next by a disc of moist pasteboard. This pile is the parent of the battery employed in electric telegraphy. A distinct shock could be felt when the highest and the lowest discs were simultaneously touched by the fingers; and it appeared that this shock could be repeated any number of times. As the result of this and further experiments, Volta set up his electrical theory of the pile, as due to the contact of each copper disc with a zinc disc, the pasteboard discs acting merely as conductors.

He recognised the existence of a continuous electric current so long as the circuit is completed by joining the highest and lowest discs. When Volta's discovery had been communicated in 1800 to Sir Joseph Banks, President of the Royal Society, Volta's experiment was repeated by Nicholson and Carlisle who, having placed a drop of water on the upper plate of the pile, in order to make the electric contact of the highest and lowest discs more efficient, observed that round the wire there was at the drop of water a disengagement of gas. They then introduced a tube of water, into which the wires from the two terminals of the pile were immersed, and observed that an inflammable gas, hydrogen, was liberated at one wire, whilst the other became oxidized.

This observation of the effect of the decomposition of the water constituted the great discovery of electrolysis, which was soon extended to the decomposition of various metallic salts in solution. It was shown by Wollaston that water could be decomposed by the discharge of frictional electricity, thus identifying the currents of the electricity of Volta's currents with those of frictional electricity. There were two views as to the mode in which the current in the pile is produced, the so-called contact theory, that it is due to molecular action between the two different metals in contact with one another, and the chemical theory, that it is due to chemical action involving oxidation of the zinc in the pile. The chemical theory was supported by Humphry Davy (1778–1829), Professor of Chemistry at the Royal Institution. He showed that there is no current when the water between the pair of plates is quite pure, and that their power of action is in great measure proportional to the power of the conducting fluid between the plates to oxidize the zinc. Davy afterwards proposed a theory of the voltaic pile in which the contact and chemical actions were both recognized as contributing to the effect; the contact of the metals disturbing equilibrium whilst the chemical changes in the liquid constantly tend to restore the conditions under which the contact energy is exerted. He was led to make the assertion that chemical affinity is essentially electrical in its nature. A comprehensive chemical theory of the electric current and of chemical combination was advanced by the Swedish chemist Berzelius (1779–1848), dependent on the hypothesis of the existence of electric charges within the atoms of matter. This was an anticipation in some respects of conceptions which have become of fundamental importance in recent decades, although the detailed theory of Berzelius did not survive him. He was inclined to regard both electricity and caloric as substances devoid of gravitation, but possessing affinity to gravitating substances.

So far as I have at present proceeded in the account of the gradual increase of knowledge of electrical and magnetic phenomena no connection between the two sets of phenomena had been exhibited. A discovery made in 1820 by the Danish physicist, Hans Christian Oersted (1777–1851), produced in its ultimate implications a revolutionary effect upon the whole future of the Sciences of Electricity and Magnetism, exhibiting as it did the closest connection between the two sets of phenomena. It had been for some time suspected that an electric discharge has an effect upon the magnetic needle, but Oersted was the first to demonstrate its existence by his observation that a magnetic needle, when placed in the neighbourhood of a continuous electric current in a straight wire parallel to the needle, tends to set itself at right angles to the wire. It thus appeared that the phenomenon of an electric current could not be localized entirely in the conducting wire, but had an effect which spread itself through the surrounding space, and produced an alteration in the orientation of the magnetic needle. Oersted's discovery was described at a meeting of the French Academy shortly after it was made, and this led to further investigation by physicists. Two of these, Biot and Savart, shortly afterwards announced their discovery of the law of force of the straight current upon the magnetic pole; that the force on a pole is at right angles to the plane through the wire and the pole, and its intensity is inversely proportional to the distance of the pole from the wire. It was shown by Arago that a magnetic field produced by an electric current may be employed in the same manner as one produced by a magnet, to induce magnetization in iron, and thus that an electric current is a magnet.

Almost immediately after Oersted's discovery had become known, André Marie Ampère (1775–1836) showed that two parallel wires carrying currents attract each other when the currents are in the same direction, and repel each other when the currents are in opposite directions. Ampère set himself the work of developing a complete theory of the pondero-motive forces which act between circuits carrying electric currents, on the basis of the conception of forces acting at a distance between pairs of elements of the two currents. He showed in particular that an electric circuit is equivalent in its magnetic effects to what is called a magnetic shell, that is a distribution of elementary magnets on a surface bounded by the circuit, with the axes of the magnets normal to that surface. He regarded magnetism as essentially an electrical phenomenon; each magnetic molecule being looked upon as having a small closed circuit within it, in which a permanent electric current flows. Ampère succeeded in developing upon this basis a complete mathematical theory of the mechanical interaction of circuits carrying electric currents, that is, of electromagnets. Of this theory, Maxwell wrote[1] several decades later:

> The experimental investigation by which Ampère established the laws of the mechanical action between electric currents is one of the most brilliant achievements in Science. The whole, theory

[1] *Electricity and Magnetism*, 3rd ed., Vol. II, p. 175.

and experiment, seems as if it had leaped, full-grown and full armed, from the brain of the 'Newton of Electricity.' It is perfect in form, and unassailable in accuracy, and it is summed up in a formula from which all the phenomena may be deduced, and which must always remain the cardinal formula of electrodynamics.

The theory of Ampère is, like the similar theories developed later by Grassmann, Stefan, and Korteweg, based upon the conception of action at a distance; no account being taken of the medium between the electric circuits. The essential assumption in these theories is that the whole effect of one circuit on another can be represented as the resultant effect of forces acting between pairs of elements of the two circuits. These theories differ from one another in respect of the nature of the forces which they assume to act between pairs of elements, but they all agree in representing the actual forces between complete circuits, which alone can be submitted to the direct test of observation. They make no explicit use of the principle of the conservation of energy, and indeed that principle would imply the existence of couples between pairs of elements, which are not taken into account in Ampère's theory. Ampère's theory cannot accordingly be regarded as a dynamical theory, although it, like the other theories of the same type, affords a sufficient representation of the interaction of complete circuits. Theories of another type, which also take no account of the medium between the circuits, were developed later by Gauss, W. E. Weber, Riemann, and Clausius. These represent the action of currents by assuming that the forces acting at a distance between electrified bodies depend upon the velocities and accelerations of those bodies. In all these theories, except that of Clausius, Fechner's hypothesis is adopted, that an electric current consists of a flow of positive electricity in one direction and a flow of negative electricity with the same velocity, and of equal amount, in the opposite direction. They differ from one another in respect of the law of force which they assume to exist between pairs of moving elements of electricity. The theories of Weber and Clausius, unlike that of Gauss, are consistent with the principle of the conservation of energy, and consequently suffice to represent the induction of currents, as well as the mechanical forcive between the circuits.

Another theory of electrodynamical forces was developed by F. E. Neumann, and extended by Helmholtz. This was a dynamical theory based upon the assumption of a law of the mutual energy between elements of currents but, like those I have hitherto mentioned, it took no account of the dielectric medium. The power possessed by different metals to conduct electric currents was investigated by Humphry Davy, but a complete theory of this conduction was formulated by George Simon Ohm (1787–1854), based upon a large amount of experimental work. Ohm compared the flow of electricity in a current with the flow of heat in a wire, of which the theory had been given in a complete form by Fourier in his *Analytical Theory of Heat*. He introduced the notion of electroscopic force as a conception which plays a part analogous to that of temperature in the conduction of heat. Tension, or difference of electroscopic force at two places in a conductor, he regarded as effective in producing a current between those places, just as conduction of heat is dependent on difference of temperature. Each voltaic cell he regarded as possessing a definite tension, which is a contribution to the driving force of a current in any circuit in which it is placed. He did not, however, relate differences of electroscopic force with differences of potential (or as we now say with electromotive force) as that conception appears in Poisson's theory of electrostatics. Notwithstanding this defect, the publication of what has since been known as Ohm's law constituted a considerable advance in knowledge of the conduction of electric currents, and much of the later development of the subject up to the middle of the nineteenth century was dependent upon it.

A complete transformation of the whole manner in which the phenomena of Electricity and Magnetism are conceived was the ultimate result of the researches of Michael Faraday, whose genius as an experimental investigator, in accordance with the inductive method, has certainly never been surpassed, and perhaps never been equalled. Only a study of his great work, the *Experimental Researches*, can lead to a just appreciation of the profound insight which led him to the discovery of a multitude of facts, in an orderly succession, under the guidance of novel conceptions radically different from those which had guided previous investigators. Upon his discoveries rest ultimately not only the modern theory of Electromagnetic Science, but also, in its practical applications, the Science of electrical engineering as we know it. The first question which Faraday set himself was to discover whether an electric current in one circuit can induce a current in another circuit, in analogy with the known fact of Electrostatics that a charged conductor induces a charge in neighbouring conductors. In 1831 he published a memoir in which he gave an account of the answer he had obtained to this question. He found that such a current was in fact induced, but that it lasted only for an instant, when the primary current was started or stopped; no induced current existing whilst the primary current flowed steadily. With a view to a formulation of the laws of the induction of currents he took the step of concentrating his attention on the dielectric, or non-conducting medium which surrounded the circuits, thus initiating the breach with the older conception that the phenomena are localized in the conductors, in accordance with the notion of action at a distance; with Faraday the phenomena are localized mainly in the dielectric. He constructed for his guidance the conception of lines of force, which he conceived to fill the space in the neighbourhood of magnets or electromagnets; the direction of each such line at any point being that of the magnetic force which would act upon a magnetic pole at that point. Every such line of force he regarded as a closed curve which at some part of its length passed through the magnet or electromagnet with which it was associated. These lines of force he conceived to form unit tubes of force such that, for any one tube, the product of its cross-section into the magnetic force is constant along its whole length. In this manner he formed an intuitional geometrical representation of the phenomena of magnetism which assisted greatly in directing his investigations. On the basis of his experiments he obtained a formulation of the law of induction of currents in circuits—that the electromotive force induced in a circuit is measured by the rate of change of the number of unit tubes which pass through the circuit. The full import of this important result was only understood later, when the theory of electromagnetic induction was formulated mathematically by Maxwell and others A few years later this fruitful discovery was followed by that of self-induction, in which the effect of the electromotive force in an electric circuit due to a change in the magnitude of a current through that same circuit was established. Faraday also completed the identification of currents, as exhibited by an electrostatic discharge, with those due to voltaic cells, by showing that the magnetic, calorific, and other effects are the same in the two cases. A very important set of Faraday's investigations were concerned with the chemical decomposition in the cells, leading to a statement of the quantitative laws of electrolysis. In this connection he made the statement, of great significance at the present day, that "the atoms of matter are in some way endowed or associated with electrical powers, to which they owe their most striking qualities, and amongst them their mutual chemical affinity."

The concentration of Faraday's attention on the dielectric media was rewarded by the discovery that such a medium has a definite specific inductive capacity, the magnitude of which is different for various dielectric substances. He regarded electrostatic induction as consisting of a certain polarized state of the particles of the medium, similar to that which precedes the decomposition of an electrolyte, into which they are thrown by the inducing surfaces or particles. This state of polarization disappears when the inducing force is removed; it can exist continuously only in insulators, because a conductor is incapable of retaining this state of its particles; an immediate discharge taking place if it be set up in the conductor.

Among many other investigations of Faraday were those connected with his discovery of diamagnetism, and his investigations of the magnetization of crystals, to which attention had been called by the discovery made by Plücker, of the University of Bonn, that certain uniaxal crystals, placed between the poles of a magnet, tend to set themselves so that the optic axis has the equatorial position.

One surmise of Faraday is of extreme interest in view of its later realization in the hands of his great disciple and successor James Clerk Maxwell. In 1851, Faraday suggested that, if the hypothesis of a luminiferous ether be admitted, that ether may have other uses than simply the conveyance of radiant light and heat. He had in 1845 made the important discovery that the plane of polarization of a ray of light is rotated when it passes through a magnetic field, when the polarized ray passes in the direction of the lines of magnetic force. This established a direct connection between optical and electromagnetic phenomena, and is of the highest interest in view of the later amalgamation of the theories of light and electromagnetism. It was found by Joule, in 1841, that the amount of heat evolved in a given time in a wire through which a current is passing is proportional to the resistance of the wire and to the square of the strength of the current, that is to the product of the current into the electromotive force on the wire. This enabled him to perfect the theory which had been developed by Roget and Faraday, that the chemical energy derived from the cell has its equivalent in work done in the outer part of the circuit. Thus the amount of energy transformed from the potential energy of chemical affinity into an electrical form has its equivalent in the heat evolved and dissipated in the circuit, and in any work done otherwise in connection with the outer circuit. It was, however, shown by Kelvin and Helmholtz that the electrical energy furnished by a voltaic cell need not be derived exclusively from the chemical energy of the cell, but may also depend upon the abstraction of energy from neighbouring bodies which is also converted into electrical energy. Helmholtz applied also the principle of energy to the case of systems containing electric currents, and showed that the phenomenon of magneto-electric induction can thus be taken into account. His theory was, however, defective in that he did not take into account the electro-kinetic energy of the currents themselves.

It was largely to the researches of Lord Kelvin that the complete theories of the magnetic and electro-magnetic fields are due. He introduced into Magnetism the distinction between the vectors afterwards named by Maxwell the magnetic force and the magnetic induction, and he extended his theory to take account of magneto-crystallic phenomena. Ohm's theory of linear conduction of currents was generalized by Kirchhoff (1824–1887) to include the case of conduction in three dimensions, the analogy with Fourier's theory of the conduction of heat being useful for this purpose. Kirchhoff showed that, in a system of conductors, the currents so distribute themselves as to produce the minimum amount of heat. Hitherto there had been no complete identification between the electroscopic force of Ohm and difference of electrostatic potential; this hiatus in electric theory was filled up in 1849 by Kirchhoff.

The theoretical investigations of Electrical Science in the last half century have in a very large degree been dominated by the conceptions of James Clerk Maxwell, and by the mathematical formulation of the phenomena in the electromagnetic field to which he was led by those conceptions. The modes of geometrical representation of the state of the electromagnetic field which were devised by Faraday, and which guided him in his researches, show that his mode of thinking was essentially of a mathematical kind, and indeed he possessed constructive mathematical power of a high order. He lacked however that command of the technique of Mathematical Analysis, the possession of which enabled Maxwell to follow out Faraday's conception of the localization of the phenomena in the dielectric medium surrounding conducting substances, and to develop in mathematical form the geometrical notion which Faraday employed of lines of force, and, by an enrichment of Faraday's ideas with conceptions of his own, to give a dynamical theory of the electromagnetic field. In this work he was appreciably influenced by the ideas and mathematical analogies contained in the work of Lord Kelvin. In particular he accepted Kelvin's idea that Magnetism is essentially a phenomenon of rotation, in the form of the conception that, in a magnetic field, there is rotation of the medium about the lines of magnetic force, and that electric currents are to be regarded as a phenomenon of translation.

One of his most fundamental conceptions is that an electrostatic field involves "electric displacement" in the direction of the lines of force, and that, when the field is varied, the variation of this electric displacement, whatever precise interpretation the term may receive, must be regarded as an electric current. This notion of electric displacement is a development of the notion of Faraday that in a ponderable dielectric there is an actual displacement of electric charge on the small conducting particles of which he assumed the dielectric to consist; whereas with Maxwell the displacement occurs even in free ether devoid of ponderable matter. Maxwell was thus led to one of his most characteristic assumptions, that every current forms a closed circuit; thus a current employed in charging a condenser is closed, being completed by the displacement-current in the dielectric between the coatings of the condenser. Another fundamental conception of Maxwell's scheme is that magnetic energy is the kinetic energy of a medium occupying the whole of space, whilst electric energy is to be regarded as the energy of a system of strains of the same medium. In his great memoir *A Dynamical Theory of the Electromagnetic Field*, published in 1864, he gave his theory in the form of equations connecting vectors in the electromagnetic field. In his treatise, published in 1871, he gave a fuller account of his theory of stresses in the ether and in dielectrics, but he was not completely successful in conceiving a mechanism by which such systems of stresses could be sustained. Space will not allow me to give even in outline an account of Maxwell's various investigations in this great branch of Physics, or of the very important developments of the subject made by his successors. I accordingly turn to the great step which Maxwell made in the unification of Science when he set up his electromagnetic theory of light. The theory of Optics has a long and intensely interesting history, but I must confine myself to a summary account of the later stages of that history, leading up to the state of the subject when Maxwell introduced his great unification.

The first Natural Philosopher who made any advance upon the crude Cartesian theory of light, as consisting of a propagation of pressure, not of motion, through a set of globules which constitute space, was Robert Hooke (1635–1703), one of the founders of the Royal Society. He appears to have initiated the wave theory of light; regarding light as consisting of a system of minute vibrations propagated in a medium, or ether. He introduced the notion of a wave surface, as a sphere with centre at the luminous point, and he made an attempt on these principles to explain reflection and refraction. On the other hand, Newton supposed light not to be constituted by the vibrations of an ether, although he regarded such vibrations as existing in close connection with light, but by streams of corpuscles emitted by luminous bodies; the various colours being due to differing corpuscles which excite vibrations of differing types in the ether. One side of Newton's views of the matter constituted the celebrated emission theory which was for a long time the rival of the wave theory. The important fact was established by Roemer, in 1675, that light requires a finite time for its transmission.

The undulatory theory was greatly improved by Christian Huygens (1629–1695), who gave satisfactory accounts, on that basis, of the phenomena of reflection and refraction, and who explained the varying velocity of light in different substances. He concerned himself with the study of the double refraction of light by such crystals as Iceland spar; and explained the phenomenon as due to the propagation of two waves in the crystal, with different fronts, a sphere and a spheroid. The ultimate triumph of the wave theory over the rival emission theory was largely due to the labours of Thomas Young (1773–1829), who showed that the former theory gives the more satisfactory explanation of the phenomena of reflection and refraction, of interference fringes in shadows, of the colours of thin plates, and of the behaviour of light in crystals. The important discovery of polarization by reflection at the surface of water, at a certain angle, was made by Etienne Louis Malus (1775–1812), who showed that such a reflected ray has the same peculiarities as one of the rays which has suffered double refraction. The discovery of biaxal crystals is due to David Brewster (1781–1868). The vibrations of the ether were for the most part conceived as longitudinal, in the direction of propagation of the light, on the analogy of sound-waves. The important step was taken by Augustin Fresnel (1788–1827) of conceiving the vibrations to be in a direction perpendicular to that of the direction of propagation of the light. His theory, the first attempt at the construction of a dynamical theory of the phenomena, was the first of a series of theories based upon the view that the ether behaves more like an elastic solid, and not like a compressible fluid. Fresnel supposed that no longitudinal wave exists, and that in a polarized pencil the direction of the vibrations is perpendicular to the plane of polarization. A celebrated instance of the power of prediction even of an imperfect theory such as that of Fresnel is the fact that it enabled Sir William Rowan Hamilton to predict the occurrence of conical refraction, in which a single ray proceeding in a crystal in a certain direction would on emergence give rise to a whole bundle of rays forming a conical surface; this prediction was verified experimentally by Humphry Lloyd, of Dublin. Investigators were confronted with the difficulty that the ether appears to behave like an elastic solid, in relation to such rapid vibrations as those of light, but at the same time to yield freely to such comparatively slow motions as those of the planets; that there is no necessary inconsistency involved in this was pointed out by Stokes, who referred to the analogy of such substances as pitch and shoemaker's wax, which have both rigidity and plasticity. The exigencies of the theory of the luminiferous ether naturally led to mathematical investigation of the vibrations which can be propagated in elastic solids. Theories of such vibrations were given by Navier, Cauchy, and Poisson, the last of whom established the existence of two types of wave, one transverse, and the other longitudinal, propagated with different velocities, both of which he determined in terms of the elastic constants of the medium. In 1828, this theory was so extended by Cauchy as to take account of crystalline substances. The difficulties in the way of conceiving a type of medium in which the vibrations propagated would accord with the known properties of light, especially the difficulties as regards the conditions which would hold at the interface of two media, led to the development of further theories of the matter by James MacCullagh (1809–1847), by F. E. Neumann (1798–1895), and by George Green. MacCullagh developed the theory of a new kind of medium endowed with rotational elasticity; this theory appears to be sound as a dynamical theory, and to accord with the properties of light; Kelvin afterwards devised a model to illustrate this kind of rotational elasticity. From the point of view of Dynamics the theory of George Green was superior to the pre-existing theories, but the vibrations of Green's type do not accord very well with optical phenomena. Cauchy, in a later theory which he advanced in 1839, introduced a type of ether so designed that longitudinal waves are suppressed; this type of ether was designated by Lord Kelvin labile ether.

At this stage the difficulties in the way of conceiving a substantial ether as the vehicle of conveyance of light, in a manner capable of being completely represented in accordance with a dynamical scheme, had become parallel to the corresponding difficulties relating to electromagnetic phenomena. The great merit of Maxwell consists in his perception that the difficulties of both departments can be concentrated upon a single scheme. He showed that, in his electromagnetic medium, electromagnetic oscillations can be propagated with a velocity in agreement with the known velocity of light. He established the fact that his equations of the electromagnetic field accord with the formulation obtained by the elastic-solid theory, and that it thus affords a general explanation of metallic reflection. There still remained various difficulties which Maxwell did not completely overcome, and these have led to a great amount of subsequent investigation; but that light is to be regarded as an electromagnetic phenomenon, to be investigated as a portion of the phenomena of electromagnetism, has been generally accepted by all recent investigators. The discovery, by Hertz, in the light of this order of ideas, of the long waves, employed in wireless telegraphy, identical in most respects, except their physiological properties, with those of light, is a striking example of the value of the unification of ideas which is due to Maxwell.

Amidst the differences of opinion which have prevailed as to the precise manner in which electromagnetic phenomena may be best conceived, Maxwell's equations remain endowed with an undoubted power of representation of what can be actually observed. It would appear that the time may have arrived at which the scaffolding constituted by notions of a substantial ether, with properties difficult to formulate precisely and consistently, may be removed. The theory of electromagnetism and of light would then have reached the stage of abstraction in which the electromagnetic field would be regarded solely as a field of vectors, distributed and changing in accordance with definite mathematical laws; the notion of a substantial ether having served its purpose as a guide, and been superseded by a more highly abstract scheme in which all such models are discarded. If this be done the theory will have reached the high stage of abstraction towards which all conceptual theories tend as they approach completion. The development of such a theory necessarily involves the previous existence of a series of attempts to represent the phenomena by means of sensuous images which always contain elements of inadequacy, and often of contradiction. The existence of a series of successive grades of abstraction is a law of the mental evolution of scientific theories. The question whether this gradual increase of abstractness in scientific theories represents a recession from, or an approach towards, "reality," in any metaphysical sense of the term, is a question which will receive differing answers from Philosophers of different schools. Natural Science has no direct concern with the answer which may be given to such a question.

# XII

## THE CONSTITUTION OF MATTER

THE notion that all the various forms of matter are constituted by, or evolved from, some primordial constituent has at all times been a speculative idea which has exercised a powerful influence upon the minds of thoughtful men. With the alchemists of medieval times the related idea of the possible transmutation of different kinds of substances led to experiments undertaken with a view to the discovery of some means of effecting such transformation, and especially of converting baser metals into gold. However, for a long time after the rise of modern Chemistry, in which the atomic theory of Dalton was fundamental, the trend of chemical investigation was in the direction of negativing the conception that all forms of matter may be conceived as ultimately one. All forms of matter appeared to be constituted by combinations of some seventy or more irreducible elements which were conceived as unchangeable and permanently distinct from one another. This conception was expressed as late as the year 1873 by Clerk Maxwell in his British Association address, in the words:

> Natural causes, as we know, are at work which tend to modify, if they do not at length destroy, all the arrangements and dimensions of the earth and the whole solar system. But though in the course of ages catastrophes have occurred and may yet occur in the heavens, though ancient systems may be dissolved and new systems evolved out of their ruins, the molecules out of which these systems are built—the foundation stones of the material universe—remain unbroken and unworn.

I propose to give some account of the steps which have led in our time to the reversal of this view.

Although speculative doubt always persisted in relation to the view that all matter must be conceived as constituted of a very considerable number of fundamentally distinct elements, until the last decade of the nineteenth century no single definitely ascertained fact was known which made untenable the current chemical conception of the irreducibility and inconvertibility of the atoms of the different chemical elements, although some facts were known which seemed to indicate certain family relationships within various groups of elements. In the last 30 years a multitude of new facts have been discovered which have made it impossible any longer to regard all the chemical elements as fundamentally distinct from one another, and which have given new life to the older conception of transmutation. Moreover they have led to a theory of the constitution of the atom which, although much remains to be done before it can be regarded as a completely verified conceptual scheme, has made remarkable advances in its power of representing a very considerable complex of phenomena. In accordance with this theory, the electron, or definite unit electric charge, is conceived as a most important part, if not the whole, of the fundamental constituents out of which all atoms are built up. Should this theory be fully established, it would appear that the goal would be reached, the instinctive desire for which had inspired all the speculations relating to the unity which may underly the diverse forms of matter.

In the years 1815 and 1816 the attention of chemists was directed by Prout to some facts which tended to show that there is a relationship between the properties of various elements and their atomic weights. He called attention to the fact that the three magnetic elements, iron, nickel, and cobalt, had in accordance with his estimate the same atomic weight, double that of nitrogen; from this, and other instances of approximate equality of atomic weights, he concluded that substances having nearly the same atomic weights resemble one another in properties, and can combine more readily with one another. From an examination of the atomic weights he made the suggestion that all the other elements have atomic weights which are integral multiples of the atomic weight of hydrogen, and he suggested that they are all compounded of hydrogen and oxygen; in his later writings he put forward the view that hydrogen is the primordial element. In this order of ideas Prout had various successors, amongst whom Newlands, who promulgated in 1865 his *Law of Octaves*, is the most prominent as a predecessor of Mendeléeff. Newlands arranged the elements in a table in which all the elements are consecutively numbered, and appear nearly in the order of their atomic weights; they are arranged in a number of groups, each group containing seven members, but in some cases two elements are coupled together so as to occupy one place in the table. The elements in each group are placed in a vertical column, and all the columns are arranged in parallel, so that the first, eighth, fifteenth, etc. elements appear in a horizontal line. In this arrangement all the elements which are in a horizontal line are analogous, having similar properties. In the table the relation between elements of the same family is the same as that which in music exists between the notes at the extremities of one or more octaves. In some cases analogous elements whose atomic weights are consecutive appear with consecutive numbers in the table, but in general the numbers attached to analogous elements differ by 7, or a multiple of 7. Various objections in detail were raised to this classification, one of which was that it left no room for the inclusion of elements which might be discovered later. The fact that caesium, thallium, indium, and rubidium had been discovered only a few years earlier made this objection a cogent one. Both Newlands' arrangement of the elements, and an earlier one embodied in the so-called telluric helix propounded by De Chancourtois in 1863, indicated a periodic variation of properties among the elements, when arranged in the order of their atomic weights.

The Periodic Law of the elements, as it is now known, is the system published by Mendeléeff, of Petrograd, in 1869. He showed that, when the elements are arranged in order of their atomic weights, elements which have a given property occur periodically; thus, for example, chlorine, bromine, and iodine have similar fundamental properties, although their atomic weights differ widely from one another. Again elements which have very similar chemical and physical properties have atomic weights of nearly the same magnitude, unless they increase regularly. Further, when the elements are arranged in the ascending order of atomic weight, they are in their proper order as regards valency. The elements, such as hydrogen, oxygen, nitrogen, etc., which are most widely distributed have the smallest atomic weight; consequently Mendeléeff called hydrogen and those elements included in the second series of his table "typical elements." The general principle on which Mendeléeff insisted may be expressed in the form that the physical and chemical properties of an element are periodic functions of its atomic weight. The table which Mendeléeff published in 1871 has been since accepted as an authoritative illustration of the Periodic Law. The gaps in his table were taken by Mendeléeff to indicate the existence of elements which had not yet been discovered, and he was enabled to predict the characteristic properties of such unknown elements. The success of these predictions was conspicuous in the cases of Scandium, Gallium, and Germanium, which were discovered a few years after the publication of the table; and this success was important evidence of the value of the Periodic Law. In the case of some elements Mendeléeff employed the law to correct their atomic weights. Thus the atomic weight of uranium had previously been regarded as 120, but Mendeléeff gave reasons, founded on his theory, for supposing that the number should be 240; and this surmise was shown by Zimmermann, in 1881, to be correct. Very important work was done by Lothar Meyer in the establishment of the theory of the Periodic Law; he expressed his results in the form of a curve in which the atomic weights are plotted as abscissae, and the atomic volumes as ordinates; in this curve there are four well-marked maxima, and the elements arrange themselves in six divisions. He considered in detail the bearing of the tabulation upon the electro-positive and the electro-negative properties of the elements, and upon magnetic and diamagnetic properties. The connection of the melting and boiling points of compounds with the places of the elements of which they are formed in the table of atomic weights was the subject of extensive researches by Carnelley, between the years 1879 and 1885. Many applications of the Periodic Law, in connection with chemical and physical properties, and especially relating to the spectra of the elements, were made by a considerable number of researchers.

In 1887, Sir William Crookes published papers dealing with the *Genesis of the Elements*. He suggested the hypothesis that all the elements, including hydrogen, have been evolved from one original substance which he proposed to name protyle. He suggested also that our knowledge of any particular element is that only of an average specimen of such element, so that, even as regards atomic weight, there may be possible variations, between limits, for one and the same element. The Periodic Law and the detailed confirmation of its value which it received from the results obtained by a great mass of research seemed to make it highly probable that some community of structure and of composition of all elements must be recognized, which would account for the law, but it was not until 1896 that a direct breach was made in the seeming fixity and inconvertibility of the different kinds of atoms.

The acceptance of the electromagnetic theory of light naturally led to the conclusion that the radiation from hot bodies must have its origin in the vibration of electric systems, and thus that the atoms of such bodies must contain electric charges capable of setting up such vibrations. This suggested an electron theory of matter, developed by Larmor and Lorentz, in accordance with which matter is composed of electric charges, and its mass is to be regarded as electromagnetic inertia. On these lines it was suggested that the movement of the electric charges would be changed by a magnetic field, and that this would displace the spectrum of the substance. That this is actually the case was shown by Zeeman in the case of the spectral lines of sodium, which were split into two under the influence of a strong magnetic field.

When an electric discharge is passed between electrodes through a glass vessel, so exhausted that the air pressure is extremely small, luminous rays, known as cathode rays, proceed from the negative electrode, or cathode. When they strike an insulated conductor they impart to it a negative electric charge. It was shown by Sir J. J. Thomson that these rays can be deflected from the rectilinear path by magnetic force, or by electric force, in the same manner as a stream of negatively electrified particles would be deflected. From this it was concluded that the cathode rays consist of a stream of negatively electrified particles in rapid motion. By measurement of the deflections, employing the highly probable assumption that all the particles are equally charged, Thomson estimated that their velocity is about one-tenth that of light, and that their mass, which appeared to be independent of the nature of the electrodes, is about one seventeen-hundredth part of the mass of a hydrogen atom. The estimate has been later the subject of revision. This amounted to a discovery of a particle of very much smaller mass than that of the lightest particle hitherto taken to exist, viz. the atom of hydrogen; and such a particle might be taken to be a component of all of the various substances which could be employed as electrodes. Since electricity in motion possesses electromagnetic energy, its effects are comparable with those of mechanical inertia; moreover the electromagnetic energy of a moving charge is increased as the velocity is increased. It appeared then that the whole, or a part, of the mass of material substances might be of electromagnetic origin. It seemed possible to regard the corpuscle as simply an electric charge of definite amount, and to identify it with the electron of Larmor and Lorentz. As the charges of the corpuscles in the cathode rays are all negative, there remained for consideration the question of the nature of positive electrons, and the part they might play in the phenomena of matter; further facts were required for the elucidation of these questions.

In 1896, a most important discovery, that of a radio-active substance, was made by Becquerel, in Paris. He found that compounds of the metal uranium, which has the highest atomic weight of all the elements, continually emit rays capable of penetrating opaque screens, and of affecting photographic plates. The intensity of this effect was measured by utilizing the fact that the rays from the uranium convert the air through which they pass into a conductor of electricity. It was thus found that the radio-activity of the substances depends solely upon the amount of uranium which they contain. It was soon afterwards shown by M. and Mme Curie that thorium, the element of next highest atomic weight to uranium, possesses the property of radio-activity; but other of the known elements failed to exhibit the same property. On experimenting with pitch-blende and other natural minerals in which uranium occurs, M. and Mme Curie found that the radio-activity of pitch-blende is greater in amount than what could be attributed to the uranium that it contains, and they then succeeded in chemically separating out of the pitch-blende compounds of a new and intensely radio-active substance, to which the name radium was given, only a minute amount of which is contained in the pitch-blende. They also found other very radio-active substances in the pitch-blende. Radium was found to be a chemical element with a well defined place assigned by the Periodic Law, and it resembles the element barium in its powers of entering into chemical compounds; it has a characteristic line spectrum, and its radio-activity is about two million times as great as that of uranium; its atomic weight was found to be next less than that of thorium. The amount of energy given off spontaneously by pure radium compounds, and capable of being transformed into light and heat, is so great that it was estimated to be sufficient to heat a quantity of water equal to the weight of the radium from the freezing-point to the boiling-point every three-quarters of an hour.

The urgent question at once presented itself as to the source of this great amount of energy which is continually being given out by the radio-active substance. It was found that, in radium and the other radio-active substances, the intensity of the radiation is independent of such physical conditions as temperature and pressure, and that the radiation cannot be inhibited by any known means. The discovery was made by Sir E. Rutherford, in 1899, that the radiation from thorium contains three distinct kinds of rays, which are known respectively by the names $\alpha$-rays, $\beta$-rays, and $\gamma$-rays. The investigation of the character of these rays was carried out by Sir E. Rutherford and Professor Soddy; from their investigations, and those of others, a large number of facts as to the radio-activity of thorium and the other radio-active substances were disclosed. Of the detailed investigations of this order, I can give no account, but must confine myself to a statement of the main facts which emerged as the result of many elaborate and difficult experiments. It was found that the $\alpha$-rays, when tested by means of magnetic and electric fields, behave as positively electrified material particles, and that they are stopped by a sheet of paper. These particles carry two atomic charges of positive electricity, and they travel with a velocity of from one-twentieth to one-fifteenth of that of light. At first there was some doubt whether these particles should be regarded as atoms of hydrogen or of helium, the latter of which is always to be found in minerals containing uranium and thorium. It was ultimately shown by a spectroscopic examination of the light from a few milligrams of radium that the $\alpha$-rays must be taken to consist of a stream of helium atoms. It was later shown that the $\alpha$-rays from uranium, thorium, polonium, actinium and other radio-elements all consist of helium atoms.

The history of the discovery of helium is a remarkable one. It was first made in 1868 by observation in the spectrum of the solar chromosphere of a bright yellow line which had not been observed to exist in the spectrum of any known terrestrial substance; from this it was concluded that there exists in the sun an element hitherto unknown, and to this the name helium was given by Lockyer, who was one of the first to observe it. Other lines were observed in the solar spectrum to accompany the yellow line, and the same spectrum was observed in the light from many stars. But it was not until 1895 that helium was discovered to exist on the earth. After the discovery, by Lord Rayleigh and Sir W. Ramsay, of the element argon in the atmosphere, a search for argon in mineral substances was undertaken, and in the course of this search a gas was found to be given off by solutions of minerals containing uranium. On examination of the spectrum of this gas by Lockyer, it was found to be identical with that of the helium discovered nearly thirty years earlier in the solar spectrum.

The $\beta$-rays have been shown to be negative electrons; they are deflected both by magnetic and electric fields, and by means of measurement of such deflections their identity with the cathode rays in the vacuum tube has been established. These $\beta$-rays travel with great velocity, approaching that of light, and much more rapidly than the cathode rays; they have great penetrative power, being able to pass through a considerable thickness of tinfoil or of glass, without losing their power of affecting the photographic plate. They consist of unit charges of negative electricity, just half as great as the positive charges carried by the $\alpha$-atoms. The effective mass of a $\beta$-electron is only a very small fraction of the mass of one of the $\alpha$-atoms; and it is the latter which contain most of the energy emitted in the radiation. The $\gamma$-rays are not deflected by magnetic and electric fields, but they have much greater power of penetration than the $\beta$-rays. They have been identified with the Roentgen rays emitted outside the vacuum-tube when the cathode rays are passing through the interior of the tube.

Besides the rays given out by these radio-active substances, radium and thorium also give out gases, called emanations, which are themselves radio-active substances. It was found that the radio-activity of these emanations suffers a gradual decay, whilst the walls of the vessels in which they are contained become radio-active. But when the walls are washed with certain acids they lose their radio-activity, and this is transferred to the acids. The inference from this is that the emanation gives rise to a new radio-active substance which is deposited on the wall of the vessel, and is soluble in the acid. It has been shown that, in a radio-active substance, chemical changes are going on which result in the production of small quantities of new substances. By chemical processes, minute quantities of these new radio-active substances have been separated out from uranium and thorium, and it is found that they have properties different from those of the original substance. At first these new substances are highly radio-active, whilst there is a diminution of the radio-activity of the original substance. After some time the radio-activity of the new substances decays, whilst that of the original substance gradually goes back to its original amount. Rutherford and Soddy, who worked together, by the examination of the phenomena connected with the radio-activity of thorium, were led to the conclusion that a whole series of new chemical substances are produced successively, the whole process being accompanied by the liberation of energy. The atom of uranium, in twelve successive changes, appears to expel 7 atoms of helium and 5 electrons, one such atom or electron at each change. The activity of a radio-active substance is found to be independent of the presence and nature of any other substance with which it may be in chemical combination; from this it can be inferred that radio-activity is to be regarded as an atomic, not a molecular, phenomenon. From the point of view of the atomic theory the changes in the nature of a substance are to be regarded as a series of successive changes in the constitution of the atoms. These atoms become dissociated into simpler parts, and the process is accompanied by the liberation of some of the internal energy of the atom. The amount of energy given out in these atomic changes is enormously greater than the amount liberated in ordinary chemical changes which only involve molecular changes, that is alterations in the grouping of atoms, whilst the atoms themselves are unaltered. It has been estimated that an ounce of radium, in the course of its average life of about 2500 years, gives out as much energy as is evolved by burning ten tons of coal.

We have before our eyes, in the case of a radio-active substance, an actual transmutation of elements, the possibility of which had, as I have already stated, often been speculatively asserted. This transmutation theory of radio-activity was formulated by Rutherford and Soddy in 1903, and soon afterwards some of its consequences were verified in a striking manner. It was shown that, when a small quantity of radium emanation was watched during its slow decay, the spectrum of the element helium could be seen; and later the presence of helium in the $\alpha$-rays was demonstrated spectroscopically. The successive products of the disintegration of radium were examined with minute care by Rutherford, who traced out a long series of such disintegration products, most of which are so small in amount that they can only be detected by means of their radio-activities. Some of these products give out only $\alpha$-rays, one gives out both $\beta$- and $\gamma$-rays, and one gives out all three kinds of rays; two of them give out no rays. They show great differences in their rates of decay; one of them decays to half its amount in three minutes, another in 143 days, another in 40 years, whilst radium itself takes about 2600 years for its half decay. Results of a similar character have been obtained as regards uranium and thorium and their radio-active products. As the result of various experiments by R. J. Strutt and B. B. Boltwood there is strong reason for believing that radium is a descendant of uranium, although probably not the direct product. An immense amount of work has been done by a large number of Physicists in the detailed investigation of the phenomena of radio-activity and of the properties of the new substances discovered as the products of disintegration.

Various hypotheses have been advanced as to the mode in which the phenomena are to be conceived; and these may be divided into two classes. In the first of these the energy emitted from the radio-elements is regarded as being obtained at the expense of the internal energy of the atom. In the second type of theory the energy is regarded as derived from external sources, the radio-elements being regarded as mechanisms for the transformation of such energy into the forms manifested in radio-activity. The evidence appears to be very strongly in favour of the theories of the first type; since no experimental evidence appears to have been obtained that the energy is obtained at the expense of any external source. The theory of Rutherford and Soddy assumes that, on an average, a definite proportion of the atoms of each radio-active substance become unstable at a given time, and when this instability occurs there is a disintegration of the atom, usually of a violent character. The disintegration is supposed to consist of the expulsion, either of an $\alpha$-particle or of a $\beta$-particle, or of both simultaneously, but in some cases the change in the atom does not appear to involve the expulsion of either kind of particle. When an $\alpha$-particle, of atomic weight 4, is expelled, the remaining atom is lighter than before, and a substance composed of such atoms has chemical properties different from those of the original substance. This new substance becoming again, as regards some of its atoms, unstable, another $\alpha$-particle may be expelled; and this process may be repeated through a series of stages. As the disintegration proceeds, the substance consists of a mixture of atoms of the original type and of the new type; that this is the case with radium is evidenced by the fact that the original spectrum of radium persists unchanged whilst the disintegration very slowly proceeds. The difference between the chemical properties of the original substance and the new substance formed by its disintegration is strikingly illustrated by the case of radium. Radium itself is an element closely allied to barium, and it has a definite spectrum of bright lines similar to the spectra of the alkaline earths; it is non-volatile at ordinary temperatures. But the emanation is a chemically inert gas which, in its spectrum, and in the absence of definite chemical properties, resembles the group of inert gases to which argon belongs; it condenses at – 150° C. This emanation is unstable, and emits $\alpha$-rays, that is atoms of helium; the residual atoms then form a new substance, called radium-*A*, which behaves like a solid, and is deposited on the surface of bodies. The question arises, what are the final substances which appear as the result of the whole process of successive formation of new substances? One of these final products is helium; this is formed by the accumulation of the $\alpha$-rays given out during the successive transmutations. The other final product is lead, which is always a constituent of uranium minerals, the atomic weight of which appears to agree with the atomic weight of the substance which would be obtained from radium by the expulsion of 5 $\alpha$-particles; one such $\alpha$-particle is known to be expelled in each of five out of the eight successive transformations of radium into lead. The whole line of descent of uranium through ionium and radium, in a considerable number of successive transformations, has been traced out; the rates at which these transformations proceed show enormous variations ranging from a few seconds to thousands of years.

It was natural that attempts should be made to devise models of the atom which should be capable of representing the facts which have been discovered relating to the instability of radio-active substances. A model of this kind was suggested by Lord Kelvin, and investigated in a detailed manner, in 1904, by Sir J. J. Thomson. The atom is supposed to consist of a uniform sphere in which is a positive electric field, and throughout which are distributed a number of negative electrons in motion. The total positive charge in the sphere is taken to be balanced by the charge of the electrons, when the atom is electrically neutral. The idea is that the properties of the atom depend upon the number of the electrons, upon their distributions in their orbits, and upon the stability of such systems. The possibility presented itself of connecting in this manner the various possible types of atoms with the Periodic Law of the elements. Sir J. J. Thomson showed that the electrons moving under the electric forces upon them must distribute themselves in a number of concentric shells of differing radii. In order to simplify the mathematical problem of determining possible arrangements of the electrons, Thomson worked out in detail the case in which the electrons move in rings in one plane, the electrons being arranged at equal angular intervals. He determined the number of rings and the number of electrons in each ring which are such that stability of the system is ensured. It appears that such an atom, containing a large number of revolving electrons, may radiate energy extremely slowly. There must however come a time when this small but continual loss of energy from the atom results in a rearrangement of its electrons into a new system, or in the expulsion of one or more of the electrons from the atom. It was suggested that the disintegration of the atoms of the radio-active elements is due to this result of the gradual loss of energy by radiation. It was shown that only a very small part of the mass of the atom was due to its electrons, the rest being presumably associated with the positive electric field. In order to account for the scattering of the $\alpha$-particles when they penetrate a thin film of matter, this theory of the atom was modified by Rutherford.

In accordance with the theory of the atom which was advanced by Rutherford, the atom contains a charged nucleus, of dimensions exceedingly minute compared with the whole volume of the atom, that is with what is called the atomic domain; and this nucleus replaces the sphere of positive electrification in Thomson's theory. The nucleus consists of a number of charged elements, some of which are positive and others negative, their resultant being a positive charge. Surrounding this nucleus there are in the atomic domain a number of electrons; that number being equal to the number of resultant positive unit charges in the nucleus, so that, in the neutral atom so constituted, the total number of unit charges, positive and negative, is zero. The resultant positive charge of the nucleus is taken to be about half the product of the atomic weight multiplied by the fundamental unit charge. The existence of the small massive nucleus was suggested by the fact that the $\alpha$- or helium-atoms when they penetrate into matter are divided through a considerable angle; the existence of an intense field of force in the interior of the atom being deduced from this fact. The properties of an atom are regarded as dependent mainly on the resultant nuclear charge, and not on its mass; this is taken to explain many facts connected with the Periodic Law of the elements. The electrons in the outer part of the atom are arranged at distances from the nucleus, and are controlled by the forces due to the resultant charge of the nucleus and their own electric fields. The nuclear electrons form a close combination with the positively charged units that make up the main part of the mass of the nucleus. It is regarded as probable that, in the region just outside the nucleus, an electron cannot be in stable equilibrium. Rutherford makes the statement as regards the electrons internal to, and external to, the nucleus that[1]:

> While no doubt each of the external electrons acts as a point charge in considering the forces between it and the nucleus, this cannot be the case for the electrons in the nucleus itself...... Under the intense forces in the latter the electrons are much deformed and the forces may be of a very different character from those to be expected from an undeformed electron as in the outer atom. It may be for this reason that the electron can play such a different part in the two cases and yet form stable systems.

In accordance with this theory, radio-active change originates always in the nucleus of the atom, the expelled $\alpha$- and $\beta$-particles coming from the nucleus. There is

[1] "Bakerian Lecture," *Proc. Roy. Soc.*, Vol. 97, pp. 377–8.

evidence that, when the rapidly moving $\alpha$-particles, or helium atoms, pass through dry nitrogen, they give rise to rapidly moving particles that closely resemble hydrogen. This is shown by observations of their brilliant scintillations when they are allowed to strike a zinc-sulphide screen. It is thus suggested that, by a breaking up of the nitrogen atoms, hydrogen has been obtained; if this inference is fully confirmed, a very important step has been taken in the theory of the transmutation of the elements. The hypothesis has naturally been made that the atoms of all the elements are built up of hydrogen nuclei and of electrons. In accordance with this view the helium nucleus is composed of four hydrogen nuclei and two negative electrons, with a resultant electric charge of two positive units. This is a modification of Prout's hypothesis to which I have already referred, that all the elements are built up of hydrogen as a fundamental constituent.

Rutherford's theory of the nuclear constitution of the atom has been further developed by Dr Niels Bohr, of Copenhagen, who studied the optical spectra of hydrogen and of helium, regarded as the simplest types of atoms. His theory takes account of, and employs, the new quantum-theory, in accordance with which radiation only occurs discontinuously in quanta, or definite unit losses of energy, at moments when one stable configuration of the atom changes into another such configuration. This theory has been further elaborated by Sommerfeld and others, and appears to have had a considerable amount of success in coordinating various phenomena connected with hydrogen and helium, and especially in accounting for the complexity of the lines of their spectra. The mathematical theory of the hydrogen atom has been worked out; the calculation of the spectral lines of hydrogen, known as the Balmer series, is in full accord with observation; and even the displacement of the lines which occurs when the hydrogen is in an electric field has been successfully determined. It is a remarkable fact that this theory of the hydrogen atoms makes use of the theory of Abelian functions, a purely mathematical theory which had been worked out without any expectation that it would ultimately have any such application. This is an example, which has many parallels in Physical Science, of the fact that the most abstract Mathematics, which of course has its ultimate roots in the physical domain, does not really lose its connection with its original source, however long the time may be before the fact of such connection attains explicit recognition.

The development of the theory of the constitution of the atom is still in progress. Complete success in devising a model of the atom which would make possible the calculation of the detailed varieties of configuration of which such a model might be capable would be a considerable step towards the goal of turning Chemistry into a deductive science, in accordance with which all possible elementary forms of matter might be ascertained, and the possible nature of compounds, with their chemical and physical properties, predicted.

The known radio-active elements, in spontaneous disintegration, are few in number, most of the elements being so stable that no such disintegration of them can be detected. We do not at present possess any fully established means of exciting artificially the process of expulsion of $\alpha$- or $\beta$-particles from the atom, and of thus liberating energy of the enormous stock which appears to be internal to the atom; but efforts are being made to accomplish this. Sir E. Rutherford, by bombarding the atoms of nitrogen with $\alpha$-rays, has, as I have already remarked, succeeded in producing hydrogen atoms from a small proportion of the nitrogen atoms. If the means of solving this problem of transmutation are in the future discovered, the practical use of such a method would lie more in the utilization of the internal energy of the atom as a source of energy, than in the solution of the ancient problem of transforming one kind of matter into another. We should be in possession of a source of energy by which quantities of energy would be liberated and made available, enormously in excess of the amounts obtainable by the ordinary chemical processes of combustion, in which only molecular energy is utilized, and in which the vast store of energy in the atoms remains unaffected.

The existence of the radio-active minerals is of much importance in connection with attempts to estimate the age of geological strata in which they are contained, as we know the rate of change of uranium and its various products into the final product, lead. As Soddy has written[1]:

> To-day we know that the radio-active minerals are in reality geological clocks, and they record more accurately than in any other way the age of the stratum in which they occur. In a uranium mineral, for example, each 1 per cent. of lead in terms of the quantity of uranium signifies the lapse of 80,000,000 years. Errors of course are possible, if lead should have been an original constituent of the mineral, but these are minimized by taking a large number of different minerals. On the other hand every cubic centimetre by volume of helium per gram of uranium in a uranium mineral signifies 9,000,000 years, and—as here helium, being a gas that forms no compounds, cannot have been initially present, and as, moreover, some will have escaped—the age of the mineral by this method is a minimum, whereas the age by the lead content may be too high. The carboniferous rocks tested by this new method appear to have an age of some 350,000,000, and the oldest Archean rocks of over 1,500,000,000 years.

It will be observed that the processes going on in radio-active substances consist exclusively of the breaking up of more complex atoms into lighter and less complex parts, and so far as our experience goes these processes are irreversible. Accordingly no light is thrown upon the

[1] *Science and Life*, pp. 100, 101.

possibility of the present complex elements having been built up out of simpler components, such as is involved in the view that a simple primeval form of material once existed, from which the present elements were evolved by processes in which there was a gradual complication of the original material. If we are to conceive that our present forms of matter were so evolved, we must imagine that such evolution took place under conditions very widely different from those which we are able to observe at present. Passage from the more complex to less complexity is what we observe at present, and the idea that change in the reverse order once took place is at present a speculation resting upon no evidence obtained from observation.

A very important discovery has been made in connection with the investigation of radio-activity, that of the existence of isotopes, the name given to elements which have identical chemical properties, and occupy the same place in the Periodic Table, but have different atomic weights. If an $\alpha$-particle, or a $\beta$-particle, is expelled from an atom, the new atom has different properties from the original one. If it is the $\alpha$-particle that is expelled, the element after this expulsion passes in the Periodic Table to the place next but one to that of the original element, the atomic weight being diminished. If it is the $\beta$-particle that is expelled, the change of place is into the next place in the table, in the opposite direction. But if an $\alpha$-particle and two $\beta$-particles are expelled, in any order, the element returns to its original place in the table; its atomic weight is diminished by four units, but its chemical and spectroscopic properties are the same as before the expulsions took place. The elements in the original and final form are in fact isotopes. The places in the Periodic Table appear to represent the integral nett charges of electricity in the atomic nucleus. The $\alpha$-particle having a charge of two units of positive electricity, when it is expelled from the atom, moves the element through two places in the Periodic Table, whereas the expulsion of a $\beta$-particle with its unit negative charge, moves the element one place in the opposite direction. Isotopes such as thorium and ionium have identical chemical and spectroscopic properties, although their radio-active properties are different. The lead obtained as the final product of the series which commences with uranium has a different atomic weight, 206, from the lead which is the final product of the thorium series, for which the atomic weight is 208. Ordinary lead has the atomic weight 207·2, and this suggests that it is a mixture of the two isotopes.

There appear to exist also isotopes which have the same atomic weight as well as identical chemical properties, although the internal energy in the atoms of the isotopes is different in the two cases. Such isobaric isotopes are obtained when, as in the case of thorium, there is a branching off of the successive disintegrations into two series, each with a final product.

By means of a new method of deflecting the rays which consist of streams of charged particles, F. W. Aston has been able to determine the masses of a number of atoms with a great degree of accuracy. By this method he has been enabled to show that some of the elements, formerly supposed to be homogeneous, consist of a mixture of two or more isotopes with different atomic weights. When the atomic weight of oxygen is taken to be 32, so that the atomic weight of hydrogen is 1·008, it has been found that the atomic weights of many of the elements are represented exactly by integral numbers. In other instances, such as chlorine, in which this is not the case, the elements have been found to consist of a mixture of isotopes, each with an atomic weight represented by an integer; the fractional atomic weight of the element as a mixture being due to the fact that the element contains portions of different atomic weights. Thus, for example, chlorine, of which the atomic weight is 35·46, obtained by chemical methods, is a mixture, in the proportion of 3 to 1, of two isotopes of which the atomic weights are 35 and 37 respectively. These investigations have led to the theory, which has been to a great extent verified, that the fractional irregularities of the atomic weights in the Periodic Table are due to the existence of isotopes.

The known facts appear to be consistent with the view that all matter is built up of electrons and atomic nuclei; all the electrons being alike, and the atomic nuclei being constructed of fundamental units all alike, and of electrons. The electrons and nuclei must be regarded as postulated concepts subject to defined interrelations, and which, from the point of view of this theory, are irreducible.

As I have already indicated in an earlier lecture, for a long period chemical and physical investigations were carried on independently of one another, and they had but few points of contact. An important effect of the great discoveries of the last quarter of a century, relating to the composite character of the atom, has been to break down the barrier which had long separated the sciences of Physics and of Chemistry. The two great departments have in our day at last joined forces; their meeting-place has been within what was long regarded as the impregnable fortress of the atom.

# XIII

## COSMICAL THEORIES

I PROPOSE in the present lecture to give some account of various theories which have been designed to describe the processes of change by which the bodies of the solar system and the stellar universe may be conceived as having reached their present configurations and general physical states. Such theories consist of attempts to trace back into the remote past the history of the solar system and of other configurations which we observe telescopically and with the aid of the spectroscope, on the assumption that the physical laws by means of which we describe the present motions of bodies, and their thermal and other physical changes, are adequate to describe what we conceive to have happened in the immense periods of time which we must assume, in accordance with those laws, to have elapsed since the primeval conditions postulated by such cosmical theories. This assumption involves a considerable element of speculation, because our knowledge of the range of applicability of physical laws is determined and limited by observations of the special physical processes which we can actually observe only during strictly limited intervals of time; and these laws can only hypothetically be extended to embrace processes going on under conditions widely different, and during immense periods of time.

An examination of the bodies of the solar system reveals certain striking uniformities within that system. In the first place, the eight great planets Mercury, Venus, the Earth, Mars, Jupiter, Saturn, Uranus, and Neptune, all revolve in the same direction round the sun; their orbits are all nearly in the same plane, and are all approximately ellipses with small eccentricities, that is, nearly circular. The same statement holds good of the minor planets or asteroids, some 900 in number, which have been discovered in the solar system. These facts are sufficient to suggest that the uniformity in the orbits of the bodies of the solar system may be represented as the result of some process of formation from a common origin. But these facts do not stand alone. The planets rotate round their axes in the same direction, and their satellites, with certain exceptions, rotate round the primaries in nearly circular orbits, nearly in the same plane, and in the same direction. Jupiter and Saturn each has a system of satellites, forming systems like the system of the sun and planets, but on smaller scales. It appears that the exceptional cases arise on the outermost parts of the solar system, and the outermost parts of the systems of Jupiter and Saturn. Thus Neptune has only one satellite, and this has retrograde motion; that is, its direction of revolution is in the opposite direction to that of the great majority of the satellites and of all the planets. Uranus has four satellites, whose orbits are all highly inclined to the ecliptic; Saturn has nine satellites, of which the outermost has retrograde motion; and its orbit has a large eccentricity. Of the nine satellites of Jupiter, the two outermost have retrograde motion. Thus in each case the deviation from uniformity attaches to the planets or satellites which are at the outermost distance from the sun or the primary.

In 1772 it was pointed out by Titius, a Professor at Wittenberg, that the distances of the six planets at that time known, from the sun, were very approximately represented by numbers which increase according to a very simple law; but that there was one interruption, the planet that should in accordance with the law have its orbit between that of Mars and that of Jupiter not being known to exist. The matter was taken up by Bode of Berlin, who suggested that an unknown planet remained to be discovered which would occupy the vacant place in the table of distances indicated by the law of Titius, which has since been known as Bode's law. The discovery of the planet Uranus was made about this time, and it appeared that its distance from the sun was in fair conformity with the law. This discovery lent further weight to the prediction that a planet would be found to occupy the vacant place, and this prediction was ultimately verified in an unexpected manner. A number of small planets or asteroids were discovered in the space between the orbits of Mars and Jupiter, the first of these being discovered on the first day of the nineteenth century, by Piazzi, a Professor at Palermo. The idea then presented itself that these asteroids were fragments of a single planet which had been broken in pieces by an explosion. It was indeed shown that three of these asteroids really conformed closely to Bode's law. Since that time the number of asteroids observed in the region between the orbits of Mars and Jupiter has steadily increased, until some 900 of them have been found, but the notion that they are all fragments of one original planet is for various reasons regarded as untenable. Nevertheless the discovery is remarkable as the first instance of a successful prediction of the existence of bodies which had at the time not been observed. The story of the still more remarkable verification of the prediction made by J. C. Adams and Leverrier, that a planet would be found outside the orbit of Uranus, has often been told. This discovery of the planet Neptune was predicted mainly as the result of observation that the motion of the planet Uranus exhibited disturbances which could only be accounted for on the hypothesis that they were due to an unobserved planet at greater distance from the sun. Whereas the prediction which led to the discovery of the first asteroids was inspired by the purely empirical rule known as Bode's law, in the prediction which led to the discovery of Neptune, that law was only employed to afford ground for a probable value of the distance of the planet from the sun; calculations based upon the law of gravitation, taken in conjunction with the data afforded by the irregularities in the motion of Uranus, forming the chief basis of the prediction.

After this survey of some of the main features of the solar system, and of the considerable degree of the regularities in its structure, I now turn to the consideration of the theories that have been proposed as to the mode in which it has been evolved, and in particular to the celebrated Nebular hypothesis, first suggested by the Philosopher Immanuel Kant, and developed more precisely by Laplace.

Kant's speculation as to the origin of the heavenly bodies was published in 1755 in his *Allgemeine Naturgeschichte und Theorie des Himmels*, and was applied by him not only to the solar system but to the stellar universe. He supposed the solar system to have been developed from an initial condition in which a cold nebula at rest consisted of a vast mass of evenly diffused particles. This nebula he supposed to commence to concentrate under the mutual gravitation of its parts, and to become hot in doing so, owing to the consequent compression. He made the quite arbitrary and unwarranted assumption that this process would give rise to a rotation of the whole mass. He assumed further that the matter would concentrate into rings which would be in rotation, and he illustrated this by the case of the rotating rings of Saturn which he regarded as giving evidence of the correctness of his views. He supposed that in course of time these rings would become unstable, and would form planets by a process of agglomeration; thus leading to a system of planets revolving round the densest and hottest core, the sun. The satellites he supposed to be formed by a repetition of similar processes on a smaller scale, as the planets gradually contracted under their own gravitation.

The Nebular hypothesis, as we now understand it, was first put forward by Laplace, in 1796, in his *Exposition du Système du Monde*. Unlike Kant, of whose speculations Laplace appears not to have had any knowledge, Laplace limited his theory to the case of the solar system. He assumed as the original form from which the solar system was evolved an immense nebula of a flat or lens-shape, consisting of extremely attenuated, but gaseous, material. Unlike Kant he assumed the nebula to be originally in a state of slow rotation round an axis, thus avoiding the dynamical difficulty in Kant's theory as to how the rotation came into being. The mass falls in upon itself, owing to the gravitation of its parts, and as this occurs, the central portion becomes hotter as the whole mass shrinks, whilst the outer parts are cooled by radiation at the surface. Since the angular momentum of the mass remains constant in amount, the shrinkage of dimensions of the nebula produces a gradual increase in the angular velocity of rotation. It was assumed by Laplace that this increase of angular velocity led to the separation of an outer ring of matter from the main mass. As the shrinkage of the main mass proceeded, this process of shedding a ring of matter would be repeated, and as the result, there would be a central core surrounded by a series of concentric rings, all in rotation about a central axis. The planets were then supposed to be formed in the manner indicated by Kant, the rings becoming unstable, and ultimately forming planets.

In accordance with this theory, the rings of Saturn represent a stage in the general process of the formation of planets and of satellites. The theory does not explain why, in accordance with dynamical theory, the successive rings should become unstable, or why the rings of Saturn should form an exception to this instability. It was pointed out by Sir G. H. Darwin that, in case a ring became unstable, the resulting planet would not be formed on the perimeter of the ring, but at its centre of gravity. The fact that a continuous rotating ring of matter would be unstable was established by Maxwell in 1857, who found that the stability of the rings of Saturn could only be accounted for by the supposition that they consist of an aggregation of separate small fragments, or a heap of stones.

The highly speculative character of Laplace's theory makes it clear that, before it can be accepted as a probable account of the manner in which we may conceive the solar system to have arisen, the various stages which it includes must be scrutinized in the light of the principles of Dynamics. The difficulties of a mathematical kind which are involved in an investigation of the forms taken up by rotating masses of tenuous gas, of the stability of such forms, and in the ascertainment of what happens when the forms become unstable, are so great that nearly all the mathematical investigators who have concerned themselves with these questions have found it necessary to treat the nebulae as consisting of homogeneous incompressible fluid. Although the results of these investigations throw some considerable light upon the dynamical possibility of the processes involved in the theory in its original form, they cannot be regarded as having an absolutely decisive weight in our judgment of the theory. The details of that theory have received considerable modification at the hands of later investigators, but the notion that the solar system originated from a nebulous mass of gas, and the conception that increased rotation of that mass owing to shrinkage has been a main factor in the change from the primitive nebula to the present system, remain as at least highly probable hypotheses; and they have not been superseded by any other theory.

By later Astronomers the nebular hypothesis has been considered more generally than by Laplace, in relation to the idea that stars in general may be conceived to have been formed out of nebulae. The enormous increase of our knowledge of the nature of heavenly bodies, due to vastly increased powers of observation, in which photography plays an essential part, to the employment of the spectroscope, not only for the determination of the chemical nature of the composition of the sun, stars, and nebulae, but also of their motions, and the discovery that double and multiple stars are extremely numerous in the stellar system, have brought Laplace's theory into relation with a number of fundamental problems of cosmology.

After the discovery, in 1842, of the mechanical equivalence of heat and motion, the attention of Physicists was directed to the question of the means by which the solar radiation of heat and light is maintained. It was pointed out by Mayer that this radiation would have long since ceased if the sun were a body simply cooling, or in a state of combustion, and without any source of energy from which heat could be obtained. As a source from which the solar radiation might be sustained he advanced the meteoric hypothesis. In accordance with this hypothesis a swarm of meteoric bodies is constantly falling into the sun, and by the stoppage of their motion an immense amount of heat is evolved. Mayer calculated that, from this source, from 4600 to 9200 times as much heat would result, as would result from the burning of equal masses of coal. The difficulty of this theory is that the amount of matter which must fall into the sun, in order to maintain its radiation, must be enormous, amounting to a mass equal to that of the moon every half year. So great an accession to the mass of the sun would materially affect the gravitation of the planets towards it, producing a diminution in the periods of their revolution round the sun. In the case of the earth, Mayer reckoned that this would involve an annual diminution of the length of the year by a not very small fraction of a second; and he postulated an emission of material from the sun to neutralize this continual reinforcement of material from the meteorites. Independently of Mayer, this theory was advanced by Waterton at the Meeting of the British Association in 1853, and was fully discussed by Lord Kelvin in the following year. Kelvin pointed out that, although the influx of meteorites "is the only one of all conceivable causes of solar heat which we know to exist from independent evidence," that source of heat is entirely insufficient. An intraplanetary supply of meteorites would be much too scanty to achieve any perceptible result, and it was estimated by Kelvin that a sufficient supply from extra-planetary space would have resulted in a shortening of the year by six weeks since the beginning of the Christian era. This theory of the maintenance of solar heat led to the hypothesis that the earth and planets had been formed by an agglomeration of meteorites, which must have been of amount much greater than they at present receive. But the evidence of geological deposits negatives this theory, since no considerable parts of these deposits can possibly have been of meteoric origin.

A theory which has been much more widely accepted than the meteoric hypothesis was put forward by Helmholtz in 1854. In accordance with this theory the maintenance of the solar heat and light is due to the shrinkage of the material of which the sun is composed. As the envelope of the sun is continually cooled by radiation, a condensation towards the centre occurs which develops heat. In order to provide for the present rate of emission, a diminution of the solar diameter by 380 feet a year would suffice, and in five million years the sun would have shrunk to less than half its present bulk. Helmholtz estimated that radiation may have gone on at its present intensity for about twenty-two million years, an estimate which geologists find insufficient. This theory is in accordance with Laplace's view that the material of which the sun is composed was once diffused through an immense portion of space, and has gradually been condensed, with the production of heat, to the present dimensions.

To that part of the nebular theory which relates to the formation of planets many objections have been raised. Among these is the failure to account for the retrograde motion of the satellite of Neptune and of the satellites of Uranus. Again, one of the two satellites of Mars revolves round the planet in a shorter time than that of the rotation of the planet round its axis, a fact which it is difficult to reconcile with Laplace's views as to the generation of satellites. An objection to the theory of Laplace is that since, under gravitation towards the sun, the velocity diminishes as the distance from the centre increases, the rings could only coalesce into globes with a backward rotation round their axes. Laplace himself was aware of this difficulty, which he endeavoured to meet by the assumption that the rings had sufficient cohesion to allow them to rotate in one piece, as if they were solid. Thus the outer part of a ring would move faster than the inner part, and would give rise to planets spinning forwards, as is actually the case. It is difficult to conceive that the materials of nebulae can have had the amount of cohesion which is involved in this explanation. To meet these difficulties a theory was propounded by Faye, in accordance with which all the planets except Uranus and Neptune were formed in a practically homogeneous nebula, before its condensation had formed the sun. This nebula revolved like a solid body, with velocity increasing with the distance from the axis. The planets were then formed by agglomeration, and consequently their rotation was direct. The sun was subsequently formed by contraction of the nebula, and Uranus and Neptune were formed at a time when this contraction was far advanced; in accordance with this theory the order of formation of the planets is the reverse of what it is in accordance with Laplace's theory.

Attention was called to a factor of great importance in connection with the origin of satellites by the investigations of Sir G. H. Darwin on the effects of tidal action between the Earth and the Moon. The moon, by producing tides in the ocean, which act as a kind of brake upon the rotation of the earth round its axis, produces a very slow lengthening of the day, and this effect must have been much greater than it is at present when the earth was liquid or plastic. A precisely similar effect was produced by the tidal action of the earth on the moon. The result of this tidal action in the case of the moon is the present state of the rotation of the moon round its axis in the same period as its rotation round the earth, so that it always shows the same half of its surface to the earth. Darwin proved that a consequence of the tidal action of the moon on the earth, in accordance with the laws of Dynamics, is that the moon must slowly recede from the earth while its time of rotation round the earth is very slowly increased. He showed that the ultimate result will be that the day and the month will become equal to one another, each being then about 1400 hours, a state of tidal equilibrium being then reached, with the moon at much greater distance from the earth than at present. If this process be traced back from the present time, at some epoch not less than 54 million years ago, the length of the day was between 2 and 2½ hours; the moon was then almost in contact with the earth, and rotated round the earth in the same period as that of the earth round its axis. The period of rotation of the moon round the earth when they were nearly in contact was, in accordance with Kepler's law, between 2 and 2½ hours. It can be shown mathematically that the most rapid period of rotation of a fluid mass, of spheroidal shape, of density equal to the earth's average density, which is consistent with stability, is two hours and twenty minutes. It was consequently suggested that the moon was generated by the separation of a mass from the earth, owing to instability due to too rapid rotation of the earth, a breakage into two masses being the consequence. That the general character of the changes in the rotation of the earth and in the length of the month is in accordance with the laws of Dynamics is not open to doubt, but serious criticism has been directed to the theory that the initial stage of this process is connected with the genesis of the moon. It was pointed out that, when the moon was very near the earth, a disruptive strain would act on the moon, sufficient to prevent it holding together as one continuous mass; to meet this objection Darwin suggested that at that time the moon may have been an aggregation of separate masses. But it would appear that the effect of tidal action would have caused a dispersion of these separate masses, and to this objection there does not appear to be any adequate answer, so that this part of Darwin's theory cannot be regarded as established.

The questions were discussed by Darwin whether satellites of the other planets in the solar system could have originated in the same manner as the moon, and whether the present relations between the primaries and their satellites could be accounted for as effects of tidal friction. The answers to both questions were in the negative. The circumstances in these cases are different from those of the earth and moon, in the important respect that the ratio of the mass of the moon to that of the earth is much greater than the ratio of the mass of any other satellite to that of its primary. Consequently tidal friction has been a much more important factor in determining the relations of the earth and moon than in any other case in the solar system. Darwin showed that satellites, such as those of Jupiter, Saturn, or Mars, in all probability never revolved round the primaries in much smaller orbits than at present; thus in these cases no such great tides ever existed as were raised on the earth when the moon's orbit was much smaller than at present. Darwin also considered the effect which the friction of the tides raised on the planets by the sun may be taken to have had in altering the periods of their orbits and their distances from the sun. He showed that no large effects upon the planetary orbits can have arisen from the solar tides, and thus that the planets cannot have arisen by separation of portions of the plastic mass of the sun, in a manner similar to that by which he supposed the moon to have been detached from the earth. This removes one possible alternative to Laplace's theory of the formation of the planets. It was, however, pointed out by Roche, in 1872, that the tides produced on the planets by the sun may have been a decisive factor in relation to the formation of satellites. According to Darwin's view, a satellite will be formed when the rotation of a fluid or plastic mass comes to exceed a certain magnitude, when the figure becomes unstable. There would then be no production of a satellite in case the solar tidal friction were sufficient to prevent the attainment of this critical angular velocity of rotation; this may be held to account for the fact that Mercury and Venus have no satellites, their nearness to the sun involving a large effect of solar tides in reducing what would otherwise have been their greatly increased periods of rotation as their masses contracted. In the case of the earth and the outer planets, the friction of the solar tides may be supposed not to have been sufficient to prevent the instability arising owing to too rapid rotation, and thus to have prevented the formation of satellites by division into two parts.

It was formerly supposed that our solar system was typical of what was regarded as the ordinary state of things in the universe; that in fact every star was a sun with its attendant satellites, forming for each star a system with a character resembling our solar system. The discovery of the existence in the heavens of a large number of binary stars has modified this view, since it has exhibited the existence of systems of a character differing widely from the solar system. When William Herschel commenced his great exploration of the heavenly bodies, his attention was early directed to double stars, that is pairs of stars apparently very close to one another. At first he shared the general opinion that the connection between such a pair was purely optical; that in fact they might be simply two stars at widely different distances, but of which the directions happened to be nearly the same. But he was later led to the conclusion that, in a very considerable number of cases, double stars really consist of binary combinations in which the two stars rotate round one another. In several cases the periods of these rotations were determined. It has since been shown that such binary combinations exist in large numbers. Of the nineteen stars which are at the present time nearest to us it is definitely known that eight, no less than 42 per cent. of the whole, are binary stars. A scrutiny of various parts of the heavens makes it probable that not less than one-third of all the stars that can be observed are binaries. As long ago as 1764 it was observed by John Goodricke, a deaf mute, that the brilliancy of the star Algol was subject to periodic variation, and he suggested that this phenomenon was due to periodic eclipses by an invisible companion star. The correctness of this theory was verified in 1889, when it was proved that the star was moving in an orbit of such a character as would result from gravitation with a dark companion star, by which it would be partially eclipsed. By means of the spectroscope it has been proved that many stars which are single in appearance really consist of two stars in orbital motion round one another. It was from the first surmised that the motions of the stars in a binary combination are such as would be in accordance with the law of gravitation, and that this is actually the case has been established in a number of instances. The first of such cases of gravitating binaries was ascertained in 1827, by Savary of Paris, who proved that, for a binary star in the Great Bear, the orbits of the components were ellipses with a period of 58¼ years, in full accord with the gravitational scheme. In a number of binaries it has been found possible to determine the ratio of the masses of the two stars which form the combination, and in all such cases the masses have been found to be not very unequal. It has been found, for example, that in the visual binaries in which the ratio of the masses is pretty accurately determined, one of the stars is never less in mass than one-third the mass of the other star, and a similar result has been obtained in the case of nineteen spectroscopic binaries. On the other hand, in the solar system, the mass of the greatest planet, Jupiter, is less than a thousandth part of the mass of the sun. It follows that the sun and Jupiter cannot be regarded as forming a binary system similar to the many binary systems which have been observed. Thus a binary star with its attendant planets, if such exist, forms a system which cannot but be radically different in character from the solar system. Moreover the existence of triple, and of multiple, systems has been observed. It has been shown that triple systems consist normally of a pair close together, with a third star revolving at a distance from the centre of gravity of the pair, about ten times the distance of the stars of the pair from one another.

A theory has been worked out by Sir G. H. Darwin which gives at least an indication of the mode in which a double star may be conceived of as generated from a single original mass. The mathematical difficulties of a theory of this kind are so great that they have to be simplified by means of assumptions which are of such a character that the conditions in the mathematical investigation are widely different from what we must assume them to be in actual cases, and yet not so widely different that they destroy all the value of the results of the investigation as an indication of what may be conceived to have happened in such actual cases. An actual star, consisting of liquid or viscous matter in rotation, radiates its heat, and shrinks gradually as it cools. Its density will be at any one time very various, as we proceed from the outer surface to the interior parts, and its average density will increase as the whole mass contracts. Darwin considered what would happen in the case of a mass of liquid which at any one time is homogeneous and incompressible. The shrinkage is assumed to be so slow as to be consistent with this assumption, although that shrinkage of course involves an increase of density of the fluid. It has long been known, from mathematical investigation, that such a mass of liquid in rotation is in relative equilibrium if it have the form of a spheroid with its axis of symmetry as the axis of rotation, this axis being the smaller axis of the elliptic sections of the spheroid through that axis. It is further known that this form is a stable one, so long as the rate of rotation of the mass is sufficiently slow. As the mass slowly shrinks, its rate of rotation increases, in accordance with the dynamical principle of the constancy of its angular momentum, and the stability of the form gradually diminishes. As it shrinks, and the angular velocity of rotation increases, its shape changes; in fact it becomes continually more flattened at the poles. When it attains a certain shape, in which the equatorial and polar axes are in the ratio of 1000 to 583, the stability entirely disappears; it has reached what has been called by Poincaré a figure of bifurcation, a point at which the series of spheroids through which it has passed during the gradual increase of its rate of rotation passes over into a new series of figures of a different kind. This new series of figures consists of a set of ellipsoids with all their axes unequal. As the increase of the rate of rotation proceeds, the axes of the equatorial ellipse on the plane perpendicular to the axis of rotation become continually more unequal until, when a certain angular velocity has been attained, and the longest axis has become about three times the shortest, a new figure of bifurcation is reached, when the stability of the motion has again disappeared. It had been shown mathematically by Poincaré that the new forms which the liquid will take up, after this form of bifurcation is passed, will consist of a series of pear-shaped figures, in which the figure is blunted at one end and prolonged into a sort of snout at the other end. The question whether these pear-shaped figures are stable is a mathematical problem of great difficulty, and the answer given to the question has been different by different investigators. Darwin came himself to the conclusion that they are stable, but Liapounoff came to the opposite conclusion, and further investigations by Jeans appear to confirm the answer given by Liapounoff. The suggestion made by Darwin was that, as the rotation still further increases, the inequality between the two parts of the pear becomes accentuated, a furrow being formed; and that in the end a new figure of bifurcation is reached, after which the mass splits into two parts which separate, the subsequent history being that of the two separate masses which represent the two stars in a binary combination. Great doubt is thrown upon the applicability of this theory to the case of the genesis of double stars by the uncertainty as to the stability of the pear-shaped figures, but even if the fact that they are unstable be finally admitted, they do not wholly lose their importance in connection with theories of cosmogony.

Various mathematical investigations have been carried out in another direction, with a view to throwing light upon the double star problem. Instead of attempting to trace out a series of changes in a single continuous mass, with a view to the ascertainment whether, and under what conditions, it may divide into two separate masses, as in Darwin's theory, the problem has been approached from the other end. In fact the problem of determining the tidal action upon one another of two separate masses rotating round one another without change of relative position has been taken as the starting point. This problem was studied in considerable detail by E. Roche, and also by Sir G. H. Darwin. By the former of these investigators the problem was simplified by considering one of the bodies to be a rigid sphere, and the series of forms of the other body, which was taken to consist of a fluid mass, was traced out. It was found that no relative equilibrium of the two masses is possible if the distance of the two bodies from one another is less than a certain minimum depending upon the ratio of their masses, so that they cannot rotate in close contact with one another. The more difficult problem connected with the double star question was discussed by Darwin, in which both masses are fluid, so that each is distorted under the tidal forces generated by the other. As the result of intricate investigations of the stability of the figures of the two rotating masses, it appears to be impossible to trace back the states of these masses to a time when they were in contact, or nearly so, with one another. In fact it appears that in those cases there are no figures which are stable, except ellipsoids and spheroids, whereas, if the masses are to be regarded as coalescing, they must pass through forms which are far from being ellipsoidal. When instability is reached it would seem that the whole character of the motion of the system must be changed. These problems are so difficult that no complete solution of them has as yet been obtained. Jeans, who has continued the investigations of Poincaré and Darwin in various directions, has expressed the opinion that it is highly probable that tidal action may produce systems such as are seen in the solar system and in the systems of Jupiter and Saturn; that increasing rotation may produce systems such as are seen in ordinary binary stars, and that the close approach of two stars revolving about one another may produce systems such as Saturn's rings, and possibly also the asteroids.

I turn now to some consideration of that part of the theory of Laplace and Kant which has to do with the generation of stars from nebulae. As discerned by the telescope, reinforced by the photographic plate, there exist nebulae of several distinct forms; some 15,000 nebulae have been hitherto investigated. The most frequent of these are the spiral nebulae which consist of a nucleus with two arms emerging from opposite points; these two arms have each a spiral form, approximately that of the equiangular spiral. The so-called planetary nebulae are comparatively very few in number; they have an ellipsoidal shape. There are also ring-shaped nebulae, and also elongated or spindle-shaped nebulae. Besides these are irregular nebulae such as the great nebula in Orion. All the knowledge we have of nebulae, apart from their apparent shapes, as disclosed by the telescope, has been obtained by analysis of their spectra. In this manner, information is obtained in the first place of their chemical constituents. In 1862, Sir W. Huggins and Miller in London, and Father Secchi in Rome, commenced the work of studying the stars and nebulae by the new means which the spectroscope provided. In 1864, a bright planetary nebula in Draco was found by Huggins to consist of a mass of glowing vapour which showed the characteristic bright line of nitrogen, and a fainter line showed itself to be the *F* line of hydrogen. By 1868, Huggins had examined the spectra of about 70 nebulae, about one-third of which turned out to be of a gaseous character, and showed the nitrogen line. It was found that all the planetary and annular nebulae, as well as the irregular nebulae in the region of the milky way, have the character of a glowing vapour. Besides the knowledge of the constitution of stars and nebulae which has been obtained by the use of the spectroscope and has built up the department of stellar chemistry, spectroscopic analysis has also been applied to determine the motions of stars and nebulae relatively to the earth, in the line of sight. In 1842, Christian Doppler, of Prague, enunciated the principle that the colour of a luminous body must be changed when the body approaches or recedes from the observer with sufficient velocity to make the change sensible. In an amended form, this principle is that the velocity of approach or recess of a star or nebula will exhibit itself by means of a displacement of the lines of the spectrum of the body from the position of the lines of the spectrum of the same substance in the laboratory, and that this velocity may be estimated by measuring the amount of this displacement in one direction or the other. This very delicate operation of measurement of displacement was first successfully carried out by Huggins, in 1868, in the case of the star Sirius, leading to the result that the star is receding from the solar system at the rate of twenty-nine miles a second. A more accurate observation made in 1872 diminished this rate to about twenty miles per second, and estimates were made by Huggins for other stars, some of which are receding from, and others approaching the solar system. Since that time the measurement of such velocities has formed part of the regular work of Astronomers. The earlier observations of nebulae did not disclose any traces of their motion in the line of sight, but later observations have shown that the spiral nebulae have very great velocities. For example, the Andromeda nebula has been estimated to have a velocity of approach of 300 kms. per second, or perhaps somewhat more. Other spiral nebulae have been estimated to have a still greater velocity of recession, exceeding 1100 kms. per second. In fact the average velocity of spiral nebulae is some twenty times greater than the average velocity of a star belonging to the system of which the sun is a member. But besides these velocities of nebulae as wholes, the spectroscopic method has been applied to detect the rotation of nebulae. A nebula in Virgo was discovered, in 1914, by Slipher, to be in rotation; and the velocity of rotation has been since estimated to be 330 kms. a second, at a distance of 2′ from the centre, the velocity increasing proportionally to the distance from the centre. In the cases of other nebulae similar observations have been made, but in one case at least the rotation does not appear to be one in which the mass rotates like a rigid body, for the angular velocity diminishes with the distance from the centre. The irregular nebulae are found to be almost at rest relatively to the stars of our system, and the planetary nebulae have, with certain exceptions, velocities much smaller than the spiral nebulae. The evidence appears to show that the spiral nebulae are quite outside our system of stars, and they are tentatively regarded as island universes, each comparable in scale with the system of stars of which the sun is a member.

The opinion that the origin of stars may be traced to nebulae has been very widely held, but there has been some considerable difference of opinion as to whether a primitive nebula should be regarded as a mass of gas or as a cloud of dust. Lord Kelvin made the suggestion that a collection of meteoric stones, vaporized by collisions, would give rise to a gaseous nebula from which a star might be generated by contraction, in accordance with the hypothesis of Laplace. The study of stars by the spectroscopic method led to a classification of them into types which were supposed to be indicative of different stages of development, after the nebular stage had been passed. Five such successive stages have been distinguished, the one first after the formation from nebulae representing the hottest and least dense stars, and the last, the red stars, being the coolest, and nearest to extinction. But recent discoveries of H. N. Russell have thrown great doubt, not so much upon the theory of the nebular origin of stars as upon the notion that these five spectral types represent the successive ages of stars. There is strong reason to think that a star of the first, or hottest, type is not at the beginning of its history as a star, but half way through it. In accordance with this later view a nebula is, or becomes, non-luminous, and remains so until the mass becomes incandescent as a giant red star. It then passes in order through the successive stages in which it becomes hotter, until it becomes of the hottest type, and then proceeds in the reverse order, until it finally becomes again a red star, previously to its extinction. It is thought that only the most massive stars ever go through all the different stages, many of them turning back before the type of the hottest star is reached. This is in accordance with the observed fact that all the stars of the hottest type are of exceptionally great mass. Russell's theory is chiefly based upon the evidence afforded by the observed absolute magnitudes of stars. The stars in the three redder and cooler spectral classes were found to fall into two detached groups. In one of these groups the absolute magnitude was found to be nearly independent of spectral type. In the other group, consisting of stars of smaller magnitude, that magnitude varies with the spectral type, being smallest in the reddest type. The stars of these two groups have been called "giant" and "dwarf" stars respectively. A mathematical investigation, by Jeans, of the changes to be expected in a mass of gas when there is a continual emission of radiation from its surface yields results which are in accordance with Russell's theory.

A new factor has been brought into the discussion of all questions connected with the production and radiation of heat by the recent discovery of the presence of radio-active substances which provide sources of heat that had previously not been recognized. It has been maintained recently that radio-active substances in the interior of the earth may have provided a source of heat of sufficient amount to form an important factor in estimates of the age of the earth based upon the length of time it may have required to cool down to its present condition. The estimates of such age formed by Physicists, such as Lord Kelvin, which have been regarded by Geologists as inadequate, may, it is held, have to be completely revised when the emission of heat from radio-active substances is taken into account. As regards the effect of radio-active substances in the sun, it has been pointed out by Lindemann that the radio-active energy in the sun must be regarded as insignificant in amount compared with the gravitational energy. It is however possible that some more effective means of production of energy, of a sub-atomic character, perhaps involving the actual destruction of matter, may have to be recognized as contributory to the solar radiation. Of the efforts made in connection with the great Science of Geology to give a conceptual account of the mode in which the earth may be regarded as having reached its present condition, I am unable to give any account. The few fundamentally important cosmical problems that I have discussed may perhaps afford a sufficient illustration of the highly speculative character of those parts of Natural Science which concern themselves with the ideal reconstruction of the physical conditions of the remote past. It is clear that the scientific theories relating to such matters are tentative and hypothetical, in a sense and in a degree which marks them off from theories which refer to short time processes, that are capable of verification or refutation of a more direct kind than is possible in the case of theories which have essential reference to immense periods of time. The investigation of cosmical theories involves a completely justifiable attempt, by an extension of our conceptions of the actual processes which we can observe, to represent such processes as but portions of processes which we conceive to have proceeded during periods belonging to a vastly greater time-span than that of any actual observer. To make the attempt, in a certain sense to understand the present by means of the past, we are impelled by ineradicable impulses of the human mind. The mental satisfaction obtained by imbedding processes which we can actually observe in long time processes beyond our direct reach is one which we shall never forego. Needless to say, when we attempt to push back physical processes ideally, as far as may be, we involve ourselves in an indefinite regress. Of absolute origins, Science knows nothing, and we can form no conception. The so-called primordial state, such as is postulated in the nebular hypothesis, presents a problem which we do not attempt to solve. With some such postulation, behind which we do not go, every attempt at an historical scientific construction must commence.

# XIV

## EINSTEIN'S THEORY OF RELATIVITY

THE theory of space, time, and gravitation, propounded by Albert Einstein, is one of the small class of scientific theories which have at once succeeded in captivating the attention, not only of the scientific world, but also, in a marked degree, of a very large number of persons whose primary interests are not scientific. The interest of Physicists and Mathematicians is sufficiently explained by the ambitious character of a theory of which the aim is to combine in a single scheme temporal and spatial measurements together with gravitational phenomena, and by the fact that the theory includes a new law of gravitation, and a new Mechanics involving a breach with certain assumptions formerly supposed to be axiomatic, if indeed they were ever explicitly recognized. But in the case of Philosophers and of others to whom Natural Science is only of mediate or of secondary interest, another factor enters into the explanation of the amount of attention they have given to the theory. This is the prevalence of the idea that Einstein's theory of relativity has implications which reach beyond the purely scientific domain; and that it may serve to throw light upon, and perhaps to lead to changes in, our general philosophical views of the nature of reality. The term "relativity" is a very general one, capable of being employed in various directions in philosophic thought. Meanings may be assigned to it, of more general scope, and perhaps of a less definitely circumscribed character, than the rigidly defined meaning which is assigned to the term by Einstein and other Physicists who have concerned themselves with the development and exposition of the theory. There has thus been exhibited in some quarters a disposition to make this theory a starting point for the development of relativistic views outside and beyond the scope of the scientific theory itself.

It is a question of general epistemological interest whether, apart from the undoubtedly great importance of the theory in its purely scientific aspect, there is anything in the nature of Einstein's theory which should properly give it, in the eyes of Philosophers and of the educated public, a unique position in relation to general thought, of a kind which other previously existing physical theories do not occupy. Is Einstein's theory not only technically, but also generically, different from earlier physical theories? Does it rest upon a philosophically different basis? The answer to be given to this question cannot be considered in complete independence of the diverging views which are held as to the true character and functions of scientific theories in general. To those who regard Science as a means of penetrating to the inner nature of reality, the theory, so far as it is regarded as established, will appear to have given new knowledge of the inner nature of the real world, at least as regards spatial, temporal, and material relations. But some of those who hold this view of the functions of Science have been disconcerted by the highly abstract mathematical form of the theory, in accordance with which matter, at least in connection with its gravitational phenomena, exhibits itself no longer as substance, but only in the guise of specializations in a spatio-temporal metric. This attitude of mind is well illustrated by the utterance of Sir Oliver Lodge, *à propos* of this theory, which I quoted (p. 60) in my third lecture. The apparent impossibility of denying the descriptive efficiency of the abstract scheme, and the admiration excited by the constructive genius manifested in its creation, struggle with a reluctance to being drawn away from what, in accordance with pre-conceived ideas, and especially the old hankering after an underlying sub-stratum of matter, is regarded as concrete reality. On the other hand, to those whose view of the character and functions of scientific theories is in general agreement with that which I have advocated, Einstein's theory of relativity, however highly its scientific importance may be estimated, will appear to be merely a new conceptual theory, much more comprehensive no doubt in its scope than older theories which it aims to replace, but epistemologically on the same footing. Its extremely abstract character will occasion no great surprise, since that must be a feature of every scientific theory which lays such a claim to precision and comprehensiveness of descriptive power as does the theory in question. This theory proposes certain changes in scientific hypotheses formerly accepted as a basis of physical theories, of so striking a character that it may without exaggeration be described as revolutionary in its tendency. But if this revolution became stabilized, and its prospects of becoming so appear to be bright, although not at present fully assured, it will, so far as I can see, be a revolution purely internal to Natural Science, and will in no sense radically affect the external relations of Science with general Thought. In fact, my answer to the question I have raised as to whether the Einstein theory is generically, or epistemologically, different from other scientific theories is that no such difference exists: that it has in fact the same independence of all special ontological assumptions and theories as has Natural Science in general.

One of the advantages which may be expected ultimately to accrue from the widespread discussion of the theory by Physicists, Mathematicians, and Philosophers, is that it will lead to a clarification of ideas as to the nature and scope of scientific theories in general. The theory of Einstein is peculiarly well suited for this purpose because, when it is clearly presented, and its foundations are scrutinized, the fact that its basal conception consists of an ideal scheme, abstract in the highest degree, incapable of representation by the sensuous imagination, and in which nearly all elements derived from perceptive intuition have been removed by abstraction, forces itself upon the attention in such wise that its essential character is less easy to disguise than in the case of many current physical theories. There can be no pretence that a four-fold ordered manifold in which the ordering is neither spatial nor temporal, with a suitably chosen metric imposed upon it, but not intrinsic to it, is anything else than a most highly abstract conception. The metrical relations imposed upon this manifold can be analysed mathematically, but the manifold has neither in whole nor in part perceptual actuality, or presentability to the sensuous imagination. It can be conceived but not imagined.

I propose to explain in general terms the main characteristics of Einstein's theory of space, time, and gravitation, so far as is possible in the short time at my disposal for the purpose. Extensions of the theory have been suggested, with a view to combining in one whole not only a theory of space, time, and gravitation, but also a theory of electromagnetic phenomena in general; but with such extensions I do not propose here to deal. I must say, however, at once that the theory is only capable of complete expression in mathematical terminology, and is of such a character that only a trained Mathematician who is prepared to spend a considerable amount of time and energy upon the detailed study of its foundations can obtain a complete grasp of it. Fortunately, however, it is possible, without making such a complete study of the theory, to obtain a general knowledge of the points in which it differs from the older physical theories. These older theories have proved adequate for the purpose of measurement of spatial and temporal magnitudes and of gravitational effects, not only in ordinary life, but for nearly all scientific purposes; and they will always retain that adequacy. It is only for certain special scientific purposes that Einstein's theory will, in case it survives the tests which are being applied to it, be applied in future; it will never be applied in any of the measurements made in ordinary life.

When it is said that Einstein's theory of gravitation has overthrown and superseded Newton's theory, that is in any case only true in a very limited sense. In all ordinary astronomical cases Einstein's theory gives the same results as Newton's; it is only in very special cases that the difference between the results they lead to is sufficiently large to be discernible by the means of measurement at our disposal. That Newton's theory has its limits of applicability will not surprise those of us who contemplate the discovery of such limits in the case of all conceptual theories. The claim is made on behalf of Einstein's theory that the limits of its applicability are wider than those of Newton's theory. Before Einstein's theory can be regarded as an established and indispensable part of our stock of scientific conceptual apparatus, it must not only be shown to be logically self-consistent and to be capable of being applied for the purpose of providing an adequate symbolical description of the range of percepts to deal with which it has been constructed, but it must further be shown that no conceptual scheme of a simpler character is adequate for the representation of the same range of physical percepts.

The first point in which Einstein's theory has made a new departure has reference to the measurement of space and time only. In order to explain the character of this new departure I must refer to the observations I made in my lecture on "Time and Space," having reference to the pre-Einstein theory of their measurements. The private spaces of individuals were, I pointed out, correlated with a single public, or physical space, in which all physical objects with which we have to deal, either in ordinary life, or in Science, are regarded as located, and in which all actual spatial measurements are made. Similarly, the private times of individuals were correlated with a single public time, in which all measurements of time, by clocks, by the rotation of the earth, and by other sufficiently approximatively equivalent processes, are made. That these constructs, of a single physical space, and a single public time, each independent of the other, are sufficient to represent, by a process of correlation, the private spatial and temporal experiences of all individuals, whatever might be their relative positions and motions, has been the universal assumption made before the rise of the theory of relativity; and it has been regarded as axiomatic. For all the purposes of ordinary life, and also for the purposes of Science, it had always been found sufficient, until attention was directed to certain facts of observation to which I shall presently refer. The first great breach of the Einstein theory with the previously universally accepted tradition consists in a denial of the sufficiency of these constructs, a single physical space, and a single public time, independent of each other, as affording the basis of a system of spatial and temporal measurements which will completely describe the spatio-temporal experiences of all observers. The measurements in the single physical space were all made in approximate accordance with an abstract geometrical scheme, of which the basis was a manifold of elements with a three-fold order, into which manifold a Euclidean metric was introduced. This three-fold ordered manifold with the imposed Euclidean metric is the space of abstract Euclidean Geometry, each element, or point, of which manifold is representable by a triplet of numbers. Its properties were developed deductively as a scheme of Euclidean Geometry, and all its metric properties correspond to facts of measurement which can be verified in physical space, with a degree of approximation dependent on the fineness of our senses when reinforced by instruments of precision. In the same manner, public time was correlated with an abstract scheme which consists of a singly ordered manifold of elements represented by the numbers of the arithmetic continuum. Thus the theory of the measurements of physical space and public time rested upon the postulation of two abstract ordered manifolds, one with a three-fold order, and the other singly ordered, entirely independent of one another. In accordance with the theory developed by Minkowski and Einstein, instead of these two abstract manifolds a single manifold with a four-fold order of its elements is postulated, to serve as the theoretical basis for all actual spatial and temporal measurements. In order that it may serve both for spatial and for temporal measurements it must be, whenever it is applied, in some manner split up into two manifolds, the one with a three-fold order, and the other with a single order. An essential element in the theory is that this cannot be done in a unique manner; indeed, if the division were unique, the system would be reduced to the former scheme, in which there are two independent manifolds, one to be correlated with physical space, and the other with public time. An essential point of the theory is that, although each element of the fundamental manifold is represented by four numbers, or coordinates, it cannot be said straight off that three of these are for the purpose of spatial, and the other for the purpose of temporal, representation. The mode in which the sets of four numbers are to be applied to represent both spatial and temporal measurements will depend upon the observer, that is upon the physical frame of reference which he employs. Thus space and time cannot be immediately separated out from one another in any absolute way, in the fundamental four-fold manifold. An element of this manifold, psecified by four numbers, may be regarded as an abstract event, that is an extensionless object at an instant of abstract time, but the actual event to which it may be made to correspond will be represented by three spatial measurements representing the position of a point, and a clock-time, both of which will be dependent upon the physical frame of reference employed by the observer. Thus the position and time of a single elementary actual event will be measured in wholly different ways temporally and spatially, when referred to different physical frames of reference. In order that the fundamental fourfold manifold may be effective for its purpose, a suitable metric must be introduced into it; and the whole value of the theory depends upon making such a choice of this metric that the manifold may serve its purpose of providing an abstract representation of spatio-temporal measurements which will succeed in resuming the actual facts in the physical domain, as measured with reference to any actual physical frame which a particular observer may employ. Einstein's success, so far as it is established, consists in his having discovered, by the employment of a refined mathematical Analysis involving the use of the Calculus of Tensors, how this choice could be made. On the observational side, the leading feature of the new scheme is that there is no longer a single physical space and a single public time, common to all observers, but that two physical frames of reference in motion relatively to one another will have different physical spaces and different time-measurements. As I have already remarked, for ordinary purposes not only of everyday life, but of Science, the differences in question are negligible; it is only for certain purposes, to which I shall presently refer, that the Einstein scheme gives results which are sensibly different from those obtained on the basis of the older theory, and thus becomes effective.

It is sometimes said that Einstein's theory involves an obliteration of all distinction between time and space, and between past and future, since their measurements cannot be disentangled from one another in any absolute way. It must, however, be remembered that the original qualitative distinctions in our spatial and temporal intuitions are untouched by Einstein's theory or by any other theory of the measurement of space and time. These qualitative distinctions, as also the qualitative intuitional distinction between future and past, are removed by abstraction, even in accordance with the older traditional scheme, when we pass from intuitional space and time to abstract space and abstract time; moreover all our measurements of public time, by means of clocks or by any other method in which some regular physical process is employed, are spatial measurements; and thus the qualitative distinction between time and space has already been removed by abstraction. This has been spoken of by Bergson as the spatialization of time. The obliteration of the distinction, in Einstein's scheme, consists in an abstraction of all qualities of intuitional space and time except the element of order, which appertains to both, a three-fold order in the one case, and a linear order in the other. The fundamental four-fold manifold involves the notion of order, that is of an abstract order, neither specifically spatial nor specifically temporal, but obtained by abstraction from both, and generalized into a single four-fold order. Both extension and duration have been removed by abstraction. That this abstract manifold is spoken of by many expositors of the theory as the "world" may perhaps have a certain convenience, but it is, I think, somewhat unfortunate, especially when it is stated or suggested that this abstract construct is the "real world," because such terminology has at least the appearance of involving a prejudgment of metaphysical theories.

The next point in which Einstein's theory has introduced new conceptions is connected not only with the measurement of space and time, but also brings those measurements into relation with physical phenomena, especially with the gravitational phenomena of matter, and with the electromagnetic phenomenon of light. The ancient idea that space and time are objects with definite metric properties of their own, independent of matter, but forming a kind of framework into which material bodies, at rest or in motion, could be fitted, and that these metric properties of space and time are independent of all physical laws, had become moribund after the investigations of the foundations of Geometry due to Riemann and Helmholtz. It was shown by them that the abstract Geometry that is applicable to describe our measurements in physical space has a metric which is not fixed *a priori*, but is dependent on the fact of experience that there exist bodies which are approximately rigid and are freely movable in physical space; the metric is then so determined that the numerical measures of the distances of pairs of points of such bodies remain unaltered during their motions. So far the choice was left open either of employing a Euclidean metric, or a non-Euclidean metric, with either positive or negative space-constant, as the basis of measurements in physical space. But, as I described in my lecture on "Time and Space," in connection with the suggested experiment of measuring the angles of a triangle with very large sides in order to determine whether a Euclidean or a non-Euclidean metric would accord with facts of observation in physical space, the interpretation of any result obtained in such an experiment would depend upon the mode in which we formulate the laws of Optics. This shows that the system of Geometry adapted to form the basis of measurements in physical space is dependent upon the form in which we state physical laws; and thus that Geometry and Physics are in our experience inseparable from one another. The choice of the particular system of metrical Geometry best adapted to describe our spatial measurements will be such as to be consistent with the simplest formulation of physical laws, especially those of Optics. This fact of the interdependence of spatial measurements and physical laws is not a convention but an unalterable datum of observation. In Einstein's theory, this interdependence is developed to a further degree, so that spatial and temporal measurements are brought into relation with the phenomenon of gravitation. In accordance with the theory, any portion of matter exhibits itself in the presence of a gravitational field surrounding the matter, in which spatio-temporal measurements must be made in accordance with a metric varying from one part to another of the field, and differing from the metric which would be applicable at places very remote from gravitating matter. This gravitational phenomenon, considered as involving a special metric distribution in the field of the body, will exhibit itself in the orbital motion of a particle, or in the path of a ray of light, in the field. Thus, for example, the orbital motion of a planet round the sun is represented, not as due to a supposed attractive force towards the sun, but simply as exhibiting the spatio-temporal metric throughout the gravitational field of the sun. Thus, in Einstein's theory, what we have been accustomed to regard as the effect of gravitational forces is included in a scheme of spatio-temporal measurements. In a sense, gravitation is included in Geometry, only that Geometry is four-dimensional, non-Euclidean, and it has a metric of a kind more complicated than in the older schemes of non-Euclidean Geometry; moreover matter, at least in its gravitational aspect, only exhibits itself in and through the spatio-temporal metric. The fact that measurements of time and space are not independent of the physical phenomena of gravitation, and of light, leads to the idea that all the laws of physics and the laws which govern spatio-temporal measurements must be regarded as belonging to one interconnected whole, and should be conceptually represented by a single unitary scheme. Einstein's theory is an attempt to attain to such a unitary scheme, at least so far as the gravitational phenomena are concerned. Extensions of Einstein's conception have been already suggested by Weyl and others with a view to embracing electromagnetic phenomena in general, as well as gravitation, in a single conceptual scheme.

In order to describe in more detail the nature of the theory of relativity it is necessary to sketch the history of its origin. The theory has been developed in two stages, the first of these culminating in what is known as the special theory of relativity, and the second in the general theory of relativity. The special theory of relativity is accepted by some Physicists and Mathematicians who are sceptical as regards the general theory, at least in the particular form in which it has been developed by Einstein. The special theory takes no account of the phenomenon of gravitation; it is applicable strictly only in localities very remote from large gravitating masses; but it applies also very approximately to optical phenomena in weak fields of gravitation such as that of the earth. It had its origin in experimental investigations undertaken in connection with the electromagnetic theory of light, in accordance with which light was regarded as an electromagnetic disturbance propagated through the ether. After various attempts to elucidate the structure of the ether and its relations with material bodies, it came to be regarded for the most part as a substance which is undisturbed by the motion of material bodies through it, and as freely interpenetrating such bodies. On this theory the ether would form a natural frame of reference with respect to which all motions of material bodies might at least ideally be measured; and it became a matter of importance, as a test of the theory, to detect by experimental observation the existence of the velocities of material bodies relative to the ether. It is known that light is propagated with a velocity approximately of 300,000 kms., or 186,000 miles, a second. If a body is moving in the direction of, or in the opposite direction to, a ray of light, with a velocity through the ether which is an appreciable fraction of the velocity of light, the ray of light would appear to have a velocity relative to the body, less in the first case, and greater in the second, than the velocity of propagation of light through the ether; the defect or excess being the velocity of the body through the ether. The earth, in its orbital motion round the sun, has a velocity relatively to a frame fixed by the sun and stars of about 30 kms., or 18 miles, a second. Although the velocity of the earth with respect to the ether was unknown, if it were assumed that it is of the same order of magnitude as the orbital velocity of the earth, it seemed certain that the velocity relative to the ether might be detected, and its magnitude determined, by comparing the velocity, relatively to the earth, of a ray of light in the direction of the earth's orbital motion with that of a ray in the perpendicular direction, provided sufficiently precise measurements were made in an apparatus devised for the purpose. The celebrated experiment of Michelson and Morley, carried out in 1887, was devised in order to establish the existence of the expected effect due to the velocity of the earth relative to the ether; the effect of such velocity was expected to be apparent if the velocity of the earth in the ether was only a quarter of its orbital velocity. A beam of light from a single source was divided by partial reflection at a mirror into two portions, one in the original direction of the beam, and the other at right angles to the first. These two portions were reflected back by mirrors, and struck the first mirror again, when portions of them would be re-united. The whole apparatus could be rotated into any position, and could be fixed so that the original beam was in any required direction. The two portions of the beam which are re-united, having required different times to pass from the first mirror to the reflecting mirrors and back again, exhibit interference fringes. A rotation of the instrument was expected to show a displacement of these fringes, owing to a change in the retardation of one portion relatively to the other. But when various positions of the apparatus were explored, no trace of such displacement of the fringes was observed, whereas the expected displacements would have been easily capable of measurement had they existed. The same negative result was reached by a still more refined experiment conducted by Morley and Miller in 1905. This and other experiments have led to the conclusion that it is impossible to detect by any observation the motion of a material body relatively to the ether. The first attempt to account for this inability to detect any effect of the motion of matter through the ether was made by Fitzgerald, and independently by Lorentz. It was suggested that, when a material body is in motion through the ether, its dimensions in the direction of the motion are shortened by an amount depending upon the square of the ratio of its velocity to the velocity of light; and thus that the whole apparatus in the Michelson-Morley experiment is shortened in this way in the direction of its velocity through the ether, thus annulling the displacements of the interference fringes. This so-called Fitzgerald contraction of a body moving through the ether could only be conceived as due to some interaction of the ether with the constituents of the moving body; the hypothesis of its existence was much discussed, and led to various difficulties.

The interpretation of the observed facts that was given by Einstein was of a radically different character; it was in this connection that he propounded the theory which is now called the special theory of relativity. The first postulate of this theory is that all physical phenomena, as observed from a material body as frame of reference, appear to be completely independent of any uniform translational motion which that body may have relatively to another physical frame of reference. The second postulate of the theory is that the velocity of light is independent of the motion of the source of light. If these postulates be accepted, the negative result of the Michelson-Morley experiment is immediately accounted for. In accordance with these postulates the velocity of light is the same when referred to any two frames of reference in uniform translational motion with respect to one another, and is independent of the origin of the light. It should be observed that the traditional Newtonian Dynamics is in full accord with the foregoing postulate of relativity, because the Newtonian equations of motion of a dynamical system referred to a given material frame of reference are unaltered when a new frame of reference is employed which is in uniform translational motion with respect to the given frame. But that is not the case with Maxwell's equations for the representation of electromagnetic phenomena relatively to a frame of reference fixed in the ether. It was shown by Larmor and Lorentz that these equations are unaltered in form when the coordinates and the time are transformed by means of a certain linear transformation, of a less simple character than the transformation of the coordinates in a dynamical system, from one frame of reference to another in uniform translational motion with respect to the first. In this so-called Lorentz-transformation, the new coordinates are expressed in terms of the old, in a form which involves the time as well as the translational velocity of the new frame of reference, and the new time-measurement involves not only the original time-measurement but also the original spatial coordinates. It thus appears that the Newtonian dynamical scheme and the Maxwellian electrodynamical scheme which represents optical phenomena, when taken together, do not satisfy the postulates of the restricted principle of Relativity; and consequently one of these must be changed if that principle is to be accepted. The bold step taken by Einstein consists of a rejection of the

Newtonian system of Dynamics, and the substitution of a new dynamical scheme in which the Lorentz-transformation is applicable, not only in the electromagnetic equations, but also in Dynamics. The consequences and implications of this step are of a far-reaching character. A complete revision of the old ideas about spatial and temporal measurements is involved in the change. When we pass from the measures of space and time which an observer with his physical frame of reference employs to the corresponding measurements of time and space employed for the same physical event by another observer with a frame of reference in uniform translational motion with respect to that of the first observer, the scales both of spatial and temporal measurements of one and the same event are different for the two observers. The measure of time for the one observer depends not only upon the measure of time of the other observer, but also upon his spatial measurements, as well as upon the relative velocities of the two observers; thus there exists no single system of measurement of time which is common to all observers. Neither can the two observers employ one and the same system of spatial measurement. If two events occur at different places, the interval of time between them will be measured differently by two observers in motion relatively to one another; for one observer the two events may be simultaneous whilst for the other observer the same events may occur at different times. The distance between the places at which the two events occur will be in general different for the two observers. It might appear that, as the intervals both of time and of space which distinguish two events depend upon the observer, there is no invariant relation between the two events; that is no relation which is common to all the observers. When, however, the complete scheme in an abstract form is set up, it appears that this is not the case. The important step was taken by Minkowski of establishing that the Lorentz-transformation is capable of simple representation and interpretation if an abstract four-fold ordered continuous manifold, which he called the "world," with a certain metric system imposed upon it, is taken as the conceptual basis upon which all actual spatio-temporal measurements are made to rest. In view of the use which is to be made of the elements of this manifold, any one of which is to be regarded as correlated with an actual observable event which for any observer is extremely small both in extension and in time-measure, these elements of the manifold may be spoken of as abstract elementary events; they are frequently also spoken of as points in a four-dimensional geometrical space, but such language ought not to be taken to imply that our intuitional notions of space and of time are applicable to this fundamental manifold. The notion of continuous order, taken in the abstract, is the sole remaining element of our spatial and temporal intuitions which is a constituent of the conception of this manifold. The order of the elements is assigned by correlating each element with four real numbers, often spoken of, analogically, as the coordinates of the element. The metric is imposed upon the manifold by means of a definition of what may be called the "separation of two elements." It is taken to be the square root of the sum of the squares of the differences of the corresponding coordinates of the two elements; but in this sum three of the squared differences are taken to have the negative sign, and the metric is not therefore strictly an extension of the ordinary Euclidean metric. If the velocity of light is not taken to be unity, the positive squared difference in the expression for the "separation" must be multiplied by the square of the constant numerical measure of the velocity of light. If now, by means of the linear Lorentz-transformation, a new system of coordinates is introduced into the manifold, so that each element is represented by a new quadruplet of numbers determined from the original quadruplet by means of the Lorentz-transformation, it will be found that the "separation" between two elements has exactly the same value when expressed in terms of the new coordinates as it had originally. It is this fact upon which the utility of the abstract scheme depends. The "separation" between two particular events is an invariant for all observers. A change of coordinates in the manifold, of the kind described, corresponds to, and represents, a change from the spatio-temporal measurements of one observer, with his own scale of measurements, to the measurements of another observer in uniform motion relatively to the former one. Thus this fundamental manifold forms the ultimate conceptual basis for Einstein's restricted theory of relativity which satisfies the two postulates I have already specified. Of the four coordinates of an element of the manifold, for a particular observer in the physical domain, that one of the coordinates which appears with the positive sign in the expression for the "separation" of two events may be taken to represent by correlation his measure of time, and the other three his spatial measurements relatively to his rectangular frame of reference. But we cannot say that, of the four coordinates of an element of the fundamental manifold, one is to be always correlated with temporal measurements and thus represents abstractly a time, and the other three represent abstractly spatial coordinates. This is the case for one particular observer, but another observer in motion relatively to the former will have to employ an entirely different set of four coordinates in the manifold for correlation with his spatio-temporal measures; this set being related with the former set by means of the Lorentz-transformation. Thus in general each one of the four coordinates in the manifold is correlated with a mixture of spatial and temporal measures in the physical domain.

A set of events in the physical domain, represented by a material particle which never impinges on any other particle, will be correlated with a continuous set of elements in the fundamental manifold, forming what may by analogy be called a straight line, and this is spoken of by Minkowski as the world-line of the particle. This purely statical abstract object in the manifold represents conceptually the whole history of the material particle, past and future; its interpretation spatially and temporally will vary according to the circumstances of the observer, involving the way in which he judges from his own relative standpoint the spatial and temporal circumstances of the particle of matter in question. The principle of relativity, as embodied in the two postulates, is incompatible with the conception of ether as a substantial medium for the transmission of light. There have at all times been great difficulties in formulating the properties of an ether which should satisfy the conflicting demands which fact and theory seemed to require. Notwithstanding these difficulties there has been much reluctance on the part of Physicists to give up a conception which was designed to afford a pictorial representation of electromagnetic phenomena, and which appealed strongly to those who regard the ether as a concrete, though not directly perceptible, reality. The idea has however gained ground that it is in the equations of the electrodynamic theory that the really effective formulation of that theory is to be found, and thus the ground has been prepared for that final removal of the formerly useful notion of the ether which is involved in the acceptance of Einstein's theory, even in its restricted form.

The restricted principle of relativity, which was completely stated by Einstein in 1905, changes radically the notions of the measurement of time and space which were employed in the Newtonian Dynamics and in ordinary affairs, but the principle suffers from the defect that the relativity is applicable only to material frames of reference in uniform motion of translation with respect to one another, and that it takes no account of the phenomena of gravitation. The decade after 1905 was spent by Einstein in an endeavour to remedy these defects; and in 1915 he found himself able to propound a complete principle, in which the relativity is applicable to all frames of reference in motion of any kind with respect to one another. The general principle of relativity includes the conception that all actual spatial and temporal measurements are dependent upon, and vary in a definite manner with, a material frame of reference employed by an observer. The scheme includes a mode of taking into account, and measuring by means of spatio-temporal measurements, gravitational phenomena as exhibited in gravitational fields. This mode of treating gravitation is fundamentally different from that of Newton; there is in it not even a suggestion of anything that could be regarded as a causal law of gravitation, in accordance with the older traditional meaning attached to the term causation. The special spatio-temporal peculiarities of a field of gravitation are taken to give the only theory of gravitation that Science requires, or can attain. Newton's law of gravitation, in accordance with which the gravitation between two material particles is represented by a stress proportional in magnitude to the product of their masses and inversely as the square of their distance from one another, being independent at any instant of the motions of the particles and of all other matter, had become quite indefinite in meaning. In the first place, the mass of a given particle has, in modern Physics, lost the characteristic property of having a constant value independent of the motion of the particle relative to an observer. In accordance with the electron theory of matter, which rests upon the observed facts of radiation and radioactivity, a material particle is constituted in part at least of electrons, of which the effective mass increases when their velocity is increased. Thus the mass of a particle may be sensibly changed if it is set in motion as a whole, with a velocity comparable with that of light, or when the motions of the electrons within it are considerably altered by the receipt of energy from without. In fact, in accordance with the special principle of relativity, the mass of a system is the total energy in it divided by the square of the velocity of light; this energy being measured relatively to the observer. It is then no longer clear how the magnitudes of the masses in Newton's law are to be fixed. Again, the distance between the two particles has no longer an absolute measure independent of their motions relative to an observer. Further, the fact that gravitational force, in Newton's theory, depends at any instant only on the positions of the particles at that instant is not consistent with the conception of the propagation of gravitation with finite velocity through a medium; and thus Newtonian gravitation was never linked up with other physical phenomena in any unitary scheme.

As I explained in my lecture on Dynamics, in the Newtonian system all actual phenomena of motion are described upon the basis of a conceptual scheme in which there is an absolute frame of reference; but in physical space, it is possible with a degree of approximation sufficient for the purposes of any special case, to determine so-called Newtonian frames of reference which can be correlated with the absolute conceptual frame. Any frame of reference which is in rotational motion, or in accelerated translational motion, with respect to a Newtonian frame can only be employed if certain "fictitious forces," among which is the so-called centrifugal force in the case of rotation, are introduced into the equations of motion. An essential element in Einstein's scheme consists in his principle of equivalence, which involves a denial of any distinction between such fictitious forces and gravitational forces. All of them alike are regarded as due to the gravitational field, and the mode in which this gravitational field exhibits itself is in the spatio-temporal measurements suitable to the particular frame of reference employed. In any sufficiently small region it is impossible by any experiment to distinguish between a fictitious and a gravitational field of force. The Einstein scheme does not depend, as does the Newtonian, upon the selection of any specially suitable frames of reference; all such frames are on a parity, and in this the completeness of the relativity consists.

This independence of any particular frame of reference is essentially connected with the fact that all actual processes of measuring space or time consist in the establishment of the coincidence of two points belonging to two material bodies; and such a coincidence is a fact which is unaltered by any change in the frame of reference employed in making the measurements.

In Newtonian Dynamics the equality of the inertial mass and the gravitational mass of a material body remained as a bare fact, derived from observation, which appeared from the theoretical point of view to be merely an accident. In Einstein's conceptual scheme the two are indistinguishable, or rather identical, in accordance with the fundamental postulates of the theory. An essential consequence of Einstein's scheme is that any phenomena which an observer perceives in his neighbourhood to be due to a gravitational field would be perceived unaltered if the gravitational field were not present, provided that the observer makes his frame of reference move with the acceleration characteristic of the gravitational field at the place at which he makes the observation. This amounts to the assertion that it is possible to eliminate a gravitational field, at a particular place, by a proper choice of the frame of reference. The fundamental four-fold ordered manifold is employed as the conceptual basis of the general theory of relativity, as in the case of the special theory, except that the metric imposed upon it is of a more complicated character, designed to take into account, by correlation, the existence of gravitational fields, and to exhibit their presence in spatio-temporal measurements. The determination of the precise mode of correlating the conceptual scheme with actual measurements would appear to be one of the most difficult points in the whole theory.

The precise mode in which the theory of this metric was developed by Einstein is of a so highly technically mathematical character, that I can only give the briefest and most superficial indication of its nature. In this matter the geometrical investigations of Gauss and Riemann were the original source of inspiration. Riemann's theory of a continuous ordered manifold, designed by him for the purpose of investigating the foundations of Geometry, led in the hands of Ricci, Christoffel, and others, to a mathematical development known as the theory of tensors, and this theory was utilized by Einstein, and developed for his purpose of determining the nature of a metric system to be imposed upon the fundamental manifold, of such a character that it could be applied to the spatio-temporal metrical characterization of observed gravitational fields. Two elements of the fundamental manifold are regarded as neighbouring elements when each of the four coordinates of one of them differs by a very small number (more exactly a differential) from the corresponding coordinate of the other. The "separation" of two such neighbouring elements is defined as the square root of a quadratic function of the differentials of the four coordinates. The ten coefficients in this quadratic function are in general functions of the coordinates of the element, and are taken to be the potentials characterizing a gravitational field. In order that this definition of the differential "separation" may be the basis of a definite metric in the manifold, these coefficients must satisfy a certain number of conditions which involve also the gradients of these coefficients. The determination of these conditions is made in accordance with the developed Riemannian theory, which had its origin in Gauss' theory of the curvature of surfaces in three-dimensional space. Einstein succeeded in overcoming the great difficulty of showing how all these conditions could be satisfied so as at the same time to make the metric available for the representation of actual spatio-temporal measurements in gravitational fields, such as those which are given us by experience.

The history of a material particle which does not impinge upon any other particle is completely represented by a set of elements in the fundamental manifold which may, by analogy, be spoken of as a geodesic in that manifold, and is obtained as an extremal of the integral of the differential separations. This geodesic is the world-line of the particular particle, and it is analogous to a curved line in ordinary geometry whenever the particle is in a gravitational field; when there is no such field the world-line is the analogue of a straight line, as in the special theory. The world-line forms the conceptual basis of a spatio-temporal description of the motion of the particle, as estimated by an observer who chooses arbitrarily his material frame of reference. All such frames may be employed indifferently, the observed path of the particle being relative to the particular frame employed; but the world-line is absolute for a particular material particle.

One of the most important applications which Einstein has made of his theory is to the motion of a planet in the gravitational field of the sun. His result contains a correction to the Newtonian law of force upon the planet which is in any actual case very small, but the effect of which in one case, that of the planet Mercury, is sufficient to render it capable of being observed. This has formed one of the tests of the applicability of Einstein's theory, and the theory appears to have proved itself able to stand the test. In accordance with Newton's theory of gravitation, a planet would move over and over again in a fixed ellipse with the sun in one of the foci, in accordance with Kepler's law, if it were not subject to the disturbing effects due to the other planets. Thus the line joining the sun to the planet, when in perihelion, that is when nearest to the sun, would preserve for all time a direction fixed relatively to the stars. However, one of the effects upon the orbit produced by the gravitation of the other planets is to give this line a very small change of direction which steadily accumulates, so as to make the change in a century sufficient to be observed. The orbit of the planet Mercury is more elongated than in the case of other planets, so that it can be observed at what times it is in perihelion more accurately than in the case of a planet whose orbit is more nearly circular. The motion of the line due to the disturbance of the other planets was calculated by astronomers to amount to 532″ in a century; but the actually observed amount was found to be 574″, so that the excess of 42″ remained to be accounted for. It was calculated by Einstein that the effect of his amendment to Newton's law would be that there would be an advance of the line of perihelion amounting to 43″ in a century, which differs only by a very minute amount from that of the discrepancy to be accounted for. It thus appears that Einstein's law leads in this crucial case to a result which is in close agreement with observation, and in which Newton's law is in default. It is not possible to apply a similar test in the case of the other planets, either because the corresponding advance is too small, or because the orbits are too nearly circular to make sufficiently accurate observations possible.

Another crucial test of the theory was provided by observation of the deviation of direction in a ray of light which takes place when the ray passes very near to the sun. It has for some time been known that radiation has inertia, as manifested in radiation-pressure; in accordance with Einstein's theory it consequently has gravitational mass, or weight. As the result of a calculation made by Einstein, the effect of the intense gravitational field near the surface of the sun would be that a ray of light passing very near that surface would appear to an observer on the earth to be deflected through an angle of $1''\cdot74$. If Newton's theory of gravitation were applicable the deflection would be just half this amount, $0''\cdot87$. It was decided to put to the test of observation the question which of these two values represents the deflection that actually takes place. At the time of a total eclipse of the sun, a deflection of the light from a star very near the sun would exhibit itself in an apparent displacement of the star from its true position in a direction away from the sun. The very delicate operation of measuring the displacements of position of stars near the sun at the time of the total eclipse on May 29, 1919, was undertaken by astronomers in two expeditions, one in Brazil, and the other in the Gulf of Guinea. Although the observations of the latter expedition were very seriously hindered by cloudy weather, in both cases the results of observation of several stars, after the elaborate process of correction for various errors of observation had been carried out, were found to be very fairly in accordance with Einstein's prediction, and definitely to rule out the correctness of the deflection as calculated by Newton's theory. Thus the result of the observations was distinctly in favour of Einstein's theory as against the Newtonian. It must, however, be remembered that such observations are of an extremely delicate character, and involve various sources of error. Accordingly it is necessary to await the results of further observations of the same kind before complete confidence may be placed upon the results of this test.

The third practical test of the theory which has been made has not as yet been brought to a decisive conclusion. It was expected that, in accordance with the theory, when the spectral lines of a chemical substance in the sun are observed, they should show a displacement from the position of the spectral lines of the same substance on the earth. The results obtained by different observers in the course of their attempts to verify or disprove this predicted effect are discordant and contradictory. It appears therefore that, up to the present time, no decisive evidence has been obtained by means of which it can be definitely decided whether the theory satisfies this test or not.

Whatever be the ultimate fate of Einstein's theory, and to whatever modifications it may in the future be subjected, it is, as it stands, of the highest interest, not only on account of its comprehensive character, and on account of the novelty of its conceptions, but also as a chapter in the history of Science. It could not have arisen apart from its two great roots, the one the physical theory of Electromagnetism, including the theory of light, and the other the highly abstract theory of Geometry in its most generalized form. The latter includes a whole series of investigations into the foundations of Geometry which reach far back into the past, culminating in the work of Riemann and Helmholtz, who were the first to perceive that Geometry, as applicable to actual measurements, is not really independent of Physics, and which have been continued in Mathematical detail by others.

All this line of mathematical work was carried out almost entirely by thinkers who had no hope that their labours would ever form an essential groundwork of a great physical theory. This piece of history illustrates in a most striking manner the fact that there is never any certainty that the most abstract mathematical theory which, although it of course has its ultimate roots in the perceptual domain, appears to have no direct relations with that domain, may not turn out to be of the last importance in relation to some unforeseen theory of physical phenomena. That the further development of Einstein's theory, so that it may embrace all the phenomena with which Physics has to deal, may be subject to limitations which, in its present form at least, it cannot avoid without undergoing at least some modification, is indicated by the rise and development of the theory of Quanta. It would appear that sub-atomic Dynamics is essentially dissimilar in character to that of systems which consist of large aggregations of atoms. This would appear to indicate that Einstein's scheme which, like the Classical Dynamics, is a continuous theory, may prove to be inapplicable as the conceptual basis for the representation of sub-atomic phenomena. Suggestions have even been made recently that a discrete manifold may be requisite as the conceptual basis of spatio-temporal measurements for this purpose, instead of the continuous manifold employed at present by the Einstein theory.

It is interesting to observe that Riemann himself made, at the conclusion of his celebrated dissertation on the foundations of Geometry, a remark which showed that he had a prophetic insight into the possibility of his conceptions being one day linked up with the physics of matter. The passage to which I refer states that[1]:

> The question as regards the validity of Geometry in the region of the infinitesimally small is connected with the question relating to the inner ground of the metric of space. In connection with this question which can surely be reckoned as belonging to the doctrine of space, the above remark can be applied, that, in a discrete manifold, the principle of the metric is contained in the conception of the manifold itself, but that, in a continuous manifold, it must come from outside. Thus the reality upon which space is based must either form a discrete manifold, or else the basis of the metric must be sought outside, in binding forces that act upon it.

[1] *Gesamm. Werke*, 2nd ed., Vol. I, p. 285.

# XV

## BIOLOGICAL SCIENCE

AMONG the manifold objects which we perceive in the physical world, those which we call living organisms, whether animals or plants, are universally recognized as forming a class, the members of which are distinguished in a variety of ways from other objects in the whole physical complex. The term Biology I here take in the most general sense, as denoting the whole group of sciences which concern themselves with the study of the physical characteristics of living organisms, their forms and parts, the various processes in which their parts are involved, and with the physical relations which they have with one another and with their environment, their genetic relations, and their geographical distribution. As all Science is a higher development of the kind of knowledge which we associate with the expression common sense, it may be worth while to inquire what are the conceptions of a living organism possessed by those persons who are not scientific biologists, but who possess sufficient powers of reflection to be able to formulate some kind of answer to the question what is to be understood under the term living organism, as distinct from objects that are not regarded as living. The concept of a living organism, as formed by such persons, cannot be expected to be capable of definition in clear cut terms, in which any single criterion of a simple character is employed as a decisive test which could be applied to any and every special instance, to determine whether a particular object is, or is not, a living organism. The descriptive account of the meaning of the term animal which might be given by a non-Biologist would refer to a group of characteristics, all of them present in ordinary cases, but some of which might be absent or indiscernible in a special instance. In fact such a definition would not be applicable to cases on the borderline. Perhaps the most general descriptive statement that might be made is that an animal is an object which appears to have an active power of self-maintenance as against its surroundings; that this involves a behaviour on its part which has at least the appearance of what can only be described in psychical terms as purposive activity, with a view to the preservation of its individuality. It would be added that, although it has a relative permanence of form, an animal takes nutriment, grows, and reproduces itself. Reference would be made to the fact that an animal is distinguished by the complex, and apparently purposive, character of its reaction to external stimuli; and that these reactions often depend upon the past history of the individual, to a degree much greater than in the case of non-living objects. In psychical terminology, in fact, an animal can learn from experience. A person who has not studied Biology would find it much more difficult than in the case of an animal to describe in general terms what he understands by a plant, because a much closer examination is required before the main characteristics of a plant can be brought to light, especially as regards its relations with its environment. At least some simulation of what appear to be characteristic properties of living organisms may be found in the inorganic domain. For example, the growth of crystals, and the phenomena of elastic fatigue, and of hysteresis, present analogies with the growth of organisms and with their dependence upon past history.

The questions now present themselves, what is the scientific definition of a living organism? What are the scientific criteria of the distinction between an animal and a plant? Is it possible to give such definitions without the employment of psychological categories? It will,

I think, be admitted that it is unreasonable to expect that these questions can receive answers that are more than tentative, until Biological Science has reached a very advanced stage. During the earlier stages, only amplifications, qualifications, and modifications, of common-sense criteria and distinctions can be expected. In fact, final answers to such questions must be regarded as goals to be attained, if at all, only when Biological Science has reached a very high stage of development. We must be prepared to contemplate as at least a possibility, the answer to the question, what is the ultimate distinction between living matter and non-living matter? to be that, as long as we remain within the categories of Natural Science, in the restricted meaning of the term which I have adopted in these lectures, that is, excluding all psychical and psychological categories, and employing only those of physico-chemical schematism, there is no such ultimate distinction. That, in fact, the difference between what is called living matter and what is called non-living material can be represented only as a difference of degree in complexity of structure and of the physico-chemical processes associated with the two types of material. In this connection it is well to remember that the range of organisms with which modern Biological Science deals has been enormously extended by the use of the microscope; and that thus, the border-line cases of organisms of comparatively simple structure and functions, inaccessible to ordinary observation, form a most important part of the study, in relation to such questions as those I have indicated.

However different the biological sciences, in respect of procedure and history, when regarded superficially, may appear to be from the inorganic sciences, the method applied in dealing with the phenomena in which living organisms are involved is fundamentally the same as that which is applicable to the inorganic domain. Based upon the observation, sifting, and classification, of the facts of perception, concepts and conceptual schemes are constructed which suffice to represent sequences of phenomena of certain classes in the domain under consideration. All that I have said in the earlier lectures as to the total inability of scientific method to attain to explanations, in the full sense of the term, of any happenings in the perceptual domain, is as valid in Biology as in Physics and Chemistry. All modern biological theories, in their descriptions, have reference to processes of the physico-chemical order, and are consequently for this reason, as well as others, subject to the same limitations as Physics and Chemistry. Thus Biology, as a great department of Natural Science, is, and from the nature of its methods must always remain, unable to discover an answer to the metaphysical question what life is, in what its essence consists. What it can do is to give conceptual descriptions of what living organisms do, of the phenomena of which the parts of the organism are the seat, and of the interactions of the organism with its environment, which includes other living organisms of similar and dissimilar kinds. It can endeavour to ascertain the physical conditions subject to which what we call life manifests itself. As in the case of the inorganic Sciences, the limitations imposed by the character of its methods have by no means always been fully recognized by those who have built up the edifice of Biology; and this has frequently led to the employment of language, in the statement of theories and laws, which, taken as it stands, implies that efficient causation has been discovered, immanent in the processes and sequences that are described. The assumptions of physical realism have been accepted by many, probably most, men of Science; but of these assumptions Biological Science is really as independent as are Physics and Chemistry.

A fundamental question in relation to the character of Biological Science is that as to the nature of the concepts which it must employ in formulating its conceptual schemes and laws. In the first place, it may be asked how far Biological Science can confine itself to the concepts employed in Physics and Chemistry; such concepts as are sometimes spoken of as mechanical, although for historical reasons it is probably better to avoid the expression. The fruitful work done by modern Physiology has been carried out subject to the assumption that the categories of Physics and Chemistry are sufficient to form the basis of that work. The striking success of investigations of this order gives ample warrant for belief in the utility of the assumption as a working hypothesis, and is such as to afford a justification for the hope that a continuation of investigations on the same lines may lead to an indefinite extension of physiological knowledge. It would however appear that these physiological investigations are confined to special processes in the organism, and require integration before they can be applied to give an account of the coordinated happenings in which the organism as a whole is involved. Moreover, it should be remarked that the concepts of Physics and Chemistry cannot be regarded as once for all fixed; at the present time they are very noticeably in a state of flux. Further, it has to be taken into account that, in accordance with the discoveries of the modern Science of Biochemistry, the chemical processes in the living organism exhibit marked peculiarities which differentiate them from those which take place in non-living matter.

It may next be asked whether Biological Science requires, in addition to the concepts of Physics and Chemistry, further concepts of a kind not specifically psychical. Attempts have been made to supplement the physico-chemical concepts by others, for the purpose of attaining a more complete and satisfactory account of what goes on in the living organism, than can be provided with the help of the former concepts alone. The difficulties encountered in such attempts have been great.

It has proved difficult, or as some would say, impossible, to define and delimit such concepts with a precision of the kind they must have if they are to perform a useful function in scientific theories. That the want of such concepts is widely felt is clear, but however desirable their construction may be, without adequate definition and circumscription, and without precise postulations as regards their relations with the physico-chemical concepts, they will represent little more than words employed to cloak ignorance. Besides the difficulties of precise definition and of correlation with existing concepts, and assuming that these difficulties can be overcome, there remains the question as to the utility of such concepts in formulating and extending physiological knowledge. It has, for example, been maintained by Dr J. S. Haldane that the concept of the organism as a whole, as distinct from all the parts and their physico-chemical relations, but having relations with those parts, is a necessary one, if we are to have any understanding of the living body. This concept of an entity which is concerned with the unification of the whole complex of phenomena in the living body is an exceedingly plausible one, and from various points of view commends itself in a high degree to the mind. But has it been formulated, and have its relations with the parts been formulated, in such wise as to enable Physiologists really to increase their knowledge of the living organism? Is such a concept, from the point of view of Physiology, more than an aspiration, an idea of great cogency? Does it assist in solving the problem of the integration of all the processes in the body? Has not all the precise knowledge which has actually been obtained by Physiologists as to these processes dispensed with this conception, because it has lacked the requisite distinctness of outline? I gather that, up to the present time, however much its need may be felt, such a purely physiological concept has not been constructed so as to satisfy the criteria I have indicated. There can of course be no *a priori* objection to the introduction of such a concept in Physiological Science, provided it is properly defined and proves itself useful in furthering the aims of that Science. Physiologists appear for the most part to think that the true line of progress of the knowledge of the organism which they seek to obtain must, at the present time, be in accordance with the methodological assumption that the living being is to be regarded, for their purposes, as a physico-chemical mechanism, to be investigated by methods in which quantitative chemical processes and measurable physical processes are alone dealt with. It may be true, and in the judgment of most of us it is true, that the physico-chemical categories will always prove to be inadequate to do more than describe conceptually greater or smaller, but certainly limited, tracts of phenomena in the living organism. Nevertheless, the Physiologist who, in the light of present conditions, accepts as his present policy that of restricting his investigations to the tracing out of physico-chemical, and consequently measurable, processes has full justification for his attitude; and he is the best judge of the prospective value of that policy. It is only when his policy is developed into a dogma that he gives an opening for legitimate criticism. It must, I think, be admitted that the problem of relating all the phenomena in the organism with a concept which represents the aspect of the whole organism, not as a mere sum of parts, but as a unified individual, has not been solved. The problem of the relation of the one and the many has proved as intractable in Physiology as in the wider domain with which Philosophy concerns itself. In recent times, the discovery that a living organism, animal or plant, contains, except in the case of the simple uni-cellular beings, an association of cells, looser or closer according to the particular kind of organism, each of which cells has its independent life, but each of which also provides its specialized contribution to the life and maintenance of the organism as an individual whole, has emphasized the urgent character of that special form of the problem of the relation between the one and the many, to which the single organism gives rise.

Suggestions have also been made as to the introduction of concepts which should be capable of employment in those parts of Biology in which the individual organism is transcended, and which should play a useful part in the formulation of racial relations. Such concepts, perhaps in even greater degree than in the case of the single organism, suffer from a vagueness of definition which unfits them for use in precise scientific theory, however great may be the need for them felt by the speculative mind.

The question whether psychical, or at least psychological, concepts are to be admitted in Biological Science is one which, besides being a methodological question of great importance, is related to general issues of far-reaching import as regards our general views of the world. The existence of a psychical side of at least the higher organisms is now universally recognized, and in the case of lower organisms some rudimentary psychical elements such as bare sensation, indistinct awareness of changes in the environment, and even some rudimentary form of memory of past experience, are most frequently assumed to be present. In so far as such concepts, of a psychical or psychological character, are employed as an essential element in the conceptual descriptive schemes of Biology, that Science does not wholly belong to Natural Science, in accordance with the meaning to which I have restricted the term in these lectures, but which restriction I have admitted to be open to very pertinent criticism, and have adopted only for convenience. Biology may, so far as such conceptions form part of it, be described as a mixed Science, partly physical and partly psychological. A survey of the history of Biological Science shows that many, or most, biological theories of a general kind have employed in their statement psychological concepts; but whether, and how far, these can be eliminated, is a crucial question which has led to much diversity of opinion in recent times. In the special department of Physiology, as I have already observed, the recent tendency has been to employ a methodological restriction to physico-chemical concepts, and thus to resume the processes within the living organism in conceptual schemes which involve these concepts only. This however does not, or should not, involve the dogmatic assumption that there are no limits to what can be attained by this method; and I shall presently give reasons which tend to prove that such limits must actually exist.

I have already, in one of my earlier lectures, observed that Psychology, of which the object is to describe conceptually the mental processes of the typical individual, has two methods open to it. The first is that of introspection, in which the observer contemplates the processes in his own mind, and accepts similar descriptions from others as regards their own minds. The second method is that of inference from the behaviour of other persons, that is from the physical phenomena exhibited by them, and assumed to be in correlation with the psychical side of such persons. In the case of the psychology of animals the second method is the only one at our disposal. The first method is the one which affords the only basis for the interpretations of physical concomitants, in man or in other animals, which we make in drawing conclusions as to the psychical processes associated with them. We have direct knowledge of the psychical processes in our own minds, and we have also knowledge of the perceptual or physical processes which we connect with the former. By what has been termed ejection we transfer the psychical side, or at least some elements of it, to other living organisms which appear to originate physical sequences of a kind similar to those which we regard ourselves as originating. Thus our knowledge of the psychical side of living organisms in general, when that side is presumed to exist, is essentially indirect and inferential. The question whether psychological concepts should be included, or not, in the conceptual laws and schemes of Biology, would appear to depend largely, if not wholly, upon whether or not such concepts are regarded as necessary for the purpose of introducing an element of contingency into the sequences to which the laws and schemes have reference; and thus of, at least to some extent, impairing the value of such laws and schemes as instruments for predicting what will happen in concrete cases. If it is believed that no such element of contingency is involved in the use of psychological concepts, it would appear that the psychical side of the living organism is relegated to the position of a mere epi-phenomenon which does not really affect the physical events; this is the position of the thoroughgoing psycho-physical parallelist. For those who hold this view, the psychical concepts can then play no essential rôle in the conceptual theories and laws in question, and might without real loss be eliminated from them.

I have at the beginning of these lectures emphasized the fact that it is not in the province of Science to deal with the purely individual; its rôle is to resume conceptually what a class of individuals have in common, to extract the universal. When a conceptual scheme is applied in any individual case, there is always some part, greater or less in amount and importance, of what happens or is observed, which lies outside the scope of the scientific generalization employed. It may even be held that in general a scientific theory is essentially of a statistical character, the individual peculiarities of particular percepts or trains of percepts, to the description of which it is applied, being left out of account, but being always in some degree present. In the case of those percepts and trains of percepts with which Biological Science is concerned, this residuum, of an individual character, is often, and perhaps usually, of much greater importance than in the cases in which no living organisms are involved. It is certainly true that, the more highly organized a living being is, the greater is the importance of the individual peculiarities which, escaping all schematic description, remain outside the purview of scientific schematism.

The contemplation of the problems presented by the living organism and its relations, especially in the case of man and the higher animals, brings us face to face with the question of how the relation between the psychical and the physical domains is to be conceived; of the relation between body and mind. It is a question which cannot be simply ignored in connection with any general view of the nature and scope of Biological Science, although it may be very properly ignored by the investigators in many special departments of that Science. That the psychical side of a human being, and his body, which represents the construct of what we directly perceive, exercise an apparent influence upon one another is a matter of common knowledge. A change in the moral character of a man is sometimes the apparent effect of a blow on his head, which may be ascertained to be accompanied by a lesion in his brain. Conversely, a psychical disturbance, such as that produced by bad news, is apparently the cause of marked physical disturbance in the body, temporary or permanent, sometimes even of death. One thing seems certain; that, in any comprehensive view of the matter, we can leave out of account neither the conceptual knowledge of the set of percepts which we call the physical side of a man or animal, as resumed in its representation by physico-chemical descriptive schemes, nor the direct apprehension which we have in our own cases of the psychical side of our being, and which we are forced to admit by inference as being present, at least to some degree of development, in other animals; although its form may be of lesser complexity, shading down to bare awareness or sensation, as we descend the scale of animal existence. For Natural Science, as applied to man and other living organisms, the question takes the form, what limitations, if any, does the presumed presence of the psychical factor introduce into the scope of that scientific method which attempts to describe, in terms of physico-chemical mechanism, the inner processes in living organisms, and their relations with the environment? One answer to this question, to which I have already referred in an earlier lecture, that given by the thorough-going psycho-physical parallelist, is that there is no such limitation, because the psychical side is a mere *Begleiterscheinung*, an epiphenomenon, of which the physical organism is really quite independent. Now it cannot possibly be maintained that the correctness of this view has been proved. The actual successes which have been attained in the representation of particular tracts of physiological phenomena as physico-chemical processes, great as they undoubtedly have been, are still at an immeasurable distance from the attainment of such proof, in relation to the organism as a whole. The assertion of this view, in its absolute form, is then merely a dogma resting on nothing but an illegitimate extension to a whole, of what may have been shown to be true of some parts. It is in direct contradiction with the immediate deliverance of our consciousness as to the real efficiency of the will, the determinations of which we regard as completely dominated by nothing outside our own spiritual nature. *Prima facie*, the validity of this primary intuitive apprehension is quite as much entitled to recognition as the more indirect conceptual knowledge provided by Natural Science. Nothing short of the most cogent evidence, far greater than any that has hitherto been adduced, would be necessary before we could properly admit that this intuitive apprehension is wholly illusory.

The attitude of the pure Idealist, or of the Panpsychist, is a quite different one. Their philosophical position is that of psychical monism, just as the system of the psycho-physical parallelist amounts in practice, though not necessarily in terms, to materialistic monism. The view of psychical monism is that there is in reality but one system concerned in the living organism, and that that one system is fundamentally psychical; that the separation into mind and body is artificial, and if made, cannot render possible the treatment of either part in complete independence of the other. Very much the same statement applies to the views of the neutral monist who does not attribute either to the physical or to the psychical elements a fundamental rôle, but regards them both as aspects or modes of some reality more fundamental than either. However much there may be to be said, from a philosophical point of view, in favour of a monistic system of a kind which refrains from completely subordinating one side of the nature of the living organism to the other, such a system has not yet been shown to be capable of being so developed in detail as to constitute an articulated monistic Science, into which all the manifold particulars of organic life, both on the physical and the psychical sides, can be fitted. It follows that, unless we are prepared to remain in a nebulous region of generalities, we must necessarily, for scientific purposes, as in ordinary life, remain content with a methodological dualism which provisionally recognizes in separation the two domains of the psychical and the physical. The investigation of these two domains are respectively the functions of Psychology and of Biology. That there are relations between the two domains is manifest, and cannot be ignored on either side, whenever a general view of the scope of either Science is to be formulated. The problem of formulating these relations is one of such great difficulty, on account of the presumed, or provisional, disparateness of the two domains, that no even tolerably complete solution is in sight. For those phenomena connected with living organisms in which mechanical or rather physico-chemical categories seemed to be insufficient, or in which it seemed to be impossible to ignore the relations between the physical and psychical sides of the organism, some method of provisionally representing those relations in a very general way has frequently been attempted in various vitalistic theories. The attempts made in the older vitalism to localize and delineate the nature of the interaction between body and mind were crude, and had the character of makeshifts introduced at various points *ad hoc*. They frequently involved the use of such vague and hybrid expressions as "vital force": and their inability to rid themselves of the nebulosity of view which the use of such expressions indicates has had the result that they have fallen into general discredit amongst Biologists. Renewed attempts have been made in recent times to deal with the problem of the relations between the physical and psychical sides of the living organism. On the one hand, efforts have been made to conceive, or rather to represent, the nature of the influence which the psychical side exerts upon the physical side of the living being; and on the other hand, the investigations which fall under the head of the department known as Psycho-physics or as Physiological Psychology have resulted in the construction of a mass of detailed empirical knowledge of the relations between physical sense-data and the sensations or the perceptions to which they are regarded as giving rise. I propose to give some indications of both these lines of thought.

A widely spread idea is that the action of the psychical upon the physical can be represented as of the nature of guidance, which must manifest itself as physical guidance. This guidance is regarded as, on the psychical side, teleological in character. It is not always assumed that the consciousness of the individual is necessarily involved in it; it is often thought of as immanent in all the parts of the organism, and manifesting itself in guidance of the motions of the molecular or submolecular constituents of the living body. Now it is clear that, if the psychical side of a living being is to be regarded as having any intelligible relation with the physical organism, and through that organism with the physical environment, the organism cannot be regarded as simply and solely a machine, or physico-chemical complex, in which all the processes can be completely described as a dynamical system in which all the changes are completely determined by the inner relations of its parts and its relations with the external physical environment. The effect of the psychical side can only be represented as an actual interference with what would be the course of the physical system if it were independent of the psychical side. It is a matter for consideration what form this interference can take, how it should be conceived as manifesting itself in the physical organism. This question can only be considered in the light of facts obtained by accurate observation of the actual working of the living body as a whole, in relation to its environment, and of the presumably connected processes in the various parts of the body. These facts of observation may be expected to throw light upon the points of similarity and of difference between a living body and a machine, or system composed of non-living materials.

A living organism is a system of complex structure composed essentially of a mixture of substances of very complex chemical constitution which fall under the term protoplasm, and also a large amount of water and some small admixture of other substances. The protein substances are relatively stable, and yet, in the living organism, they are continually breaking down into chemical substances of simpler atomic constitution, a process known as katabolism; and they are constantly being built up, this latter process being known as anabolism. To the totality of physico-chemical changes in the living substance of the organism the term metabolism is applied. In an anabolic process energy is required for the purpose of transforming substances of comparatively small chemical energy into more complex substances of much higher chemical potential. On the other hand, in katabolic processes energy is liberated by chemical changes of the reverse kind. In the higher animals, the proteins, carbohydrates, and fats, contained in the food, are broken up into simpler constituents, in the process of digestion, and are then synthesized into more complex constituents by an anabolic process, and become capable of being transported by the tissue fluids to all parts of the body. In the animal body an important part is played by the purely physical processes of osmosis and diffusion of liquids in the circulation of food materials, secretory, and excretory substances, from blood to lymph, and from lymph to cell-substance, or to glandular cavities. All such processes, anabolic, katabolic, and mechanical, may be regarded from the point of view of Energetics, of which the concept of energy and its transformations is the basis; and the question arises whether, or how far, the principles of Energetics apply to the living body and to its various parts? As regards the law of Conservation of Energy, known also as the first law of Thermodynamics, it appears to have been established, as the result of various investigations conducted with great care, that the law holds good for the living body of an animal. The energy of the food and oxygen absorbed by the animal is shown to have its equivalent in the mechanical work done by the animal, together with the energy lost from its body by conduction and radiation. In general, the metabolism of the animal as a whole, and the chemico-physical changes in detail, conform to the law of Conservation of Energy. The forms which energy takes in the living body are mechanical, chemical, thermal, and electrical; in fact the same forms as are found in the inorganic domain. The existence of no form of energy peculiar to living matter has to be postulated as a condition that the law of Conservation of Energy should hold good for the organism and its parts. The idea that some such special form of energy, often denoted by the term Biotic Energy, is required to account for the phenomena in the living organism can thus, in all probability, be rejected. The whole weight of evidence goes to show that the influence of the psychical side upon the physical side of the organism cannot be conceived of as taking the form of a supply of energy, and it must consequently be such as to be consistent with the law of Conservation of Energy; the various forms of the energy in the living organism being the same as in the non-living domain. But it by no means follows from this that the physical side of the living being must be regarded as independent of the psychical side. For the changes in a system are not determined by the law of Conservation of Energy alone; the nature and amounts of the transformations of one form of energy into equivalent amounts in other forms have to be known before the changes in the system are completely determinate. It is just in these transformations from one form of energy to another that striking differences are found between the cases of the living organism and of the non-living domain.

With a view to the consideration of the nature of these differences, let us compare, in respect of the transformations of energy, the processes which go on in the working of a steam engine and in a warm-blooded animal. In a steam engine, the source of the energy from which the work done by the engine is derived is chemical potential energy of the coal. When the coal is burned in the boiler-furnace, this energy is transformed into heat which vaporizes water in the boiler, producing steam at high temperature. By the expansion of the steam in the high-pressure cylinder, mechanical work is done in driving forward the piston. The steam is then cooled in the condenser. During the whole process a certain amount of heat is converted into an equivalent amount of mechanical work. The high chemical potential energy of the coal and oxygen is transformed in large part into the energy of heat, the remainder being the low chemical potential energy of the residual products. Of the heat produced, a large part is dissipated, or rendered unavailable for use, by radiation from the boiler, steam-pipes, and other parts of the machine, and by friction, or remains as low-temperature heat in the water of the condenser. The nett result is that only a comparatively small part of the energy derived from the chemical transformation of the coal and oxygen has been made available by the engine for the performance of mechanical work. In the warm-blooded animal, in which the temperature is maintained at a constant level, usually higher than that of the medium in which it lives, the source of energy is the chemical energy of the food and oxygen which is taken into the body. These substances undergo chemical transformations in the alimentary canal and in the tissues of the body, in the processes of digestion and assimilation. A certain proportion of the food taken into the body has become a part of the muscles, nerves, etc. of the body. At the same time, portions of the substances of high chemical potential, the proteins, carbohydrates, and fats, are transformed into water, urea, and carbon-dioxide, which are substances of relatively low chemical potential. The energy taken into the animal reappears in large part as mechanical work, that used in bodily movements, in the motion of the heart, lungs, blood, etc.; and a part appears as heat, sufficient to compensate the loss of heat by radiation and conduction from the body; also a portion is employed in the formation of digestive juices and in the propagation of nerve currents. A chief point in which this process is distinguished from that which goes on in the steam engine consists in the much smaller part played by the production and dissipation of heat. It appears that the chemical transformations proceed without the production of anything approaching the amount of heat developed when chemical transformations occur in connection with non-living matter. The changes from chemical to mechanical energy appear to take place more directly than in inorganic processes; a comparatively small amount of heat is produced, and the dissipation of energy is accordingly greatly reduced. This feature of the transformation of chemical energy, with only an insignificant production of heat, is still more marked in the case of the cold-blooded animals, whose temperature is nearly the same as that of the medium, and rapidly adjusts itself to changes in the temperature of the medium. In their case the rate of the metabolic changes is dependent on the varying temperature of the environment, whereas the corresponding changes in the warm-blooded animal are, within limits, almost independent of the temperature of the environment. It thus appears that the living animal, considered as a chemico-physical mechanism, in which energy is received from outside and is converted into mechanical work done by the organism, is very much more efficient than is the steam engine or any other machine that can be constructed for doing mechanical work; in the sense that the proportion of the energy supplied that is transformed into such work is much greater in the former case than in the latter.

The transformation of energy which goes on in the cells of the green leaves of plants is of a still more peculiar character. Whilst the animal receives substances of high chemical potential in its food, the green plant receives carbon-dioxide and water, substances of relatively low chemical potential, and transforms them in the tissues of the leaves into starch, a substance of much higher chemical potential. The energy required to produce this transformation is derived from the solar radiation; in some manner, the details of which we are unable at present completely to describe, the chlorophyll in the cells of the green plant absorbs these radiations and utilizes their energy in the anabolic process of building up starch. This starch is converted into soluble sugar which is circulated through the vessels of the plant. The plant at the same time absorbs nitrates from the soil, and the soluble sugar together with the nitrogenous salts is employed in building up and accumulating proteins and other organic substances in the plant. During these processes there is no dissipation of heat, and only a little mechanical work is done by the plant in connection with the circulation of protoplasm, in the movements of the tendrils, and in other ways. The general effect of the processes which go on in the living plant is to build up, retain, and accumulate, a store of energy in a form in which it is available for doing mechanical and other work when it is liberated. This effect is produced by transforming energy from the sun which would otherwise be dissipated and become unavailable. In the general economy of living organisms the energy of the organic materials in vegetable foods, built up in the manner I have described, is essential for the existence of animal life. Moreover the stores of energy in wood, and the energy derived in past ages from the solar radiation and stored up in coal, form the chief sources from which we draw the thermal energy which we require for the practical purposes of human life.

The transformations of energy which occur in machines, and generally in the inorganic domain, are irreversible, always in one direction tending to the diminution of available energy. This fact of observation has been formulated in the second law of Thermodynamics, the most precise form of which is the law of Clausius that, in any isolated system, the entropy tends to a maximum. The question arises whether a living organism, considered as a physico-chemical mechanism, is such that its transformations of energy are in accordance with this law? It is probably difficult to show that there is in the organism a direct infringement of the law; that the processes are reversible; but it is clear that, as a matter of degree, the amount of dissipation of energy in the processes connected with living organisms is very much less than in any corresponding processes connected with a non-living mechanism alone. The second law of Thermodynamics is usually regarded as a statistical result which holds for the vast swarms of particles, in motion in all directions, of which material systems are conceived to consist. From this point of view, either the law would not hold good, or the consequences to which it leads would be modified in degree, in a system in which the number of molecules involved in a particular transformation of energy was so small as to make the purely statistical method inapplicable. A similar default in, or modification in the effect of, the law, might occur if we conceived that, by some selective process applied to the individual particles, without any alteration in the energy of their motions, but affecting the directions of these motions, the statistical result obtained by considering all directions as alike in reference to the motions, were invalidated. In either of these ways it might be possible to account for the striking difference which I have described between the transformations of energy in the living organism and in the mechanisms to which the second law of Thermodynamics applies in its most precise form. These considerations give rise to the suggestion that the peculiarities of the processes which occur in the living organism may be regarded as due to some control of the transformations of energy in the organism which is not present in the case of the processes of a corresponding kind which occur in the inorganic domain. As we have already seen, unless the supposed influence of the psychical side of the living organism is merely an illusion, there must be somewhere or somehow what can only be represented as an interference manifesting itself within the physical processes themselves; and it would appear that some regulation or control of the transformations of energy is the way in which this interference can be best conceived; for the facts of observation appear to negative the idea that any supply of energy which would nullify the law of conservation is involved. The processes in the living organism may thus be represented as controlled by an agency, of which the effect is to favour the anabolic building up of stores of chemical energy of high potential, and, by release of such energy, to permit of sudden transformations of it into mechanical work.

A view of the character of the processes in the living organism in this order of ideas has been developed, both on the biological and on the philosophical side, by Prof. Driesch, in his theory of entelechy. He regards the mode of operation of entelechy to consist in suspension of such transformations of energy as would be possible on the basis of pre-existing differences of intensity, for example of chemical potential, and further, in the relaxation of such suspension. He regards entelechy as non-material and non-spatial, but acting, so to speak, into space. Although it regulates the transformations in the organism, its effects are strictly limited by the possibilities afforded by the structure of the organism; its action must be subject to given, preformed, conditions of a physical kind, and it does not itself involve the supply to, or the extraction from, the organism, of energy in any form. A full account has been given, in his Gifford lectures, by Driesch of the "proofs" and arguments upon which his vitalistic theory rests. I must rest content with having indicated, in the briefest manner, a possible mode in which the purposive activity of a living organism may be held to manifest itself in the physical processes in the organism, and consequently indirectly in some degree in physical sequences in the physical world in general.

It is no doubt possible to exaggerate the part which the mental side of an organism should be regarded as playing in what happens in the body. In this connection, the experiments of Loeb, and his deductions from them, leading to his theory of tropisms, are of much interest; he has endeavoured with apparent success to account for such phenomena as the turning of organisms of certain species towards the light, or in other cases away from the light, on purely physico-chemical principles, without taking account of the existence of sensations, or of will, as mental factors in the phenomena. He attempts to show that, in such cases, neither the notion of will nor that of blind instinct prompting the organisms to their actions is necessary. The following utterance shows that he is sanguine as to the extension of his substitution of physico-chemical description for psychical factors, to a degree which but few will regard as warranted by any evidence which we possess at present. He writes[1]:

> Our wishes and hopes, disappointments and sufferings, have their source in instincts which are comparable to the light instinct of the heliotropic animals. The need of the struggle for food, the sexual instinct with its poetry and its chain of consequences, the maternal instincts with the felicity and the suffering caused by them, the instinct of workmanship and some other instincts, are the roots from which our inner life develops. For some of these instincts the chemical basis is at least sufficiently indicated to arouse the hope that their analysis, from the mechanistic point of view, is only a question of time.

However far Loeb's mode of viewing the responses to stimuli, in the case of lower organisms, may be justified, in the case of man and higher animals it very soon reaches limits which cannot be passed. The lack of uniformity

[1] *The mechanistic conception of life*, p. 30.

in the responses of men to external stimuli is no doubt in part traceable to differences in their organic constitutions, dependent on differences of past history, but, at least in the case of the more complicated complexes of stimuli, when all possible allowance has been made for individuality of physical constitution, there remains a residuum which renders futile the hope, or the fear, that the notion of an individual psychical character, exhibiting itself in the nature of the response to such stimuli, can ever be dispensed with. So far as this individual psychical element is operative, it escapes, not only physical analysis, but also schematic psychological analysis. In it consists the freedom of will of the individual, not necessarily to be regarded as ultimately purely arbitrary, but as not completely determined by, or capable of being linked up in a determinate scheme with, anything foreign to itself. The recognition of this fact does not make all phenomena in which that element is concerned wholly inaccessible to scientific description; for a determination of the will is preceded and succeeded by physical sequences of such a character that they, or at least portions of them, are representable by psychophysical conceptions; and on the psychical side a somewhat similar statement holds as regards psychological sequences.

The physiologists of the eighteenth century, especially von Haller, were the first to demonstrate the importance of the properties of irritability and sensibility in the nervous system, and to emphasize the function of the central organ of the nervous system, the brain, in synthesizing the elements which represent sensation on the physical side. But the emergence of the Science of Psycho-physics or Physiological Psychology as a distinct department may be regarded as due to Cabanis (1757–1808), who laid its foundations in his *Rapports du Physique et du Moral de l'Homme*. His leading idea, related with the Philosophy of Locke, was that the function of the brain in relation to thought is parallel to the physical functions of other organs of the body; as he expresses himself:

> In order to arrive at a correct idea of those operations from which thought arises, we must consider the brain as a particular organ, destined specially to produce it in the same way as the stomach and the intestines are there to produce digestion, the liver to filter the bile, the parotid, maxillary, and sublingual glands to prepare the salivary juice.

This amounts apparently to a practical subordination of thought to its physical concomitants.

At the end of the eighteenth, and the beginning of the nineteenth, centuries, many vague and fanciful theories arose in the shape of speculations connected with animal electricity, animal magnetism, mesmerism, and Phrenology, which retarded the progress of the Science, and for a time produced a deflection from those really fruitful lines of investigation which were approached by the school of Johannes Müller, and were later put on a firm basis by Helmholtz and Emil Du Bois-Reymond. The distinction between sensory and motor nerves, that the anterior nerves of the spine as efferent, or motor, nerves carry the nervous stimulus to the different organs, whilst the posterior nerves, as sensory or afferent nerves, carry the peripheral stimuli of the senses to the nervous centres, was established by the labours of Charles Bell, Magendie, and Johannes Müller. The last of these investigators, in his doctrine of "specific energies," introduced another important distinction into the theory of the sensory nervous apparatus. This theory was afterwards adopted, and its importance emphasized, by Helmholtz, who brought it into connection with the theory of colour first advanced by Thomas Young. In accordance with the doctrine of specific energies, by the stimulus of any single nerve-fibre, only such sensations can be produced as belong to the order of one definite sense, and every stimulus which affects this nerve produces only sensations belonging to this definite order. For example, any effective stimulus of the optic nerve-apparatus produces always the sensation of light, whereas the same stimulus would, if effective, produce in the auditory nerve-apparatus the sensation of sound. Thus the quality of our sensations depends, not on the stimulus, but on the nervous apparatus. A prodigious amount of physiological and anatomical work has been done with a view to elucidate the functions of the external or terminal organ relating to a particular sense, of the connecting nerves, and of the central organ, situated in the brain, which we must regard as related, on the physical side, with perception of the particular kind. Of first rate importance, in this order of investigation, is the work of Helmholtz in Physiological Optics and Physiological Acoustics. He showed, for example, that the ear, when subjected to anatomical and acoustical analysis, exhibits itself as a delicate resonator which absorbs the different elementary periodic movements into which musical sounds can be analysed harmonically, and that different nerve-fibres carry them separately to the central organ in the brain with which we must regard perception of sound as connected. He was enabled to relate the quality of musical notes known as "timbre," which is different for the same note produced by different instruments, with the physical production of the notes. Helmholtz accepted Young's hypothesis that there exist in the eye three distinct kinds of nerve-fibres, related to three distinct modes of colour sensation, corresponding to the three simple colours, red, green, and violet, of which all colours are compounded. As regards sound, he regarded differences of pitch and character in notes as dependent upon the differences of sensitive nerve-fibres; each nerve-fibre exhibiting only difference of intensity of the stimulus. That this should be so is in accordance with the doctrine of specific energies.

The investigations of the brothers Weber, and especially of E. H. Weber of Leipzig, were commenced even before those of Johannes Müller. They had as their object the establishment of the relation between the subjective side of sensation and its physiological side. This experimental work was continued by Fechner, to whom the term Psycho-physics is due. His treatment of the subject, as represented in his *Elemente der Psycho-physik*, published in 1860, was an investigation of the relations of mind and body involving the measurement of psychical quantities and the establishment of their correlation with physical quantities. The philosopher Herbart had already attempted to subject psychical phenomena to exact methods of calculation, and Lotze had suggested the existence of a definite and constant connection between sensation and stimulus. The fundamental difficulty of such quantitative correlations consists in the fact that sensation is not an extensive magnitude consisting of equal units, although a particular sensation may have intensive magnitude; and this fact has naturally led to much criticism of all attempts at correlation in a mathematical form between the psychical and physical sides of sensation. Of the investigations of Wundt and many other investigators in this important field it is impossible for me to speak. A flood of light has been thrown by these investigators upon the physical processes within the organisms which are related with sensations and perceptions. But great as is the scientific importance of the detailed investigations of Physiological Psychology, and invaluable in their application to practical problems of life as the results obtained, and to be obtained, may prove to be, there remains a gap between the physical and psychical sides of the organism, unbridged, and perhaps from its very nature impassable for our intelligence.

In many respects the progress of the Science of living organisms has been much slower and more difficult than in the cases of those branches of Science which deal with inorganic processes. This can be accounted for in a large measure by the fact that the extreme complexity of structure and of function of the living organism makes the process of isolating those elements which should be examined as separate objects or processes a much more intricate and difficult one than in the class of phenomena with which only Physics and Chemistry are concerned. In the latter Science a particular phenomenon which it is desired to examine can be more readily submitted to laboratory experiments in which the requisite conditions of isolation from other, and disturbing, elements are artificially produced, than in biological Science, where the phenomena to be examined occur in connection with an animal or plant simply as it is given as a whole, with all its complex inter-relations, and in which the complete isolation of a particular process is either impossible or can only be made with some approximation by the employment of a very high degree of technical skill. In Biology, the problems of mere classification are more complex, and occupy a much larger place in the history of the development of the Science than is the case in the inorganic Sciences. That stage of Biological Science which is resumed under the term Natural History is one of vastly greater extent than what corresponds to it in such sciences as Physics and Chemistry; moreover it must be regarded as still far from complete.

The method of abstraction and generalization by which scientific laws and theories, descriptive of processes, are set up is of slower and more difficult application in Biology than in the inorganic Sciences. Great as have been the difficulties of reaching conceptual laws which resume extensive tracts of phenomena in the inorganic Sciences, and however incomplete the results of this process may still be, in the case of the Biological Sciences the difficulties of such achievement have been still greater, and the stages at present reached must be in all, or most, departments regarded as less advanced than in the sciences which are not concerned with life. The degrees of abstraction and concreteness are very various in different departments of Science, but the ultimate aims of all branches of Biological Science are parallel with those of inorganic Sciences. Much of Biology remains purely descriptive, with a low degree of abstractness, and a large amount of detail remaining to be filled up.

Many of the great generalizations of Biological Science have been due in large measure to the breadth of view attained by those who examined living organisms, their habits, distribution, and environment, on a large scale, as travellers; and who thus avoided the narrowing effect of too exclusive occupation with work in the laboratory. Such work in the laboratory is largely of an anatomical and morphological character in which the dead organism is examined; and even in the physiological work the living organism is torn asunder from the environment in which its race was developed.

In the next three lectures I propose to sketch the origin and development of some of the great general theories and special departments of Biological Science.

# XVI

## THE LIVING ORGANISM

FROM the earliest times in which animals and plants formed objects of study, a certain bifurcation in Biological Science is discernible which corresponds to two different points of view in relation to such objects. This cleavage of the Science into two divergent lines of advance has persisted throughout its history, and has manifested itself in the division of Biology into several special departments, each of which is connected with one or other of these two modes of regarding the organism. Many investigators have concerned themselves exclusively or mainly with one or other of these two divisions of the subject; and many others who have not exclusively cultivated either side of the subject have shown a tendency to emphasize the greater relative importance of one or other of the two aspects, especially in relation to the problem of classification of animals and of plants. In the first of these modes of regarding the organism the attention is directed to the study of "form"; this is the point of view of the Anatomist; and the term Morphology, invented by Goethe, is now employed to designate the study of form in the most general sense; the term being not even always confined to the study of the forms of living organisms. Throughout the history of the Biological Sciences, Morphology has advanced from the study of the forms of animals and plants as wholes, to that of the structure, connections, and spatial positions, of the various organs of which they are composed, and their comparison in different animals or plants; then to the structure of the tissues; and, in the nineteenth century, to the study of the forms and structure of the cells of which the body of the organism is built up; and finally to the study of the chemistry of protoplasm, consisting of a mixture of complex substances, which is the fundamental material that distinguishes the living organism from non-living matter.

Whilst Morphology has to do with the statical aspect of the organism, it is the province of the other great division of the Biology of the single organism, Physiology, to deal with the kinetic aspect, as expressed by "function." Taking as its starting point the activities of an animal as a whole, Physiology has advanced, through the study of the functional activities of the various organs, to that of the activities of the tissues; later to the study of the active life of the individual cells, and finally to the metabolism of the protoplasm. It seems as if the two great departments of Morphology and Physiology had found their meeting point in the study of the protoplasm, of which the structure and metabolism appear to represent the physical aspect of what we call life. To the department of Morphology there belongs that study of the anatomy and histology of extinct species which is denoted by the term Palaeontology. The study of Embryology, that of the early stages in the growth of the organism, its organs and tissues, has belonged chiefly to Morphology, but in recent times the Physiology of the processes at work during the development has also been studied. Both Embryology and Palaeontology are in close relation with those racial and evolutionary aspects of Biology of which I shall speak more directly in the two following lectures. The distinction between the Morphological and the Physiological points of view has exhibited itself in the history of the Science in relation to the problem of the classification of animals and plants. It is in fact clear that the element of arbitrariness which appertains to such classification leaves ample room for differences in the emphasis laid upon form and upon function, in relation to the similarities and divergences upon which the classification is based.

Although there existed some knowledge both of the anatomical and of the functional aspects of animals, and especially of man, before the time of Aristotle (384–322 B.C.), as recorded in the writings of Hippocrates and his school, Aristotle may be regarded as the founder of Comparative Anatomy. That great thinker and observer was much less of an Aristotelian, and reached much more nearly a true conception of scientific method, than the medieval thinkers who based their views upon his Philosophy. He had a knowledge of over 500 different animals, and had an extensive acquaintance with the structure of many of them. He studied not only the more ordinary beasts, birds, and fishes, but also cuttlefish, snails, oysters, crabs, crawfish, lobsters, sea-anemones, sponges, fish-lice, and even intestinal worms. Extensive as was his anatomical knowledge, his interest in it was however secondary to his interest in the functions of the various organs and parts of animals. He appears to have been the first to study in any detail the development of the chick; and he made a commencement of the study of comparative Embryology. The results of his study of form are contained in his great work the *Historia Animalium*, but that work also contains the results of his observations in comparative physiology, and in the distribution and behaviour of many species of animals. His later book *De Partibus Animalium* deals with what he regarded as the causes of the form and structure of animals, and thus amounts to a discussion of the functions of their parts, of the relations of form with function, and of the adaptedness of structure. In the comparison of an animal of one species with that of another, a clear distinction has been made in later times between homologies and analogies. By homology is understood the correspondence between the organs and parts of animals of different species, in relation to a common spatial order; whereas analogy denotes a correspondence between organs and parts which have the same function in the two animals. Aristotle made use of both these notions, although he did not state quite clearly the distinction between them. He made the first attempt at a scientific classification of animals into a number of groups based upon similarities of structure. He distinguished between back-boned and back-boneless animals, but held the erroneous view that only the former have blood. He recognized various modes of classifying animals, not only in accordance with their structure, but also in manners dependent on function, such as their manner of life, their mode of reproduction, their food, etc. Aristotle recognized clearly that animals belonging to the same one among the great groups into which he divided them have a unity in the plan of their construction. In his *Historia Animalium*, taking man as the standard, after describing his external and internal parts in detail, he compared viviparous quadrupeds with man, and traced out the unity of plan in the structure of man and all such quadrupeds. Although this constitutes a definite contribution to Morphology, what he really sought for were not homologies, but parts with the same functions; his interest being mainly in functioning organs, and not in merely spatial relationship of parts. Aristotle foreshadowed various ideas and distinctions which came to be of great importance in Science in later times. Among these is the principle of division of labour amongst different organs, a point which was emphasized in modern times by Milne-Edwards. Thus he writes in *De Partibus Animalium*: "Whenever, therefore, Nature is able to provide two separate instruments for two separate uses she does so, instead of acting like a coppersmith who, for cheapness, makes a spit and lamp-holder in one." He recognized the distinction between tissues and organs, the homogeneous and heterogeneous parts of the body. Although he was unable to describe the structure of tissues, as does the modern Histologist, he described their distribution in the body. Aristotle also foreshadowed the notion of a scale of beings, to the development of which in modern times I shall presently refer. It is of interest to observe that, in Aristotle's view, the gradation of organic forms is the consequence, not the cause, of gradation in their activities. Thus, in the *Historia Animalium*, he writes:

Plants have no work to do beside nutrition, growth, and reproduction; they possess only the nutritive soul. Animals possess in addition sensation, and the sensitive or perceptive soul.

Again he writes in *De Partibus Animalium*:

Plants, again, inasmuch as they are without locomotion, present no great variety in their heterogeneous parts. For where the functions are but few, few also are the organs required to effect them.... Animals, however, that not only live but feel, present a great multiformity of parts, and this diversity is greater in some animals than in others, being most varied in those to whose share has fallen not mere life, but life in high degree. Now such an animal is man.

It thus appears that, in Aristotle's teleological view, the pre-determined character of the activities of organisms is to be regarded as the cause of their structural characteristics. Aristotle's Physiology was to some extent based upon observation, but contained a large element of erroneous assumption. All the functions are connected with animal heat, associated with the blood, and centralized in the beating heart, which is the seat of the soul. The brain is bloodless, and produces mucus, and the sense-organs are in the head, so that they may not be overheated by the blood.

After the time of Aristotle, and before the commencement of the modern epoch, the only progress of importance in knowledge of Physiology was that made by the celebrated Physician Galen (132–200 A.D.), who recognized that the art of medicine should rest upon the basis of a physiological knowledge which must be dependent upon a groundwork of Anatomy. By his experiments on monkeys and swine, he showed that the arteries contain blood, not air; and he attained to an understanding of the meaning of the brain and nervous system. He was the first to show that the nerves connected with sensation are different from those which have to do with motion, and that they form separate parts of the nervous system. He elaborated a pathological theory which dominated the theory and practice of medicine until the sixteenth century.

The inauguration of the modern period, involving a breach with the dominant physiological tradition of Aristotle and Galen, is due to Harvey (1578–1657). As in other branches of Science, the fuller recognition of the importance of observation and experiment, as the basis of all theory, differentiates this period from the medieval. Harvey's introduction of precise methods of observing and experimenting into physiological Science had an importance, in directing the attention of others to the true line of progress, as great as his own great discovery of the circulation of the blood. The results obtained, owing to the new impetus, were collected and systematized by Albrecht von Haller (1708–1777) in his *Elementa Physiologiae Corporis Humani*. Among von Haller's own researches were those on respiratory movements, the contractility of muscles, and the irritability of nerves.

The origin of all modern mechanistic and physico-chemical theories of the living organism may be traced back to Descartes (1596–1650). The astronomical discoveries of Copernicus, Tycho, and Kepler, the mechanics of Galileo, and Harvey's discovery of the circulation of the blood, suggested to him the idea that mechanical laws could be applied to the purpose of explaining the phenomena of life in the bodies of man and other animals. His Physiology was mechanistic, in the strict sense, and included no chemical conceptions; everything was made to depend upon heat, hydraulics, tubes, and valves. He believed it possible to account, on these lines, for all the phenomena of organic life in animals and in man. His physiology was based upon that of Galen, supplemented by Harvey's discovery of the circuital motion of the blood, but he did not accept the idea that the heart acts as a propulsive apparatus. The food in the intestine was absorbed by the blood, and carried to the liver, where it became charged with the "natural spirits," and then passed to the heart, which charged it with "vital spirits," in virtue of the innate heat of the heart and the action of the lungs. The blood became rarefied, owing to the flame of the heart, fed by the natural spirits; and this expansion of the fluid produced the circulation, when directed by the valves of the heart and great vessels. The finer and rarefied parts of the blood pass off in two directions, one to the organs of generation, and the more important to the cavities of the brain, where they not only serve to nourish that organ, but also give rise to a fine ethereal flame or wind, through the action of the brain upon them, and thus form the "animal spirits." From the brain these spirits are conveyed by means of the nerves, which are regarded by Descartes as pipes, to various parts of the body, where they act upon the muscles. The impressions of the sense-organs are also conveyed by these tubular nerves to the brain.

In Descartes' view, all animals except man are pure automata, simple mechanisms, devoid of any element of feeling or consciousness. "The animals," he says, "act naturally by springs, like a watch." In the case of man, there is added to the bodily mechanism the rational soul, spiritualistic and immortal, which is located in the pineal gland, situated in the middle of the brain. By this assumption, Descartes attempted to reconcile his mechanistic conceptions with his idealistic philosophy, but it is safe to assert that he failed to give any intelligible account of the relations between the rational soul and the purely mechanistic body, a failure which is characteristic of all the later theories of which his may be regarded as the parent. In the latter part of the seventeenth century, Physiology continued for some time to be purely mechanistic, but in the modified sense that chemical discoveries were utilized in the description of the life of the organism.

The eighteenth century was a period in which vitalistic hypotheses dominated physiological conceptions, especially owing to the influence of the chemical and vitalistic ideas of Stahl and his followers. During this time the progress of Physiology was much retarded by the lack of progress in Chemistry, which was probably the result of Stahl's phlogistic hypothesis. Not until the great chemical discoveries of Lavoisier and his successors were adequate chemical conceptions made available in Physiological Science. Physico-chemical conceptions, without vitalistic hypotheses, became again dominant in Physiology, and lasted throughout the nineteenth century, culminating in the writings of Huxley and Max Verworn.

But little progress was made in the classification of animals during the eighteen centuries subsequent to the time of Aristotle. During the fifteenth and sixteenth centuries of our era, considerable additions were made to the list of known animals, but hardly any improvement was made in their classification. The first to make any such advance was John Ray (1628–1705), the predecessor of Linnaeus. He was the first to define the use of the term "species," as denoting a group of similar individuals exhibiting constant characteristics from one generation to another; and he was the first to emphasize the anatomical characteristics as a basis of classification. The greater systematizer Linnaeus (1707–1778), who has been described as "a classifying, coordinating, and subordinating machine," in his great work, the *Systema Naturae*, introduced a system of classification of plants and animals which formed the starting point of modern Systematics. Linnaeus employed a binomial system of nomenclature, and graded his classification into classes, orders, genera, species, and varieties. He recognized six classes of animals,—Mammals, birds, amphibians (including reptiles), fishes, insects, and vermes; this classification was afterwards made more precise by Lamarck, who established sixteen classes instead of six. Linnaeus believed that each species was descended from a pair originally created, each expressing an idea in the divine mind. Moreover, in his doctrine of continuity, taken in a loose sense of the term, he maintained that the species could be arranged in series with no hiatus between two consecutive series. Thus, unlike the present view, he recognized no genetic relations between closely allied species, and left no room for the occurrence of the discontinuous variations which many persons at the present time believe to be an important factor in evolution.

The advances made in human anatomy from the time of Aristotle until the end of the sixteenth century were accompanied by but little advance in the knowledge of comparative Anatomy. A great impetus was given to anatomical studies by the invention of the microscope, at the beginning of the seventeenth century. One of the first effects of the use of the microscope was the discovery of the complex structure of tissues. Up till that time they had been regarded as little more than inorganic substances, possessing, however, some organic properties such as contractility. Thus the study known as Histology came into being, and in particular, important discoveries relating to muscle fibres were made. Among other applications of the microscope was the study of the comparative Anatomy of lower animals, and that of the metamorphoses of insects, with which the name of Swammerdam is associated. One of the first to make extensive use of the new instrument was Malpighi, who studied by its aid the development of the chick. He also ascertained the minute structure of the lungs, and demonstrated the connection of the arteries with the veins; further he described the histology of the spleen, the kidney, the liver, and the cortex of the brain; he showed that the liver is really a conglomerate gland, and discovered the Malpighian bodies in the kidney.

The biological studies of the eighteenth century were largely in the direction of general natural history and, under the influence of Linnaeus, of the problem of classification. The minute study of insects was continued by Réaumur and Bonnet, who attached, however, more importance to their habits and physiology than to their anatomy. The general conception, to which I have already alluded, of a scale of beings was first put into a detailed and systematic form by Bonnet (1720–1793), and was in fact extended by him to all the objects in the Universe. He constructed a long table, headed "Idée d'une Échelle des êtres naturels," which begins with Man, the Orang-outang, the Ape, Quadrupeds, and ends with earth, water, fire, and more subtle matter. The scale is not based upon any definite principle, either morphological or functional, but is supposed to represent a gradation involving all possible orders of perfection. This conception of a gradation of beings was also held by Buffon (1707–1788), but in his hands it takes more consistent form as a functional gradation. He pointed out the fact that the groups of Invertebrates are very different in structural plan from those of the Vertebrates, and had a clear conception of the unity of plan which is characteristic of the Vertebrates. Moreover he, for the first time, expressed the idea that community of origin might be at the base of the unity of plan, although he was far from being a consistent Evolutionist. He pointed out the difficulty of supposing that one species may arise from another by a process of degeneration, and oscillated between the ideas that species are definitely discontinuous with one another, and that they can be united in larger groups.

Xavier Bichat (1771–1802), who was mainly a human

anatomist, worked out in detail the Aristotelian distinction between the animal and the vegetative parts and functions of animals, to which Buffon had also drawn attention. The animal life, which does not appertain to plants, he described as the order of functions which connect the animal with its environment; these organs are the afferent and efferent nerves, the brain, the sense-organs, and the voluntary muscles; the brain being the central organ. The organic or vegetative life has for its central organ the heart, and includes the processes of digestion, circulation, respiration, exhalation, absorption, secretion, nutrition, and calorification. He regarded the plant and animal as standing for two different modes of life. The only relation which the plant has with the environment is that involved in nutrition; the animal has in addition to this organic life, a life of active relation with surrounding things. He observes[1] that:

One might almost say that the plant is the framework, the foundation of the animal, and that to form the animal it sufficed to cover this foundation with a system of organs fitted to establish relations with the world outside. It follows that the functions of the animal form two quite distinct classes. One class consists in a continual succession of assimilation and excretion; through these functions the animal incessantly transforms into its own substance the molecules of surrounding bodies, later to reject these molecules when they have become heterogeneous to it. Through this first class of functions the animal exists only within itself; through the other class it exists outside; it is an inhabitant of the world, and not, like the plant, of the place which saw its birth. The animal feels and perceives its surroundings, reflects its sensations, moves of its own will under their influence, and, as a rule, can communicate by its voice its desires and its fears, its pleasures and its pains. I call organic life the sum of the functions of the former class, for all organized creatures, plants or animals, possess them to a more or less marked degreé, and organized structure is the sole condition necessary to their exercise. The combined functions of the second class form the "animal" life, so named because it is the exclusive attribute of the animal kingdom.

[1] *Recherches Physiologiques sur la Vie et la Mort*, 3rd ed., p. 2.

Bichat contrasts the symmetry of the nerves and muscles of the animal life with the asymmetrical arrangement of the visceral muscles and the sympathetic nerves which belong to the organic life. He points out that habit is all-important in the animal life, but denies that habit has any influence upon the organic life. He states that the organs of the organic life attain their full perfection independently of use; whereas the organs of the animal life require education in order to reach perfection. These views as to the independence of the organic life of habit and use would probably no longer be accepted without considerable modification.

A very similar distinction between animal and vital functions was emphasized by the great Comparative Anatomist Cuvier (1769–1832), who studied both structure and function, and even regarded the latter as the more important in that it determined the former. Following Aristotle he asserted that a plant is an animal that sleeps. He was aware, owing to the recent progress of Chemistry, that the material of the body is principally composed of combinations of Carbon, Nitrogen, Hydrogen, and Phosphorus, forming albumen, fibrine, etc. Although the discovery of the cellular nature of tissues was not made until after his death, he observed that the organism can be resolved into small flakes and filaments which form a "cellulosity." Cuvier was the first completely to recognize as a definite principle the harmony between structure and function. "It is," he writes, "on this mutual dependence of the functions and the assistance which they lend one another that are founded the laws that determine the relations of their organs; these laws are as inevitable as the laws of metaphysics and mathematics, for it is evident that a proper harmony between organs that act one upon another is a necessary condition of the existence to which they belong." We have here an attempt to form a concept of the coordinated organism as distinct from the sum of the parts considered either in relation to structure or function; but by the conditions of existence he meant adaptations of functions and organs within the organism, and he hardly considered the external conditions or the environment. In accordance with his well-known principle of correlation, from one part of an animal, having given a model of the group to which it belongs, the whole may be constructed, for as he says: "All the organs of an animal form a single system, the parts of which hang together, and act and react upon one another; and no modifications can appear in one part without bringing about corresponding modifications in all the rest." From the shape of one organ the shape of the other organs can be inferred, having given sufficiently extensive knowledge of functions, and of the relation of structure to function in each kind of organ. The functional dependence of the parts he interpreted in terms of what is later known as the general metabolism of the organism, that is the constant chemical changes in all the parts, and the accompanying interchanges with the outside.

One of the most important results of Cuvier's work is his division of the animal kingdom into four principal types of form, the Vertebrates, Molluscs, Articulates, and Radiates. The first three have bilateral symmetry, and the last radial symmetry. In formulating these four divisions, each of which is built upon one plan, Cuvier was influenced by the idea that the characters of the two sets of organs, the vegetative and the animal, and the correlations within each, must form the basis of the classification.

In contradistinction with Cuvier it was held by his adversary Geoffroy Saint-Hilaire (1772–1844) that the structure of all animals may be referred to a single type. He went so far as to assert that "There is, philosophically speaking, only a single animal." By that single animal he meant an abstract generalized type to which, by the principle of homology, all actual animal structures could be made to conform. With him, homology was not so much homology of organs as of parts and connections. His principle of connections was his guide in tracing an organ through all its functional transformations, for as he says: "an organ can be deteriorated, atrophied, annihilated, but not transposed." As a pure Morphologist for whom "form" was everything, his difference from Cuvier's attitude, in which the stress is laid on "function," was fundamental. His attempts to prove, by means of what appear to be very far-fetched homologies, that animals, as far apart as Vertebrates and Cephalopods, conform to the same fundamental type, were criticized and demolished by Cuvier. Cuvier showed that, although Saint-Hilaire had discovered many hidden homologies, especially by his important discoveries concerning foetal structure, the unity of plan and composition, as conceived by Saint-Hilaire, does not exist in actuality. Cuvier further maintained that the whole principle of homology, so far as it is valid, is subordinate to the principle of the functional coordination and adaptation of the parts.

Biologists in Germany, and also to a large extent in France, were, during the early part of the nineteenth century, under the influence of a number of ideas which formed part of what is known as the Philosophy of Nature. The principal conceptions of this school were the existence of a unique plan of structure, the idea of the scale of beings, and the notion of parallelism between the stages of individual development and the stages of the scale of beings, which latter has for us an obvious connection with theories of evolution. A further theory of this school was that of the repetition or multiplication of parts within the individual, of which the vertebral theory of the skull is the most striking example. The law of parallelism, first laid down by Kielmeyer, and afterwards, in a more developed form, by Oken (1779–1851), asserted that the embryo of every animal passes during its development through all stages of the animal kingdom, or at least through the stages of one or more classes lower down in the scale. It was held that the animal kingdom is a dismemberment of the highest animal, man, and thus that animals are only the persistent foetal stages or conditions of man, who contains within himself all the animal kingdom. The embryo of higher animals was compared with the adults of lower animals. It was stated, for example, by Tiedemann that "Every animal before reaching its full development, passes through the stage of organization of one or more classes lower in the scale, or, every animal begins its metamorphosis with the simplest organization." A detailed account of facts which support this theory was given by Meckel, but his treatment contains very imaginative comparisons between organs of animals of widely differing groups, and involves a mixture of morphological homologies with physiological analogies. Meckel admitted that man does not pass in his development through the whole animal series, but asserted that, at least as regards single organs or organ-systems, the embryo of man passes through many animal stages. Although he was not a thoroughgoing evolutionist, he held that the higher animal, in his gradual evolution, passes through the permanent organic stages which lie below it. As he says: "The development of the individual organism obeys the same laws as the development of the whole animal series." An adherent of this school, K. G. Carus, asserted as a general law of Nature that the higher formations include the lower; that the animal includes the vegetable; that it is by a rational necessity that the development of a perfect animal repeats the series of antecedent formations. The theory of the repetition or multiplication of parts within the organism was pushed to the absurd extreme of attempting to demonstrate that the whole organization is repeated in certain of its parts; for example Oken asserted that in the head the whole trunk is repeated, the upper jaw corresponding to the arms, the lower to the legs, and that in each jaw the same bony divisions exist as in the limbs. Nevertheless the recognition of serial homologies was a real contribution to morphology, although many absurd or arbitrary homologies were maintained. It was maintained by Carus that the whole skeleton is only a repeated vertebra. The influence of the Philosophy of Nature, as an *a priori* theory, is indicated in the statement of Carus that, in respect of the formation of the skeleton throughout the animal kingdom, he wishes to know "how such and such a formation is realized in virtue of the eternal laws of reason." He held that all forms of skeletons can be deduced from the hollow sphere, so that every skeleton can be represented schematically by a number of hollow spheres suitably modified in shape and suitably arranged; and he endeavoured to work out this idea as applied both to vertebrates and invertebrates. Although he was strongly under the influence of the *à priori* ideas of the German anatomists of the period, he was in many respects less a disciple of Saint-Hilaire than of his great opponent Cuvier; he held that the connections of bones and muscles change in accordance with functional requirements, and he did not accept the "law of connections" in its rigorous form.

The leading ideas of Cuvier in relation to functions, of Saint-Hilaire in relation to the principle of connections, and Oken's notion of the serial repetition of parts, were combined, modified, and reduced to clearer forms, by Richard Owen. The main idea developed in his work *On the Archetype and Homologies of the Vertebrate Skeleton* is that the vertebrate skeleton is composed of a series of segments, each of which he calls a vertebra. His archetype is a scheme of what is usually constant in the vertebrate skeleton, both the number and arrangement of the bones in any actual vertebrate being subject to variation. He defined the object of his work to be to deduce the relative value and constancy of the different vertebral elements, and to trace the kind and extent of their variations within the limits of a plain and obvious maintenance of a typical character. He accepted in a modified form Oken's vertebral theory of the skull, which was afterwards demolished by Huxley. In the determination of homologies he followed Saint-Hilaire's principle of connections, and rejected the method of their determination by the mode of development. In his view, comparative anatomy explains embryology, and not the reverse. Relations of homology he analysed into three kinds, special, general, and serial homology. Special homology consists of the correspondence of a special part or organ, determined by its relative position and connections, with a part or organ in a different animal; thus involving reference to a common type. Owen's general view of the nature of living things was that organic forms result from the competitive working of two principles, the first of which brings about a vegetative repetition of structure, while the other, involving a teleological factor, shapes the living organism to its functions. The first of these principles, "a general polarizing force" illustrating the archetype of the vertebrate skeleton, is the same principle which produces repetition of the forms of crystals in the inorganic domain. The second principle, the "adaptive" or "special organizing force," produces the diversity of organic beings.

Although observations of the development of the embryo had been made from the time of Aristotle, Embryology attained its position as a Science in the hands of von Baer (1792–1876). The first volume of his great treatise on the subject, published in 1828, contains a full and adequate account of the development of the chick, as obtained by minute and accurate observations, and a discussion of the laws of development in general. In the *Scholia* at the end of his description of the development of the chick he refutes the notion of preformation; and defends the idea that the essential nature of the animal determines its differentiation, and that no stage of development is solely determined by the antecedent stage. He holds the vitalistic conception that a guidance, involving the idea of the completed whole, is active at each stage of development. He shows that the process of the development of the embryo is one of differentiation, by which the germ becomes increasingly individualized, and the developing animal increasingly independent. In describing the stages of development, he lays down his theory of germ-layers. In his account of the process he states that first of all the germ separates out into heterogeneous layers, which with advancing development acquire ever greater individuality, and on their first appearance show rudiments of the structures which will characterize them later. In the germ of the bird, at the beginning of incubation, there can be distinguished an upper smooth continuous surface and a lower, more granular, surface. The blastoderm then separates into two layers, of which the lower develops into the plastic body-parts of the embryo, the upper into the animal parts; the lower shows clearly a further division into two subsidiary layers, the mucous-layer and the vessel-layer; the original upper layer also shows a division into two, which form respectively the skin and the muscle-layer, the latter of which contains, in an undifferentiated state, the skeletal and muscular systems, the connective tissues, and the nerves belonging to them. Thus the process of determination results in the formation of four layers. The uppermost layer will form the outer covering of the embryo; from it there differentiates out at an early stage the rudiment of the central nervous system, forming a more or less independent layer. Below this outermost layer is the one from which the muscular and skeletal system is developed, then the vessel-layer which gives rise to the main blood vessels. The innermost layer will form the mucous membrane of the alimentary canal and its dependencies; it is originally, like the other layers, a flat plate. From all these layers tubes are developed by the bending round of their edges; the process of primary differentiation is then complete. He then describes the later histological and other morphological differentiations which occur concurrently. Through morphological differentiation the various parts of the fundamental organs become specialized through unequal growth, first into the primitive organs, and then into the functional organs of the body. At the same time there is a differentiation in the substance of the layers, whereby cartilage, muscle, and nerve separate out, and a part of the mass becomes fluid. Through histological differentiation, the texture of the layers and incipient organs becomes individualized. Originally the germ consists of an almost homogeneous mass, with clear or dark globules in suspension. This homogeneity gradually gives place to heterogeneity, the structureless mass becomes fibrous to form muscles, hardens to form cartilage and bone, liquifies to form blood, and differentiates in a large number of other ways.

Von Baer showed that the extreme views of those who were dominated by the law of parallelism, as stated by Meckel and Serres, that the higher animals repeat in their development the adult stages of the lower, went far beyond the facts. He pointed out, for example, that the developing chick is at a very early stage demonstrably a vertebrate, and does not recapitulate the organization of a polyp, a worm, or a mollusc. By means of various examples he showed that the recapitulation is never that of the whole organization of a lower animal, but only that of particular parts or simple organs. Like Cuvier, von Baer recognized four main types of animals, the vertebrate, the molluscan, the longitudinal, and the radiate. He held that the manner of development is determined by the type of organization; the type being observable in the very earliest stages of development. In his view the development of the individual is always from the general to the special; the general characters of the group to which an embryo belongs appear earlier in development than the special characters of the particular animal. There is, during the development, a progressive gradation of generality, the less general structural features always appearing after the more general. So far from endorsing Meckel's law of Parallelism, he held that the embryo, instead of passing through the states of other definite forms, separates itself from them; and that the embryo of a higher animal never resembles the adult of another animal form, but only its embryo. Thus, not even within the group, is there a real scale which the higher forms must mount. The apparent resemblance between the embryos of the higher, more differentiated, members of the group with the lower, less differentiated, adult forms arises from the fact that these latter diverge but little from the generalized type. Embryology is of great assistance to comparative anatomy, of which the aim is to discuss the general type, or the common plan of structure of the animals of a group. For, as the embryo develops from the general to the special, the state in which each organ first appears should represent the typical state of that organ within the group. Thus the true homologies of parts will be best determined by studying their earliest developments. Homologies should be restricted to a single type; organs with similar functional relations belonging to members of different types should not be regarded as homologous, on account of their different modes of development. The classification should not depend on the adult structure but on the characters shown in early development, because these latter show the characters of the type to which an animal belongs in their more generalized form.

Von Baer, like Cuvier, rejected the notion of the scale of beings, not only by recognizing the existence of four totally different types, but by showing that, even within one type, any serial arrangement only held good for separate organs or sets of organs, and not for the whole complex of organs; so that there is no actual scale of beings even within one and the same type. It has been said that man is only the highest animal in respect of his nervous system.

The publication by Theodor Schwann, in 1839, of the discovery that the tissues in the animal body are composed of cells, constituted a most notable and far-reaching advance upon the discovery of the complex structure of the tissues. In accordance with the great conceptual scheme, the cell theory, the cell is both a morphological, or statical, unit, and also a physiological, or dynamical, unit, in the formation of the complete body. It is a statical unit, inasmuch as all plants and animals are built up of cells and modifications of cells. It is a dynamical unit, inasmuch as the life of the organism is dependent upon the activities of the cells of which it is composed. In the words of Schwann: "The whole organism subsists only by means of the reciprocal action of the single elementary parts." It was in the domain of Botany that the existence of cells was first observed, but Schwann was the first to announce the identity of the cellular structure of all living beings, animal or vegetable. In the seventeenth century, cells in plants had been discovered by Robert Hooke, Malpighi, and Leeuenhoek, but they were then not regarded as living independent structural units. In the nineteenth century, when great improvements in the microscope had been made, it was established by the work of various observers that the tissues of plants are composed of elements which can be reduced, for the most part, to spherical closed cells. By the coalescence and elongation of the cells the vessels and fibres of the plants are formed. At first, attention was concentrated on the cell-walls, but in 1838 Schleiden pointed out the importance of a discovery, that had been made a few years earlier, of the existence in every cell of a small body which he called the cytoblast, and which is now usually called the nucleus; this nucleus is embedded in a gummy substance called the cytoplasm. Schleiden showed that plants are built up of cells and their modifications, and that a single cell, or ovum, is the original form of the plant embryo; further he gave a theory of the mode in which, by cell division, new cells are formed. The conception that the cell is a partially independent living unity was expressed by him in the statement that "Each cell carries on a double life; one a quite independent and self-contained life, the other a dependent life, in so far as the cell has become an integral part of the plant." Here we have, in a distinct form, the view, in relation to the plant organism, that it is to be regarded as a multiplicity in unity, a notion which was extended by Schwann to the animal organism. The occurrence of small globules in animal tissues had been observed earlier, for example by von Baer in the embryo of the chick; and that cells were to be found in animal tissues was well known to Johannes Müller and other investigators, but a complete theory of cells in animal tissues was the work of Schwann[1]. In the first part of

[1] Professor J. Arthur Thomson has drawn my attention to an article entitled "The Dawn of the Cell Theory," by Professor J. H. Gerould, published in *The Scientific Monthly*, vol. XIV (1922), in which it is pointed out that the cell theory is older than has generally been supposed. The beginning of the theory is traced back to Lamarck, who wrote in his *Philosophie Zoologique* published in 1809: "No body can possess life if its containing parts are not a cellular tissue, or formed by a cellular tissue....Thus every living body is essentially a mass of cellular tissues in which more or less complex fluids move more or less rapidly; so that, if this body is very simple, that is, without special organs, it appears homogeneous, and presents nothing but cellular tissue containing fluids which move within it slowly; but, if its organization is complex, all its organs without exception, as well as their most minute parts, are enveloped in cellular tissue, and even are essentially formed by it." In the same year, 1809, Mirbel, in the second edition of his *Exposition de la Théorie de l'Organisation Végétale*, reaches the conclusion that "The plant is wholly formed of a continuous cellular membranous tissue....Plants are made up of cells, all parts of which

his book *Microscopic Investigations Concerning the Agreement in Structure and Growth of Animals and Plants* he dealt with cartilage cells and with the cells of the notochord, showing that the nucleus, the nucleolus, and the cell-walls, play the same part, and behave in the same manner, as in the plant cells, in accordance with Schleiden's theory. It was already known that all animals are developed from an ovum; the fundamentally important discovery of the existence of the ovum in mammals had been made by von Baer, in 1827. The position of the ovum in relation to the cell theory was a subject which Schwann carefully investigated; he came to the conclusion that the ovum is itself a cell, but not one of the simplest kind. Every individual organism begins its life as a single cell; and in the simplest or uni-cellular organism, such as the Amoeba, it remains uni-cellular; in other cases the gradual formation of the body is due to the division and successive re-division of the fertilized ovum into a coherent mass of cells. Although there are considerable differences between the cells in any one organism, and in the cells of different organisms, there appears to be a considerable amount of agreement in the structure of all cells of animals and plants. The cell theory has been the basis of the modern theories of heredity of which I shall speak in the next lecture. The study of the cleavage of the fertilized ovum, and the gradual formation of cells by segmentation, leading to the formation of tissues, was commenced by Schwann, and carried on by various investigators, of

form one and the same membranous tissue." Both Mirbel and Lamarck were associated with the Musée de l'Histoire Naturelle. With Mirbel and Lamarck the emphasis was laid upon the membranous cellular tissue as fundamental, rather than upon the cell as a vital unit. The idea of the individuality of the cell is due to Dutrochet, in 1824. Up to this point the existence of the nucleus had not been discovered; the general occurrence of nuclei was recognized by Robert Brown, in 1833, and his work inspired that of Schleiden. It would thus appear that the cell theory had been developed to a considerable extent many years before the time of Schwann.

whom von Kölliker was the most eminent, on account of the vast amount of detailed embryological and anatomical work he accomplished. Although the cell theory has been for the most part accepted as a representation of the facts, a certain amount of criticism has been directed against its all-sufficiency, based upon some exceptional cases which have been observed, in which cell-walls appear to be absent, the nuclei dividing without cellular division, the result being a mass of protoplasm containing many nuclei, but without cell boundaries. The fact that, both in animals and plants, intercellular filaments, composed of protoplasm, have been discovered connecting the cells, may also not be without significance. The complete generality of the process of cell-formation in organic life has been challenged; and that process has been held by some investigators, such as Sachs on the botanical side, and Sedgwick on the zoological side, to be a secondary process, the growth of the protoplasmic mass itself being the primary factor in the development of the plant or animal. It has in fact been held that the plant forms cells, but the cells do not form plants. However this may be, the cell theory, as a working scheme, has been of inestimably great assistance in the development of biological science.

# XVII

## HEREDITY

THAT some living organisms arise by what is called spontaneous generation from non-living matter was the universal belief of men of Science until about three centuries ago. That invertebrate animals at least are generated spontaneously was held by Aristotle, who remarked the occurrence of small animals in putrefying substances. Until the seventeenth century this belief was firmly held both by zoologists and botanists; even Bacon believed in the spontaneous origin of plants, such as thistles, from dead earth. The first doubter of the truth of this conception of abiogenesis appears to have been Francesco Redi (1626–1697), the Florentine contemporary of Harvey, who retained the traditional belief. Redi showed by experiment that no grubs or insects appeared in the flesh of a dead animal if it was protected with sufficient care from intrusion from outside. With the problem of the origin of parasites he had less success. By some it was supposed that parasites were generated spontaneously from the juices of their hosts, by others that they had been originally created with their hosts; Adam being supposed to have harboured all the human parasites from the beginning. Not until the nineteenth century was the germinal origin of internal parasites established by careful experimenters.

During the nineteenth century various experimenters demonstrated the possibility of preventing the appearance of animalcules in infusions, by a process of sterilization which prevented the entry into the infusions of living germs from the external air. But even as late as 1858, Ponchet asserted, before the Paris Academy of

Sciences, that he had proved that microscopic organisms arise apart from pre-existing germs. However, the investigations of Louis Pasteur, who showed that all putrefaction and many kinds of fermentation are due to microscopic living organisms, and the work of Tyndall, who showed that absolute sterilization of infusions could be attained by intermittent applications of heat, led to the general acceptance of the law of biogenesis. As at present accepted, the law of biogenesis is of a negative character, asserting that there is no known evidence that living organisms at present arise from non-living matter. As regards the further question whether we should conceive that, under conditions widely differing from those which obtain at present, there has been continuity between living and non-living matter, the answer must be entirely a matter of speculation, because, from the nature of the case, the only mode of obtaining a decisive answer, that of observation and experiment, is not available. It is somewhat remarkable that Huxley, in stating before the British Association his conviction that the law of biogenesis had been fully established, expressed as his opinion that he would have witnessed the origin of protoplasm from non-living matter if he could have been present at the beginning of organic evolution.

There have been various speculations made as to the origin of life on the earth; among these is the suggestion of Lord Kelvin that living organisms reached the earth in meteorites. The persistence of the conception of the continuity of living with inorganic matter is exhibited in the views of prominent men of Science such as Haeckel, Ray Lankester, the Botanist von Nägeli, and the Physiologist Pflüger, all of whom accept the idea that this continuity has at some former time been effective in relation to the origin of living matter.

The term "heredity" implies the fact that living organisms can produce their like, the resemblance, though never absolutely perfect, extending to the most minute details of construction and function. The modern theories of heredity have been constructed upon the basis of an enormous amount of work in the collection and sifting of facts. Their aim, in accordance with the view developed in these lectures, is to represent in one or more conceptual schemes, abstract in some greater or less degree, the main facts of the general likeness of parent and offspring, of the occurrence of variations in the offspring, that is of characters not exhibited in the same degree by the parent; of specific similarities in the offspring with characters in one or both parents; of the fact that characteristics may occur in the offspring which are not to be found in either parent, but which were exhibited by a grandparent or by a remoter progenitor; and lastly, of the fact that characters acquired by a parent in the course of his or her life, as the apparent result of interaction with the environment, seem, in some cases, to reappear in the offspring. The last of these facts has given rise to much controversy; and its proper interpretation is one of the most crucial questions which have arisen in connection with theories of heredity and in the more general theories of evolution.

I have pointed out in my earlier lectures the fact that a conceptual theory frequently employs, not only concepts to which there are directly corresponding percepts, but also concepts which have not, or at any given time are not known to have, any percepts to which they directly correspond. The modern theories of heredity make use of concepts which belong to both these classes. Not only concepts of the first class, but also those of the second class, have almost invariably been regarded by those who have originated or developed the theories as denoting really existing entities, in the sense attributed to the term by physical realists. But, as in the case of the theories of Physics and Chemistry, the value of these theories may be estimated quite independently of the acceptance of the views of physical realists; they may be judged exclusively in relation to their coherence as conceptual schemes, and to their effectiveness in representing the perceptual facts and sequences to which they are applied.

Most of these theories have as their starting point the group of perceptual facts resumed in the cell theory of the construction of the living organism, applicable to both animals and plants. There are, in the case of most multi-cellular organisms, amongst the cells of which they are constituted, special kinds of cells known as germ-cells, distinct in various respects from the somatic, or body-cells, which constitute the greater part of the organisms. Prolonged investigation has shown that the embryo of a new individual organism appears to begin, in the case of the majority of animals and plants, in the union of two minute cells, the sperm-cell and the ovum, or egg-cell, which combine to form a single cell, the fertilized ovum. The embryo is then observed to arise by a process of successive formation of new cells obtained by division of the fertilized ovum into two cells, followed by a series of successive divisions of cells into two parts each of which is a complete cell. Although this sexual mode of reproduction holds for the great majority of living organisms, it is not the only process by which reproduction occurs. In some organisms, particularly amongst those which are uni-cellular, there is no fertilization of the ovum, which divides without previous fertilization into two cells; so that the reproduction is asexual; and in some cases there are no special germ-cells. It will be sufficient for our purpose to leave aside such special modes of reproduction which occur in various lower organisms, and to confine our attention to the sexual mode of reproduction which holds good for the higher animals and for many plants.

Since the seventeenth century, the ovum and the sperm have been regarded, or at least one of them has been regarded, as in some sense the physical basis of the new organism. A type of theory called preformationist was held from the seventeenth century until the beginning of the nineteenth. In accordance with preformationism the ovum, or the sperm, contained a complete preformed miniature organism which only required to be unfolded or evolved, and increased in size, in order to become the new animal. Moreover, either the ovum or the sperm, according to the views of two contending parties, not only contained the miniature copy of the animal of the next generation, but this contained a still more minute copy which would ultimately develop into an individual of the second generation; and this contained a still more minute copy, and so on for all future generations. This theory of emboîtement, sometimes known as the pill-box theory, contained as its essence the principle that nothing new arises, that nothing is generated, but that everything pre-existed in an invisible form; as it was expressed by Albrecht von Haller, "Es gibt kein Werden"; there is no such thing as becoming, there is only an unfolding or evolution. The opposite theory of Epigenesis, that the evolution of the animal consists of a gradual increase of complexity from what at first appears to be comparatively simple, and that thus something essentially new arises, was shown by Wolff to be a fact of perception in the case of the embryonic development of the chick, and ultimately got the upper hand over preformationism. However arbitrary the preformationists' theories may appear to be, we must, I think, recognize, when it is remembered that the successive miniature organisms were never supposed to be discernible to the senses, that they have some of the essential elements of a genuine scientific theory, considered as a purely conceptual representation of what can be observed to happen. The crudity of preformationism as a conceptual scheme exhibits itself in the very restricted range of the perceptual facts to the representation of which it lends itself. It fails, for example, to distinguish the specific characters which may be inherited from either parent, and gives no suggestion as to how the origin of congenital variations is to be represented. That it did not account for the preformation is not a decisive objection to the theory, for every scientific theory necessarily suffers at some point or other from a similar defect. Moreover, some of the modern theories, especially that of Weismann, are closely akin to preformationism.

The next type of theory designed to represent facts of inheritance is the pangenetic theory. The general notion of pangenesis was adumbrated by Democritus and Hippocrates, and in later ages by Maupertuis and Buffon. The general idea of pangenesis is that the germinal cells contain samples contributed by all parts of the body, and that these samples give rise in the embryo to parts similar to those from which the samples came. The modern forms of the pangenetic theory are exhibited in the theory of Physiological units, propounded by Herbert Spencer in 1863, and in the well-known theory of Pangenesis, suggested as a provisional hypothesis by Charles Darwin in 1868. The essential principle of Darwin's theory is that every cell of the body, not too highly differentiated, throws off at each stage of its development characteristic gemmules, or small particles, which can later multiply by fissure, and give rise to cells like those from which they originated. These gemmules are continually given off and conveyed in the blood; they become specially concentrated in the germ-cells of both sexes, or in buds. In the development of the embryo these gemmules unite with others like themselves, and being aggregated in the germ-cells, they invest these latter with the power of developing into a new complete organism; but some may remain latent during the development of embryos through several generations before they become active. Each gemmule reproduces the cell from which it was derived, and they become active in the same order as that in which the corresponding cells follow each other in the ontogeny of the part to which they belong.

If Darwin's theory be regarded as a purely conceptual theory, in which the gemmules are not regarded as representing perceptual objects, it possesses the essentials of a genuine scientific theory. The suggestion that the gemmules might be replaced by a theory of transmission of force, as in the theory of Herbert Spencer, would seem to lead to no improvement on the theory in its original form, because it would not increase the capabilities of the theory as a scheme for the representation of observed facts. For force-centres have no advantage over conceptual gemmules in depictive power, and the importation of the term "force," in an undefined sense, is open to serious objection. As a scientific theory Darwin's seems to be a marked advance, in point of effectiveness for its purpose, over the preformationist theory; because it lends itself to the representation, not only of the simpler facts of heredity with which the latter theory was alone concerned, but also to that of the transmission of characters which may remain latent for several generations, and further to the transmission of characters special to the male offspring through the female parent, without becoming manifest in her. Also, as has been pointed out by Sir Ray Lankester, the theory can be made to account for the appearance, at a particular period of life, of characters inherited and remaining latent in the young organism.

It is interesting to quote some remarks made by Weismann, in reference to Darwin's theory of Pangenesis. Weismann[1] wrote in his *Keimplasma*:

I certainly consider even now that Darwin's theory must be looked upon, and that he probably considered it, rather as an inquiry into the problem of heredity than as a solution of the

[1] See *The germ-plasm*, p. 5.

problem. His assumptions do not, properly speaking, explain the phenomena. They are to a certain extent a mere paraphrase of the facts, an explanation of a purely *formal* nature, based on speculative assumptions, which were made not because they seemed possible, or even likely, but because they provided a formal explanation of all the phenomena on one principle. If we suppose that each cell arises from a special gemmule, and that these gemmules are present wherever they are wanted, it is easy to see how that structure, the origin of which we wish to explain, may appear in any given position. Further, when a large number of cells is to arise in regular succession from *one* egg-cell, the desired sequence of cells must of course result if we assume that the gemmules present become active in the required order. But this supposition does not really explain the phenomena. Even at the present day our explanations are imperfect enough, and are far from going to the bottom of the matter, but they differ from Darwin's provisional hypothesis in that they attempt to find out the actual facts concerned in the process, and to arrive at a *real*, and not a merely formal, solution of the problem.

This criticism is, I think, based upon a radical misapprehension of the nature, scope, and possibilities, of a scientific theory. Weismann's own theory, of which I shall presently give an account, founded, as it no doubt is, on a much more extensive range of observational knowledge than that of Darwin, is open to criticism on similar lines, more especially in its latest and most developed form. Neither Weismann's theory, nor any other theory, does or can "go to the bottom of the matter," or "really explain the phenomena," or find a "real" solution of the problem. Modified theories of pangenesis have been propounded by Francis Galton, and by W. K. Brooks; also a general theory of intracellular pangenesis has been developed later by De Vries; but it is not necessary for our purpose to consider these in detail. As in the case of Darwin's theory, the scope of the first two of these is limited, because they do not come to close quarters with the known detailed facts relating to the inner constitution of germ-cells.

The later theories, especially that of Weismann, make full use of such knowledge, and the scope of their power of representing facts of heredity is correspondingly increased. A most important part of Weismann's views on the mechanism of heredity is contained in his theory of the continuity of germinal protoplasm. This is not the only theory of the kind that has been suggested; the theories of Jäger, Galton, and Nussbaum are of a somewhat similar character, although in some of them the continuity asserted is rather that of germ-cells than of the protoplasmic contents of such cells. As Weismann's system has been worked out in great detail, and was gradually developed by him into a form of increased complication, designed to increase its scope and to meet objections raised to earlier forms of it; and as it has become not merely a theory of heredity but also of racial evolution, I propose here to consider it in some detail, as an example of a biological theory developed with consummate skill and with most ambitious aims, especially as it has become very prominent in relation to various matters connected with the most crucial questions of Biology, in particular the question relating to the inheritance of acquired characters.

Weismann's theory, like other theories I have mentioned, is based upon the idea that the continuity of characters in heredity is to be depicted by means of a continuity of material between parent and offspring, that continuity holding through a complete series of generations. In order to understand the nature of Weismann's theory it is necessary to state a number of observed facts, accumulated by many workers in the subject, relating to the fertilized ovum, its organization, and what happens in it on maturation, and in the events leading up to the development of the embryo. Every cell in the body of an animal of given species contains a general cell-substance, or cytoplasm, consisting in part of protoplasm, that is of living matter, and partly of non-living material of various kinds. Within the cytoplasm is a small body called the nucleus of the cell, and within this nucleus there is contained a substance known as chromatin, because it can readily be stained and thus distinguished from the rest of the material contained in the nucleus. In many animal cells there are small bodies known as centrosomes, which appear to play a very important part in the operation which results in the division of the cell into two cells; in higher plants there appear to be no centrosomes. The chromatin, contained in the nucleus, in certain conditions of the cell, consists of a number of separate masses, called chromosomes or idants; these are rod-like, looped, or granular bodies, and in any one species the number of these chromosomes is a definite even number, the same in all the cells. In man this number is now said to be 48; but it was for some time thought to be 24; in some other organisms the number is smaller, and in some larger. It was shown, in 1883, by Van Beneden, and has been since confirmed, for the cases of many animals and plants, that the male sperm-cell and the female ovum, when prepared for conjugation, each contain only half the number of chromosomes which each somatic cell of the species contains. This process of maturation, which results in the halving of the number of chromosomes in a germ-cell, either male or female, is carried out in a complicated manner; and, in the case of higher animals, it usually commences two cell-generations earlier than the formation of the gametes, that is of the matured sperm-cell and egg-cell which conjugate. When a sperm-cell reaches an ovum, it bores its way into it, and leaves behind its cytoplasmic substance; the nuclei of the sperm-cell and of the ovum then move towards each other and unite as a single nucleus which contains the chromosomes contributed by the two cells. The number of the chromosomes is thus brought up to the full number characteristic of the cells of the species. The walls of the ovum then become harder, and this prevents the entrance of any other sperm-cell. There is no centrosome in the mature ovum, but the sperm-cell brings its centrosome with its nucleus. This centrosome then divides into two, which appear to play an important part in the subsequent division of the now fertilized ovum into two cells. The ovum is very much larger than the sperm-cell, and thus provides the bulk of the material necessary for the initial stage of the development of the embryo. Both parents contribute alike the chromosomes, and in Weismann's theory it is these chromosomes upon which rest the foundations of the inherited organization of the offspring; the male parent contributing the centrosome, which appears to play the part of directing the process of division of the fertilized ovum. In accordance with the view of Weismann and his school, in any given species, the nuclei of the two gametes contain the hereditary substance; and more particularly this is contained in the chromosomes; the hereditary substance contains the primary constituents of the whole organism. After the single nucleus of the fertilized ovum has been formed, which contains the paternal and maternal chromosomes in equal numbers, the process of sub-division of the cell begins. Each chromosome divides longitudinally into two halves, and these halves ultimately arrive at the poles of the nucleus, one half at each pole, where one of the paternal centrosomes is situated. From each centrosome there radiates a system of rays which seem to be associated with the direction by the centrosomes of the movements of the half-chromosomes. Near each pole there comes to be a group of semi-chromosomes, half of paternal, and half of maternal origin. Each of these groups forms a new nucleus; the cytoplasm of the cell divides into two halves, across the equatorial plane, and thus two new cells are composed, each of which has a nucleus, of which the chromatic material is half of paternal, and half of maternal, origin. In certain cases this process of division of the cell into two, each containing a nucleus, half of paternal, and half of maternal, origin, has been observed through several successive divisions. The development of the embryo then proceeds by continual cell division, starting with the pair of cells into which the fertilized ovum divides. The swarm of new cells which is thus developed consists partly of somatic cells, of various constructions, which form the various parts and organs of the developing embryo, and partly of undifferentiated cells which ultimately become the germ-cells of the new individual. Investigation has shown that, in certain cases, these young germ-cells contain equal numbers of paternal and maternal chromosomes; in later stages it is thought probable that there is a separation of the paternal and maternal chromosomes, such that each cell contains some of each, but not necessarily in equal numbers. This is regarded as giving rise to the possibility of various combinations of the paternal and maternal hereditary characters, as represented in the final germ-cells of the new individual.

We are now in a position to consider, in some detail, Weismann's theory of inheritance; a theory which, it should be observed, takes for granted the essential phenomena of life, nutrition, assimilation, and growth. In the first place, the assumption, based on evidence of observation to which I have already alluded, is made, that there is a definite hereditary material located in the chromosomes of the nuclei of the sperm-cell and the egg-cell. Weismann next assumes, also on evidence based upon observation, that the germ-cell contains not only the primary constituents of a single individual of the species, but also those of several, often even of many, individuals. The whole material of which the chromosomes, or idants, are composed, is called the germ-plasm; as Weismann writes: "I call the idioplasm of the germ-cells, germ-plasm, or the primary constituent substance of the whole organism, and the complex of primary constituents necessary to the production of a complete individual I call ids." Thus a chromosome, or idant, is made up of several ids, each of which is the bearer of a complete inheritance. The fundamental conception of the theory of the continuity of the germ-plasm, as stated by Weismann in 1885, is that, in each development of the new embryo, a portion of the germ-plasm contained in the fertilized ovum is not used up in the formation of the offspring, but is reserved, and remains unchanged in the body of the offspring, for the formation of the germinal cells of the following generation. Thus what is continuous is usually not germinal cells but the germ-plasm "of definite chemical and special molecular constitution." This idea of a physical nexus between the successive generations of a race, in the handing on from one generation to another of the germ-plasm, has sometimes been imaginatively described as the immortality of the germ-plasm. As an embryo is developed by continual bifurcation of cells, starting from the fertilized ovum, gradual differentiation of these cells sets in, and various body-cells, with specialized structure and function, appear; these form the various parts and organs of the developing embryo. But some cells remain undifferentiated, and gradually lead up to the germ-cells of the new organism. Ultimately, on the maturity of the new organism, these germ-cells become liberated, and each of them contains chromosomes both of paternal, and of maternal, origin. It is then assumed, as a generalization of a number of observed facts, that each such chromosome contains germ-plasm derived from the ancestors of both parents; and thus there comes to be in the chromosomes an accumulation of material derived from earlier ancestors, both on the paternal and maternal sides. It should be observed that, up to a certain point, namely the existence of ids, the concepts employed are of the kind which correspond to perceptual objects.

But when, as Weismann assumes, each id is regarded as containing in itself, in some sense, all the generic, specific, and individual characters of a new organism, a concept is created to which no perceptual element corresponds; in fact the theory passes beyond the stage in which it is exclusively descriptive of perceptual facts that can be *directly* observed. But Weismann does not stop at this point; since the dissimilarity of parts of the organism must be represented in the ids, he regards each id as consisting of many invisible constituents, which he calls determinants. Each determinant is concerned with the formation of some special organ in the embryo. It is not necessary to assume that there are in the germ-plasm as many determinants as there are cells to be determined in the individual at every stage of its development. There must however be as many of these as there are regions in the fully formed organism capable of independent and transmissible variation, including all the stages of development. Thus a single determinant can represent a group of cells which can vary *en bloc*. In order to account for the mode in which the determinants give rise to cells and tissues, the special characteristics of which are represented by the different determinants, Weismann regards them as composed of groups of biophors, which are to be regarded as the minutest vital units. These biophors exhibit the primary vital characteristics, assimilation, metabolism, growth, and multiplication by fissure; each such biophor is supposed to have a molecular constitution. Whilst the id represents the complete individual, the determinants represent its different parts and groups of cells; the biophors represent characters. A germ-cell contains as many biophors as the individual which this cell is to produce possesses of elementary individual characters; and each biophor may vary independently, so as to produce a corresponding modification in the character it represents. The biophors are supposed to be liberated in the cytoplasm of the cells of the embryo. As the fertilized ovum divides and re-divides, the blastomeres, that is the resulting cells, become heterogeneous, if not at the very earliest stages of re-division, at all events at an early stage; that is they become suited to form certain parts only of the embryo. This fact is represented in Weismann's theory by the conception that the determinants, bearing various characters, become gradually distributed amongst different cells during the process of successive segmentation of cells. This process of distribution must be regarded as of an orderly character; and thus the various determinants must be looked upon as being definitely localized, in accordance with a definite structure of the germ-plasm. In order that this distribution may take place so that each determinant may reach the proper locality in the new individual, there must be qualitative differences, as regards the distribution of determinants, between the first two cells which appear on the first division of the ovum. The blastomere which contains the determinants corresponding to one class of organs will then at the next division split into two cells which again differ from one another in respect of the determinants they contain, and so on; the ontogenesis depending at every stage upon cleavage of a cell into dissimilar cells, and producing in the end dissimilar structures. As the organs and tissues become differentiated, the germ-plasm becomes less complex, owing to the release of continually more of its determinants, and ultimately transforms itself into the idioplasm to be found in each cell, and which only contains determinants of one cell and of parts of it. The biophors then break off from the determinants, and these scatter themselves through the cytoplasm, and thus impart to the cell its specific character. Some cells may contain determinants which do not break up into biophors, but which remain in reserve until a later time when the cell may become more differentiated. The biophors require, for the actual production of the characters they represent, the co-operation of the cytoplasm which constitutes the body of the cell in which they operate.

Weismann speaks of the determinants as being kept in relation with one another by "vital affinities"; and this can only be taken to mean that it must be accepted as a postulation of his theory that they do keep in relation with one another, and that he does not attempt to set up a definite theory as to the modes of such relations. When "vital affinities" and the like make their appearance in a biological theory, it is a sure sign that a point has been reached beyond which the descriptive power of the theory cannot pass. The whole secret of the phenomena may be regarded as having been concentrated in these determinants and biophors with their vital affinities, and in principle, the distance from an "explanation," in any ultimate sense, is precisely where it was before the theory was developed. The spatial scale of the phenomena has been reduced, and that is all. This does not however imply that such a theory is useless; quite the contrary. The determinants and biophors do not represent perceptual objects, although definite properties are assigned to them. They are regarded by Weismann as having a definite chemical constitution, and this may differ in different kinds of biophors; or there may be a difference of the arrangement of the atoms in the matter even with one and the same chemical constitution. The biophors may play a similar part in a conceptual scheme to that which atoms or electrons play in physico-chemical schemes, and may be equally indispensable. Thus, I take it, Weismann's scheme, when regarded as a conceptual scheme, is not open to objection, any more than was Darwin's theory of Pangenesis, merely on the ground that it employs concepts such as biophors. It should however be remarked that Weismann himself was far from accepting the view that biophors are purely conceptual entities.

His physical realism appears clearly in his statement: "The biophors are not, I believe, by any means mere hypothetical units; they must exist, for the phenomena of life must be connected with a material unit of some kind."

The actual value of Weismann's theory, as a descriptive scheme worked out in great detail, and with consummate skill, is one which only experts can estimate; and the opinions among them seem to be sharply divided. Besides the objection that the primary constituents, determinants and biophors, are purely ideal, the further objection has been raised that these concepts are unnecessary, and introduce an undue complication. The weight of this latter objection can perhaps best be estimated by contrasting Weismann's theory with other theories of the germ-plasm, such as that advanced by Delage. This does not employ such units as determinants or biophors, but relies upon the variety of chemical substances contained in the germ-cells. In Semon's mnemonic hypothesis, the basis of heredity is the "unconscious memory" of the organism, transmitted in the germ-plasm.

Very serious objections to the correctness, or at least to the generality, of the assumptions of Weismann's theory, have been made upon the basis of experimental observations. Some experiments tend to show that the importance of the chromatic matter of the nucleus as the bearer of heredity is exaggerated in Weismann's theory; that, in some cases at least, the cytoplasm plays a larger part than that merely of nutrition of the embryo. Experiments carried out by Delage on the eggs of sea-urchins have shown that it is possible to fertilize a fragment of the ovum which contains no nucleus, and that some measure of development may occur in this fertilized cytoplasm. Doubt has also been thrown upon the essential character of the effect of the sperm-cell in fertilizing the ovum. It was shown by Loeb that the eggs of the sea-urchin may be fertilized by adding to the water in which they live certain chemical substances. This artificial parthenogenesis suggests that, in conjugation, the essential part played by the sperm-cell is that of introducing some chemical substance which stimulates the ovum to activity. That part of Weismann's theory which has to do with the segregation of the determinants was supported by experiments carried out by Roux. He found that the first division of the fertilized ovum of a frog marked out the right and left halves of the body, the one blastomere developing into the right half of the embryo, and the other into the left half. Roux succeeded in destroying one of the blastomeres, and found that the other one developed into a half-embryo. This seemed to be decisive proof that, in the segmentation, there was a differentiation of the two blastomeres as bearers of heredity. But a later experiment of Driesch showed that if, after one of the blastomeres has been destroyed, the egg be turned upside down, the uninjured blastomere develops into a whole embryo, only smaller than if the whole egg had been allowed to develop. He showed also that, if the uninjured egg, when in the two-cell stage, be turned upside down, two whole embryos are developed. This and similar results obtained appear to show that, up to the two-cell stage, there can be no such segregation of the determinants as is indicated by Weismann's theory. Upon these experiments Driesch has based one of his proofs of the existence of entelechy.

In close connection with Weismann's theory of heredity, and with other theories of a more or less similar type, is the great controversial question as to the inheritance of acquired characters. Not only on account of its importance in Biological Science, especially in connection with theories of Evolution, but also on account of the consequences which any answer to it entail in relation to social questions reaching far beyond the narrower sphere of Biology, this question has given rise to acute and lasting controversy. By the term "acquired character," or "somatic modification," is to be understood any structural change in the body of a multi-cellular organism, of a kind which involves some change from the normal structure of the species to which the individual belongs, and which is acquired and remains permanent during the lifetime of the individual, and can be shown to be traceable to a change of environment, such as climate, or to functional use or disuse, such as is involved in specialized habits, or in mutilations. From the point of view of any theory of germ-plasm, the question whether a somatic modification of this kind is heritable or not is equivalent to the question whether such modification is accompanied by a specific change in the germ-cells, such that the offspring will inherit, in some degree or other, the modification which the parent acquired. It is quite clear that this question can only be answered empirically, after an abundant amount of observation and experiment specially directed to elucidate the matter. No answer can be accepted as decisive which depends upon a theoretical deduction from a special theory of heredity such as that of Weismann; on the contrary, a decisive answer to the question, obtained by means of fully sifted observations, would be a most crucial test of the value of such a theory as descriptive of the facts of heredity. In the empirical investigation of the matter, it must in the first instance be shown that what purports to be an inherited modification is really an acquired character, in accordance with a precise definition of the meaning of the term. The mere fact that a bodily peculiarity reappears in several generations is not sufficient proof that it has been inherited as such an acquired somatic modification; it may be a congenital germinal variation which has not been originally produced in the manner described. It must further be shown that the apparent recurrence of an acquired modification in a later generation is not connected with a recurrence of the environmental change, or of the specialized habits of life, which were the origin of the parental modification. Lastly, it must be shown that the somatic modification of the parent is not accompanied by a modification of the germ-cells. If such modification of the germ-cells of the parent can be shown to occur, as may happen for instance when the whole body, including the reproductive cells, is poisoned, there will be some effect produced upon the offspring, not necessarily owing to the somatic modification of the parent, but owing to the accompanying modification of the germ-cells. This last restriction naturally introduces an element of difficulty into the investigation, because it cannot be easily possible to ascertain the circumstances in which a somatic modification is accompanied by a change in the germ-cells. The negative answer given by Weismann and others to the question whether acquired somatic modifications are heritable depends upon the assumption that, at least in the case of the more ordinary somatic modifications, there is no corresponding specific modification in the germ-cells.

In accordance with the theory that the basis of inheritance is the germ-plasm which is separate, and remains segregated, from the somatic cells, everything turns upon the relations between the body-cells and the germ-cells. Weismann's view is that an acquired modification in general affects only the somatic cells, and has no influence on the germ-cells, at least in the direction of producing in them such specific modification that the acquired somatic modification becomes heritable. It should be observed that Weismann does not assert that the germ-cells remain absolutely unaffected by the modification in the body-cells, but only that they are not so specifically affected that the offspring will thereby exhibit the same modification that was acquired by the parent, or even a tendency to it. Some of the later developments of the theory of Weismann I shall have occasion to refer to in connection with theories of Evolution.

The theories of heredity to which I have referred all depend upon the procedure of employing facts obtained by observation and experiment to suggest a descriptive scheme, or as some would say a mechanism, which will enable us to trace out in detail the processes which lead up to the facts that are observed, and to use such descriptive scheme as an instrument for predicting occurrences, not yet observed, which would follow in appropriate circumstances, on the assumption that the descriptive scheme is sufficient for this purpose. The verification, or lack of verification, of the potency of the scheme in such predictions is then to be regarded as a test of the value and scope of the particular theory in question.

There exists however another method of procedure which may be described as the statistical method, that has been in recent years applied to matters relating to the study of heredity. The essence of this method consists in the ascertainment, by direct observation, of the frequency of occurrence of a certain character or group of characters in a large number of individuals of a particular species, compared with the occurrence of the same character or group of characters or related characters, or of certain conditions, in the parents, or in the remoter ancestors, of those individuals of the species in which they are found. The purely statistical facts obtained are then analysed by the mathematical methods of statistics with a view to the determination of correlations, the existence of at least some of which may have been previously unsuspected, and in particular of obtaining numerical estimates of the average frequency with which parental or ancestral characters reappear in the offspring. The application of this method does not require the employment of any theory as to the genetic modes in which the correlations between different generations of a race are set up; but the results of statistical theory may be used to assist in the setting up of such theories, or in the work of discriminating between alternative theories that have been suggested by workers who proceed by the non-statistical method.

The first person who seems to have fully grasped the possibility of applying the statistical method to problems of heredity, and to the general problem of variation in connection with evolutionary conceptions, was Francis Galton, who applied the method to a variety of questions in this order of ideas. The method has been followed up by a band of researchers, of whom the most prominent has been Professor K. Pearson, and their work has been incorporated in the department known as Biometrics; this department of research is represented by an important periodical appearing under the name *Biometrika*. The particular characters amenable to this method are often of a measurable character, and are thus capable of being correlated with numbers. In a given race, the average measure of a particular character of the kind considered has, for a very large number of individuals, at any one time an average value known as the mean value of the character for the race. It may happen that the mean value of the character may change from one generation to another, but in point of fact many characters preserve their mean value unchanged for many generations. When the mean value changes, it cannot be concluded that the character is necessarily heritable; the change may occur without any individual heredity in the particular character. Galton himself conducted a most careful statistical inquiry, partly by using the records of about a hundred and fifty families, relating to stature, colour of the eyes, some kinds of disease, and artistic faculty. It will be observed that the last two characters are not measurable in the same sense as the others, but their occurrence in different members of the families can be counted; moreover Galton applied his method, as exhibited in the last case, to mental as well as physical characters. He also made observations on characters in sweet-peas and moths, as well as various measurements in his anthropometric laboratory. One of the simpler problems which can be treated by this method is that of ascertaining whether a deviation of a particular character, such as stature, from its mean value in the race is heritable or not. On the basis of a large series of observations, it was shown by Galton, and more precisely by Pearson, that the average stature of a son is, reckoned in inches, 31·1 inches plus nine-twentieths of the number of inches in the stature of his father. This warrants the conclusion that deviation of stature from the mean of the race is heritable. Thus if a father differs in stature from the average of the race, the same is *on the average* true of his son. It also shows that, on the average, the deviation of stature of the son is in the same direction as that of the father but, on the average, smaller in amount. This last fact is a particular case of a generalization propounded by Galton, after careful statistical inquiry, and known as the law of filial regression. This law is most simply illustrated by the case of stature, although it has been verified in its general features in the case of other characters. If, for example, fathers whose height is 72 inches be taken, the mean height of their sons is 70·8 inches; these are still, on the average, taller than the average of the general population, but differ less from it than the fathers, and thus they regress towards the mean of the stock. A similar result holds in the case of fathers who have some height less than that of the average of the stock; their sons are, on the average, taller than the fathers, but less tall than the average of the stock; they have progressed towards the mean. Galton also showed by his statistical studies that the average of human stature is very constant from generation to generation, although there is statistical evidence that there is no correlation between the statures of people who marry. These facts Galton attempts to formulate by means of a principle of organic stability of the race, in accordance with which there is a stable type, or average of the particular character, which is preserved unchanged through successive generations. He does not however connect this principle of organic stability with natural selection, in virtue of which the stability of the average value of the character and the fact of regression towards it might be accounted for on the principle that the average value was more suited to the environment than a different average value would be. The explanation of this specific stability given by Galton is that each child inherits in part only from the parents, and in part from the more remote ancestors, and since what he calls the mid-parentage is, on the average, nearer than the exceptional parents to the mean for the race, the children of selected parents are, on the average, more mediocre than their parents. The fact that the mid-parentage is nearer to the average of the stock than the exceptional parents is supposed to be due to the fact that, when the ancestors are counted back for many generations, they consist of so many and such varied elements that they become, as regards their average characters, indistinguishable from the general population. The theory has been criticized by W. K. Brooks on the ground that, while the child is descended from a long line of ancestors, it inherits from none but the parents, and that it can only be said in a figurative sense to inherit from more remote ancestors. To estimate the force of this criticism, it should be observed that Galton establishes at most merely the bare fact that there is a correlation between the characters of the child and those of its more remote ancestors, and that when this has been done the power of his method, at least as regards this particular application, is exhausted; the method establishes no intermediate nexus between the remote ancestors and the child. Galton's opinion that the correlation is due to the organic stability of the race is in no sense an explanation of the correlation, but consists merely of the introduction of an expression to denote the ascertained fact of constancy of type through many generations. Only such a theory as that of Weismann's theory of the continuity of the germ-plasm is in a position to exhibit by a pictorial representation how the nexus between the child and its more remote ancestors is to be regarded as being through the parents; in that the germ-plasm of the parents contains elements derived from the ancestors. It would thus appear that Galton's ascertainment of the fact of correlation between the characters of the child and those of its more remote ancestors affords some confirmation of the descriptive potency of the theory of the continuity of germ-plasm. The criticism of Galton's theory seems then to be valid only as pointing out a necessary limitation in the scope of the method, so long as that method is considered, as it should be, only as a mode of establishing facts of correlation, and not as providing a description of any mechanism of the correlation. Criticism has also been directed against the view that the remote ancestors are so numerous that, in the bulk, they may be taken to be equivalent, for the purposes of discussion, to the general population. A man has for example theoretically 4096 twelfth grandparents, but it is argued that this number must, in practically all cases, be markedly diminished, since the descendants of the largest part of a more or less isolated population, as it exists and intermarries at any one time, die out after a few generations; it is thus argued that a tolerably isolated population must actually be descended from a relatively small number of individuals who existed and intermarried at some former period, and that, this being the case, the force of Galton's reasoning is much diminished. But this consideration would seem merely to point out that caution is required in pushing Galton's argument too far, rather than to provide a complete refutation. In any case the ascertained facts of correlation stand on a much firmer basis than any attempt to explain them theoretically.

Galton went further than merely to ascertain the fact that there is a correlation between the characters, such as the stature, of an individual and the corresponding characters in his ancestry. The law of ancestral inheritance, which he based upon his observations and statistics, includes a numerical estimate of the amount of correlation. The law is that:

> The two parents between them contribute *on the average* one half of each inherited faculty, each of them contributing one quarter of it. The four grandparents contribute one quarter, or each of them one sixteenth; and so on, the sum of the series $\frac{1}{2} + \frac{1}{4} + \frac{1}{8} + \frac{1}{16} + \ldots$ being equal to 1, as it should be. The pre-potencies or sub-potencies of particular ancestors, in any given pedigree, are eliminated by a law that deals only with *average* contributions, and the varying pre-potencies of sex in respect to different qualities are also presumably eliminated.

It must always be remembered that such a law is purely statistical, dealing only with averages in a large population; its correctness cannot be refuted by production of single or a few individual cases in which it can be shown not to represent the facts; the only refutation possible must itself be based upon statistics dealing with a very large number of individual cases.

A very different class of investigations relating to the laws of inheritance in hybrid varieties have been of great prominence in the biological work of the last two decades. These investigations are dependent upon the original discoveries of G. J. Mendel, Abbot of Brünn, who, after many years of experimental investigation, chiefly of varieties of peas, published them in 1866. The law known as Mendel's law was rediscovered by several botanists in 1900, and the attention of Biologists was drawn to Mendel's results by Mr Bateson, who extended them by a series of his own experiments. The theory of the Mendelian school is based upon statistical results relating to the descendants of hybrid varieties obtained by crossing two varieties of a species which differ from one another in respect of some particular character. Mendel found that, in investigating the effect of crossing two varieties, for example of peas, differences in certain specific characters could be separately investigated, and this gave rise to the conception that the difference between the two varieties is compounded of differences in respect of separate unit characters. When a variety with one particular character, such as tallness, is crossed with a variety which differs in respect of the corresponding character, such as dwarfness, in the offspring, one of the two characters, in particular tallness, is found to occur in all the offspring. That character, tallness, is said to be dominant, and the character which does not appear in the offspring is said to be recessive. The hybrids, all of which possess the dominant character, were then crossed with one another, and it was found that amongst the offspring both characters occurred, but the number with the dominant character (tallness) was three times the number of those with the recessive character (dwarfness). When the recessives are crossed with one another, they give rise only to recessives, and they breed true for any number of generations. But when the dominants fertilized themselves, they produced one-third pure dominants, and two-thirds cross-bred dominants. The former of these breed true, and the latter give rise to a mixture of dominants and recessives in the proportion of three to one.

The law of dominance, that is of the fact that one of the two characters is dominant, in that all the hybrids possess it, is the first part of Mendel's scheme. The second part is the law of segregation, which expresses in numerical form the facts, in relation to the successive generations, as to the segregation of the two characters in sets of individuals which thereafter breed true. Mendel suggested as an explanation of his experimental results a theory of the existence of two kinds of germ-cells which exist in all the hybrids in equal numbers—the theory of gametic segregation; and he made an hypothesis on this basis as to how the segregation of the characters is effected. This theory of gametic segregation and combination has been applied by Mendelians to represent and to predict, in accordance with the law of averages, the results of crossing varieties, and it has been extended to cases in which the characters considered are not distinguishable into dominant and recessive characters.

In certain cases the laws of Mendel have been verified, but in the long series of experiments which have been made in this order of investigations various complications have been introduced in the interpretation of the results. Important practical applications of the theory have been made, especially in the production of wheat which possesses the quality of immunity to the disease known as rust. Hopes have been entertained that investigations on Mendelian lines might throw light upon the general theory of evolution of species, but it appears to be doubtful whether there is at present much prospect that such hope will be fulfilled.

# XVIII

## THE EVOLUTION OF SPECIES

THE general notion of evolution, that in some sense the present state of things has arisen from earlier states by some kind of development, is as old as recorded thought; and traces of the application of the idea to the case of the evolution of living organisms, by a process of growth and modification of primordial germs, are to be found in the utterances of the earliest Greek thinkers. Even the idea of the survival of the fittest was indicated, in a crude form, by Empedocles (495–435 B.C.), who depicted the gradual origin, in accordance with abiogenesis, first of plants, and then of animals, through the chance play of combining and separating forces acting on the elements fire, water, earth, and air. After elimination of the earliest forms, fitter ones were, he thought, produced, but still fortuitously. Both the atomists, typified by Democritus, and their rivals, such as Anaxagoras, who entertained teleological conceptions, had some general ideas as to the occurrence of fitness or adaptation. With Aristotle, the notion of evolution attained to greater precision than with his predecessors. He followed Plato in regarding the creation of the cosmos as a process of descent from the more to the less perfect, but his dualistic interpretation of this process permitted him to hold a teleological theory of organic evolution. Although he regarded types as the realization of an original formative principle, and in this sense fixed, he appears to have entertained the possibility of the spontaneous generation of the lowest organisms. He regarded organs as fashioned by nature in the order of their necessity, those essential to life coming first. Aristotle clearly stated the conception of the survival of the fittest, but only to reject it. He definitely refused to admit that adaptation is due to the elimination of the unfit, but regarded the process of successive adaptation as due to an immanent principle striving to attain a certain end; and that end he believed to be the production of man.

During the centuries between the close of the Greek period and the Renaissance, under the influence of the Church, the theory of special creation of species, resting upon a literal interpretation of the Mosaic cosmogony, became dominant. But there were not wanting, among the Fathers of the Church, and later among the Scholastics, those who, like St Augustine, by accepting a less literal interpretation, combined it with the general conception of evolution. The following utterance[1] of St Augustine in favour of freedom of scientific thought is of much interest, as showing how far he was in advance of the later attitude of the Church for many centuries in this matter.

> It very often happens (he writes) that there is some question as to the earth or the sky, or the other elements of this world ...respecting which one who is not a Christian has knowledge derived from most certain reasoning or observation, and it is very disgraceful and mischievous and of all things to be carefully avoided, that a Christian speaking of such matters as being according to the Christian Scriptures, should be heard by an unbeliever talking such nonsense that the unbeliever perceiving him to be as wide from the mark as east from west, can hardly restrain himself from laughing.

Unfortunately this more liberal interpretation of the Mosaic cosmogony died out amongst theologians, and the doctrine of special creation became completely dominant until the nineteenth century. When it was discovered that many species had become extinct, and been replaced by new ones, the doctrine was extended to embrace a whole series of special creations of species.

[1] The passage is quoted by Osborn in his work *From the Greeks to Darwin*, p. 19.

In the period immediately after the Renaissance the attention of scientific workers was devoted rather to detailed work in Anatomy and Physiology than to the working out of so general a conception as that of organic evolution. It was not until the seventeenth century that, largely in the hands of Philosophers, the notion of organic evolution, as a speculative idea, again came into some prominence in the modern world. One of the first to suggest the transmutation of species by accumulated variations, and to advocate the experimental investigation of the subject, was Francis Bacon (1561–1626). Descartes (1596–1650), although he was much hampered in the expression of his opinions by what he regarded as the necessity of paying lip-service to the orthodox doctrine of special creation, clearly showed, in his *Traité de l'homme*, and in his essay *Sur les passions*, that he believed himself to have found an explanation of the universe, and in particular of the phenomena of life, on purely physical principles, and also of the mode in which the universe had been evolved. Leibniz, in his doctrines of monadism, and of continuity, developed the view that each monad is the focus of an endless process of evolution and involution; and that all natural orders of beings form a single chain along which progress is made continuously. Spinoza also expressed a belief in evolution in accordance with invariable laws. Immanuel Kant should also be mentioned as having, at least in his earlier period, advanced the conceptions of Selection, Adaptation, Environment, and Inheritance; but he appears later to have abandoned his evolutionary views, as unconfirmed by observation.

The first Naturalist who clearly expressed the idea that the unity of plan in the structure of animals may be due to community of origin was Buffon (1707–1788). His statements on the subject are vacillating, perhaps owing to difficulties which he felt in breaking with the orthodox conceptions, and he had no complete or consistent theory of evolution; but some of his utterances are quite explicit as to the possibility of the descent of all the species of animals from one stock. For example, in his account of the Ass, he argues that: "Once admit that within the bounds of a single family one species may originate from the type species by degeneration, then one might reasonably suppose that from a single being Nature could in time produce all the other organized beings." Elsewhere he speaks of the reduction of the two hundred species of quadrupeds which he described to a small number of families "from which it is not impossible that all the rest are derived." Within each of the families the species branches off from a parent species. Buffon anticipated various ideas, such as pangenesis, the struggle for existence, artificial and natural selection, and geographical isolation; all of which became later of fundamental importance in the theory of evolution. Being originally in accord with Linnaeus, his views gradually changed, but there were, in his statement of them, continual indications of indecision as regards the opposed conceptions of special creation and evolution.

The great systematist Linnaeus (1707–1778) reckoned at first "as many species as issued in pairs from the hands of the Creator," but later he modified his views so far as to admit that although "all the species of one genus constituted at first one species they were subsequently multiplied by hybrid generation, that is by intercrossing with other species."

It is remarkable that Cuvier (1769–1832), although he recognized very clearly a succession of forms in time, was a most resolute opponent of the evolutionary view of descent, mainly on account of lack of sufficient evidence for it. The course of the changes in his opinions was diametrically opposite to that of Buffon and Linnaeus. Starting with the most advanced views of Buffon as to mutability of species, he gradually arrived at a point such as that from which Linnaeus started, insisting on the fixity not only of species but also of varieties. Although evolutionary ideas had some influence over the minds of such men as Charles Bonnet (1720–1793), Robinet (1735–1820), Oken (1779–1851), and Geoffroy Saint-Hilaire, none of them developed a definite and comprehensive theory of the matter. Some of them, such as Bonnet and Oken, appear to have regarded evolution as a purely ideal conception of the development of types of structure, and not to be regarded as an historical account of the genesis of species. Others, such as Robinet, Treviranus, Tiedemann, and Meckel, held the view that it described an actual historical process. Under the influence of such ideas as that of the scale of beings, evolution was most frequently regarded as taking place through a linear series, as it were along a straight ladder, and not as figured by a genealogical tree, as it came to be by Lamarck and Darwin, and by all later evolutionists. The necessary preliminaries of a theory of evolution based upon scientific induction, and not merely upon philosophical speculation, consisted in the development of the studies of Embryology, Palaeontology, Comparative Anatomy, and Distribution. This preliminary work was carried out with much vigour in the eighteenth century, but not until the nineteenth was it sufficiently completed to provide a sufficiently solid basis of facts and ideas.

Jean Baptiste Pierre Antoine de Monet, known as the Chevalier de Lamarck (1744–1829), first a soldier, then a medical man, commenced his investigations as a Botanist, and afterwards passed on to Zoology. He did important work in systematic Botany and Zoology; and these studies led him to the conception of the mutability of species, and to the theory of the origin of species by descent. In this domain he may be regarded as the most important figure previous to Darwin; and certain of his opinions which differ from those of Darwin have not lost their importance in relation to the more recent discussions of the problems of evolution. In one of his earliest works, written in 1766, presented to the Paris Academy of Sciences in 1780, and published in 1794, he affirmed his belief in the immutability of species, and his disbelief in abiogenesis; he asserted that all organic individuals descend from other individuals entirely similar, which taken together constitute the whole species. His change of opinion did not become manifest until 1802, when he sketched out his evolution theory in his *Recherches sur l'organisation des corps vivants*. About the same time he proposed the term "Biology" for the sciences of life; it is remarkable that in the same year an identical proposal was made by Treviranus, also one of the pioneers of the theory of evolution, largely from the point of view of the Philosophy of Nature. The highest point of development of Lamarck's views was exhibited in his *Philosophie Zoologique*, published in 1809; in which his whole scheme of Evolution is gradually built up in logical form. His scheme includes the main ideas that species vary under changing external influences; that there is a fundamental unity in the animal kingdom; and that there exists a progressive and perfecting development. Lamarck's theory of evolution is closely allied in some points with the theory given somewhat earlier by Erasmus Darwin, the grandfather of Charles Darwin. The fundamental assumption made by Lamarck is that changes acquired by means of the functional reactions of an animal with its environment, and changes produced in a plant by means of changes in its environment, are transmissible by heredity to the offspring. He believed that changes in the environment produce no direct changes in the animal organism, but that the indirect changes are due to induced changes of habit, or of functional reaction, of the animal. The changed habits, involving the new or changed use of parts, result in structural changes which are transmissible to offspring, and thus form the main factor in evolution. In the case of plants, changes of soil and climate have a direct influence upon the structure; and the changes so produced are transmissible.

The essence of Lamarck's theory is contained in four laws; the law of growth; the law of functional reaction; the law of use and disuse; and the law of use-inheritance. In accordance with the *law of growth*, the size of any living body, and the dimensions of its parts, are increased by the body's own activities, up to some limit imposed by the nature of the living body itself. The *law of functional reaction* asserts that the production of a new organ results from a new need which continues to be felt, and from the new movements originated and sustained by this need. The *law of use and disuse* asserts that the development and power of action of organs are in proportion to the use made of those organs. The *law of use-inheritance* asserts that all that has been acquired, imprinted, or changed, in the organization of an individual during the course of its life is transmitted to the new individuals that descend from the individual so modified. The difficulty of understanding the exact nature of Lamarck's conceptions is much increased by the fact that, although he propounded the most thorough-going materialistic ideas, his view of life and evolution is expressed in psychological terms; and thus contradictions arise within his theory, owing to the irreconcilability of his materialism with his vitalism. His apparently materialistic conceptions are expressed in the statements:

No kind or particle of matter can have in itself the power of moving, living, feeling, thinking, nor of having ideas; and if, outside of man, we observe bodies endowed with all or one of these faculties, we ought to consider these faculties as physical phenomena which Nature has been able to produce, not by employing some particular kind of matter which itself possesses one or other of these faculties, but by the order and state of things which she has constituted in each organization and in each particular system of organs...Every animal faculty, of whatever nature it may be, is an organic phenomenon, and results from a system of organs or an organ-apparatus which gives rise to it and upon which it is necessarily dependent. . . . The more highly a faculty is developed the more complex is the system of organs which produces it and the higher the general organization; the more difficult also does it become to grasp its mechanism. But the faculty is none the less a phenomenon of organization, and for that reason purely physical.

Lamarck regarded evolution as a process of gradual complication of organisms, by which new organs and therefore new faculties arise. On the other hand he divided animals into three groups; insensitive animals; sensitive animals; and intelligent animals. The first group have no principle of reaction to external excitations, but passively prolong them into actions, in accordance with purely mechanical principles; they possess merely irritability. The second group have, in addition to irritability, a power which Lamarck calls the "sentiment intérieur." This is a power of reaction to external stimuli which involves the feeling of a need, and results in instant action; it is usually called instinct in animals, and does not imply either consciousness or will, but acts by transformation of external into internal excitations. The third group, consisting of vertebrates, have the faculties of intelligence and will, in addition to the irritability of the other two groups and the "sentiment intérieur" possessed by the second group. In Lamarck's opinion, intelligence and will have little or nothing to do with evolution. The "sentiment intérieur" produces not only instinctive actions, but also the formation of new organs as due to needs experienced by this inner feeling. It is not easy to apprehend the precise meaning which Lamarck attached to his first law, that of growth. The extension is not produced simply by nutrition, an immanent power is required to produce it; and this power he regards as dependent on a subtle fluid, somewhat in accordance with the ancient conception of a soul consisting of a very subtle form of matter. In his second law, that of functional reaction, the psychological conception of a felt need is the essential factor; and this it would appear cannot be reconciled with his materialistic conceptions, unless indeed it is capable of some other interpretation; for, although the need is not a conscious one, it is *felt* by the "sentiment intérieur." Moreover, at this point, a definite hiatus appears between the case of animals of the lowest group, and those of the two higher groups. For animals of the insensitive kind possess no "sentiment intérieur" which can experience needs; in their case, their behaviour, and consequently their evolution, depends upon the purely mechanical action of fluids set in motion by the direct physical action of the environment; whereas as Lamarck states "this is not the case with the more highly organized animals which possess feeling." As soon as a need is felt, the inner feeling directs the fluids and forces to the part of the body which can, by its action, satisfy the need. If the requisite organ exists it is stimulated to action; if no such organ exists, and the need is of a sustained character, the required organ is gradually produced and developed in accordance with the third law, that of use and disuse.

It is in the third and fourth laws, which admit of verification or refutation by observation, that the main present interest of Lamarck's theory lies. The third law asserts the priority of function to form, and the fourth law asserts the heritability of acquired characters, a proposition which has become in our day one of the most keenly debated questions. Great as was the work of Lamarck as a Naturalist, his reputation as a man of Science has suffered, owing to his unrestrained tendency towards speculation of a philosophical kind.

In the decades immediately preceding the new epoch which commenced with the publication of Darwin's *Origin of Species*, the idea of the derivation of species from one another, or from primordial forms, was familiar to Biologists, but no general theory of the evolution of species was developed. By such men as Cuvier and von Baer, the evidence in favour of these current notions of evolution was regarded as insufficient; although, in 1834, von Baer expressed his belief in a limited amount of evolution. The facts of variability and those of palaeontology led him to believe that many species have been evolved from parent stocks; but the absence of sufficient evidence led him to reject any comprehensive doctrine of descent from a primordial stock. Both in England and in France the fixity of species was the creed of the great majority of Biologists, including Richard Owen, who never accepted Darwinian views; they spent their energies in the detailed study of special departments of Biological Science, without taking much interest in any wide generalizations. Those of them who had read the works of Lamarck were for the most part unconvinced. The French Naturalist Geoffroy Saint-Hilaire, a colleague of Lamarck, who was by nature more a Philosopher than a Naturalist, belonged more to the Buffon-Lamarck school of thought than to that of Cuvier, the great opponent of the evolutionary idea, although Saint-Hilaire held views very different from those of Lamarck as to the principal factors in evolution, and was less radical in his views. Saint-Hilaire denied the inherited influences of habit, which Lamarck regarded as all-important, and considered that the transformations of organisms were due to the direct influence of the environment, the rôle of the organisms being comparatively passive. In the celebrated debate between him and Cuvier which was held in the Academy of Sciences in 1830, the superior knowledge of Cuvier gave the victory to the anti-evolutionist side; the basis of fact necessary to build up a solid theory of evolution by induction being at that time insufficient. It is interesting to remark that Goethe, who was himself a philosophical evolutionist, expressed much greater interest when he received the news of this discussion than he did as regards the news of the Paris revolution which arrived at the same time.

Robert Chambers published his well known work *Vestiges of the Natural History of Creation* in 1844. The evolutionary views which it contained led to its being badly received by reviewers, who charged the author with irreligious tendencies. Like Saint-Hilaire and Buffon, Chambers regarded modifications of animal structure as due to the direct action of the environment, in opposition to the opinion of Lamarck; but he combined with this the Aristotelian conception of a perfecting principle.

The state of opinion in the period just before the Darwinian has been described by Weismann in his *Descendenz-Theorie*; he writes[1]:

> It is impossible to estimate the effect of Darwin's book on the Origin of Species unless we fully realize how completely the biologists of that time had turned away from general problems. I can only say that we, who were then the younger men, studying in the fifties, had no idea that a theory of evolution had ever been put forward, for no one spoke of it to us, and it was never mentioned in a lecture. It seemed as if all the teachers in our Universities had drunk of the waters of Lethe, and had utterly forgotten that such a theory had ever been discussed, or as if they were ashamed of these philosophical flights on the part of Natural Science, and wished to guard their students from similar deviations.

It should however be observed that there were some exceptions to this attitude of indifference to general conceptions. In 1830 there appeared Lyell's *Principles of Geology*, in which the author rejected the notion of Cuvier that sudden catastrophes have been an important element in producing terrestrial changes. Various palaeontological discoveries, from the time of Cuvier, paved the way for the conception of evolution. From a philosophical point of view, Herbert Spencer, in 1852, advocated the transmutation theory. But neither the cell-theory nor the great advance of knowledge of

[1] See *The Evolution theory*, Vol. 1, pp. 27–28.

Embryology succeeded in overcoming the hostility of the Naturalists of the period to such ideas, and indeed to the discussion of any general conceptions.

The inception of the idea of theory of Descent in the mind of Darwin was due to the varied observations he made during the five years of the voyage of the *Beagle* to South America, commenced in 1831. He took with him Lyell's *Principles of Geology*, in which Lamarck's doctrines are fully discussed. He himself stated in later years that three classes of facts had brought the matter strongly before his mind; the manner in which closely allied species replace species in going southward; the close affinity of the species inhabiting the islands near South America to those proper to the continent; and the relation of certain living species to the extinct species. From 1835 onwards Darwin devoted his time to meditation on the theory and mode of transmutation, and to the collection and sifting of an enormous mass of facts, especially relating to domestic animals and cultivated plants, until then mostly ignored by scientific men. The idea of the survival of the fittest was first suggested to him by a study of Malthus' *Principles of Population.* Malthus had pointed out that a competition amongst individuals arises, in relation to the food supply, from the fact that man tends to increase in geometrical ratio, whilst the increase of the food supply is only in arithmetical ratio; thus bringing about the disappearance of individuals less suited than others to sustain the contest. Darwin saw that an extension of this idea might explain the adaptation of all living organisms to the environment. Until 1858, Darwin published no account of his new theory; he was determined to hold it back until the verification from fact should afford evidence of irresistible weight. In that year, Alfred Russel Wallace, who afterwards became his friend, had independently reached a similar theory, and the result of a communication by Wallace of his own manuscript embodying his views to Darwin was that two short papers were published in the *Journal of the Linnean Society*, on June 30, 1858; the first, by Darwin, consisted of an abstract of manuscripts written in 1839 and 1844: "On the variation of Organic Beings in a state of Nature; on the Natural Means of Selection; on the Comparison of Domestic Races and True Species"; together with a letter written in 1857 to Asa Gray. The second paper, by Wallace, consisted of an essay written in February 1858, "On the Tendency of Varieties to depart independently from the Original Type." There is much similarity, together with striking differences, in the views therein expressed by the two investigators.

The publication of the *Origin of Species* in 1859, followed in 1871 by the *Descent of Man*, had an effect upon Biological Science, and ultimately upon other departments of Science, such as Sociology and Anthropology, which can only be described as revolutionary. This was not due mainly to the novelty of Darwin's conceptions, for, as we have seen, not only had the idea of organic evolution been from the earliest times amongst the speculative conceptions of Philosophers, but it had also been developed in detail, especially by Lamarck and Erasmus Darwin. Even the special conception of Natural Selection as a factor in Evolution, although it arose quite independently in Darwin's mind, had been suggested by several earlier writers, as Darwin himself acknowledged. The vast influence of Darwin's work was due in the first place to the fact that he, for the first time, firmly established the *fact* of the transmutation of species, as an induction resting upon a vast accumulation of facts obtained by observation and experiments of very various kinds. In the second place he, for the first time, worked out the special theory of Natural Selection as a prime factor in Evolution, and adduced a great and varied array of facts illustrating the modes of its operation. The first of these elements in Darwin's work was ultimately decisive in its effects. After a period of bitter strife, the fact of organic evolution has been accepted by all Biologists, and by the educated part of the general public. After Darwin's work, it was no longer a speculative hypothesis, but a well attested deduction from observation. As regards the position of Natural Selection as the chief factor in Evolution, it is not possible to speak so positively. On this matter the opinions of Biologists have been, and still are, much divided; a fact to which I shall refer later in some detail.

The arguments employed by Darwin in support of his first main thesis, the relation of species to one another by descent, are drawn from various departments of Science; and he shows that they all point to the same conclusion. In Darwin's opinion the most conclusive evidence is drawn from the facts disclosed by Embryology. The resemblance between the embryos of various animals is much closer than the resemblance between the adults; the fact that the embryos of vertebrates, such as birds and snakes, are almost indistinguishable from one another at the earliest stage of their development; the similarity in embryos of homologous parts which become later on differentiated; all these point back to an ancestor common to a whole group of different animals. A strong confirmation is afforded by the survival of vestigial organs. From the morphological point of view, Darwin deduces that unity of type represents actual relationship between the species which possess that unity. From Palaeontology he produced cogent evidence dependent on the close similarity of the fossil remains of two successive geological formations. He shows that more highly organized forms of life have been developed successively and gradually from parents closely resembling them. The facts of geographical distribution of species and fauna form an important part of his argument. The fact that the faunae of regions in which the geographical and climatic conditions are not very dissimilar diverge widely from one another is taken to indicate the local development of those faunae. The various barriers to migration are to be regarded as an important factor; valuable indications being given by an examination of the fauna of islands. Thus, when an island is sufficiently far from a continent, it can be observed that certain genera of animals which exist on the continent are entirely lacking on the island. It will be observed that the theory of descent, or of such a relationship involving the transmutation of species, is a descriptive theory of the most general kind; a wide generalization starting from the perceptual fact of the relation between individual animals and plants and their offspring, rising to the conception that the existence of the most diverse species of animals and plants can be unified by universalizing this one kind of relationship. It takes the fact of heredity as the fundamental fact, but assumes, on the basis of induction from a great mass of observed facts, that the relation of similarity between parent and offspring is not so close but that it is consistent with extreme dissimilarity between individual organisms, which are to be conceived as connected with one another by a complex of relationships (not a single chain) of which the unit is the single relation which holds between parent and offspring. Subject to this restriction, the theory is independent of any special conception, such as that of germ-plasm, as to the details of the relation between parent and offspring; that is as to any mechanism by which this relation of heredity can be described. It is also independent of any theory of the origin, nature, or magnitude, of the variations, or dissimilarities, between parent and offspring, which, by their accumulation in the whole complex of which the relation between individuals of different species consists, amount to the actual dissimilarity between such individuals.

The second main thesis of Darwin is that the chief factor in Evolution is natural selection. Setting out from the existence of small heritable spontaneous variations in the individual organism, that is of small deviations from the average of the stock to which the individual belongs, the structural variations of a particular character being in the directions both of excess and defect, in accordance with the theory, those of the variations which give some advantage to the individual over his fellows in the struggle for existence lead to the preservation of the individuals possessing and propagating them, in preference to those who do not possess them. This selective process going on through many generations, in which the originally small variations are increased by accumulation, at every stage giving an advantage to those individuals in which the variations are in the profitable direction, leads ultimately to the survival of a stock which differs from the original stock in respect of the particular characters; thus a new stock more fitted to the environment than was the original stock comes into being. In Darwin's own words: "Natural Selection acts only by the preservation and accumulation of small inherited modifications, each profitable to the preserved being." It should be observed that the doctrine does not include any account of the mode in which the variations arise or are inherited; their existence as variations, of small but varying magnitude, in every direction, and both in excess and defect relatively to the mean, and heritable, is simply accepted as a datum derived from observation. These variations are frequently described as "fortuitous," or as "spontaneous," which can only be taken to express the fact that they are not assumed to be variations in any particular direction, or to be the outcome of any assigned process.

From a general point of view, this theory is remarkable on account of its anti-teleological tendency; for it is designed to account for what has *prima facie* the appearance of a process of progressive purposive adaptation to the environment; it substitutes an account of the mode in which such adaptation can be conceived to arise which does not involve, at any stage, any element which requires the notion of guidance towards an end, due either to conscious or unconscious agency. When such an expression as "struggle for existence" is employed in connection with this theory, the expression must be taken in a very broad sense. The struggle may last for many generations, and does not necessarily imply violent individual contests in which the less favoured individuals succumb. A slow racial elimination of the less fit, owing to their lesser ability than the more fit to cope with their fellows, with their racial enemies, to obtain sustenance, and to propagate their kind, must be included in the denotation of the expression. Natural selection provides, not only for the progressive adaptability to the environmental conditions, but for the transmutations of the adaptations when they have reached a maximum.

It is impossible, and indeed for my purpose unnecessary, to give any account of the wealth of the evidence which Darwin gave in support of his theory, or of the many illustrations and applications of it which he gave. Many objections to it were seen and dealt with by Darwin himself; others have been since advanced in the course of the ceaseless discussion of the factors in evolution which has taken place since the theory was advanced, and during which the attention of researchers has been largely devoted to the origin and mode of inheritance of variations, which remained entirely open in the original theory. In the course of his work, Darwin paid a great deal of attention to the artificial selection of varieties amongst domesticated plants and animals, and especially in the breeding of pigeons. In this case the purpose of the gardener or breeder plays an effective part in selecting for propagation those individual plants or animals which "spontaneously," according to a common expression, exhibit those variations which the gardener or breeder selects as according with his purpose. This is quite consistent with the anti-teleological character of Darwin's theory, for the purposiveness is not immanent in the plants and animals, but is in the mind of the gardener or breeder. The selection in the case of undomesticated animals and plants differs from this only in the fact that no gardener or breeder interferes in its free play.

Darwin believed that natural selection is sufficient to account for the evolution of the most complicated organs, but he at all times admitted the existence of other contributory factors of evolution, especially of what may be described as the Lamarckian factor, the dependence on inherited effects of use and disuse. Thus he writes[1]:

> It is generally acknowledged that all organic beings have been formed on two great laws—Unity of Type, and the Conditions of Existence. By unity of type is meant that fundamental agreement in structure which we see in organic beings of the same class, and which is quite independent of their habits of life. On any theory, unity of type is explained by unity of descent. The expression of conditions of existence, so often insisted on by the illustrious Cuvier, is fully embraced by the principle of natural selection. For natural selection acts by either now adapting the varying parts of each being to its organic and inorganic conditions of life; or by having adapted them during past periods of time; the adaptations being aided in many cases by the increased use or disuse of parts, being affected by the direct action of the external conditions of life, and subjected in all cases to the several laws of growth and variation. Hence, in fact, the law of the Conditions of Existence is the higher law; as it includes, through the inheritance of former variations and adaptations, that of Unity of Type.

It has been however pointed out by E. S. Russell that Cuvier meant by "Conditions of Existence" not environmental conditions, as Darwin assumed, but the coordination of parts to form the whole of an organism.

It has been held that Darwin gives little weight in his theory to, and does not account for, the correlation

[1] *Origin of Species*, 6th ed., Pop. Imp., pp. 260–1.

of variations in different parts of the organism. In the course of time Darwin somewhat modified his belief in the relative weight of the factor of natural selection as compared with the factor due to the direct action of the environment. Thus, in 1862, in a letter to Lyell, he writes: "I hardly know why I am a little sorry, but my present work is leading me to believe rather more in the direct action of physical conditions"; again in 1876, he wrote to Moritz Wagner: "When I wrote the *Origin*, and for some years afterwards, I could find little good evidence of the direct action of the environment; now there is a large body of evidence." In view of the great influence which Darwin's theories have exercised upon modern views of mental evolution, and upon Sociology, it is important to observe that Darwin recognizes psychological factors as contributory to organic evolution. In the *Descent of Man*, he advocated the view "that there is no fundamental difference between man and the higher mammals in their mental faculties." In this matter he differed from A. R. Wallace, who held that the higher mental faculties of man had been derived from a special stream of spiritual influence at some period in the evolution of the race.

Darwin had no theory of mental evolution, but took into account the fact that conscious experience, with concomitant physiological processes, occurs in connection with some kinds of animal behaviour. The close correlation of psychological and physiological processes he accepts without entering upon the difficult and highly controversial philosophical discussions which arise in this connection. If the theory of psycho-physical parallelism be accepted, a psychological description can be regarded as a merely convenient form in which the precisely corresponding organic processes in the brain and nervous system can be denoted. There is no evidence that Darwin accepted this theory in its complete sense, but he appears to have accepted, at least implicitly, the methodological hypothesis that mental evolution is correlated with organic evolution through heredity. Only by the assumption of this regulative idea can psychological factors be recognized as within the domain of Natural Science; without it no account of the behaviour of the higher animals would be in any degree adequate which was not based in large part upon a developed mental science as an independent department. The nature of the responses of an organism to stimuli, arising from the environment, depends upon the detailed structure of the organism; Darwin held that, in the main, natural selection was the mode in which the organism had been adapted through heredity to respond fittingly to such stimuli. Especially what is known as instinctive behaviour he regarded as the result of racial preparation, transmitted through organic heredity, principally in accordance with natural selection, but in many cases arising partly or wholly from the inheritance of modifications due to use and disuse. But Darwin recognized that instincts can be modified in the course of the individual life. Intelligent modification of behaviour Darwin regarded as due to the combination of incipient variation and acquired modification; under natural selection this combination has survival value. A factor of evolution which Darwin considered to have been of importance in certain cases is sexual selection; and this involves the *prima facie* recognition of a psychological element, since it includes the conception of the choice or preference of the females for males with certain characteristics, as the origin of secondary sexual characters other than weapons of offence and defence; these latter being of importance in this connection only in the struggle between males for the possession of the females. Darwin states as regards the female that "it is not probable that she consciously deliberates; but she is most excited or attracted by the most beautiful, or melodious, or gallant, males." This part of Darwin's theory has been the subject of much criticism on the part of later Naturalists.

The great cleavage of opinion as regards the relative importance in evolution of the two principles, that of natural selection, and the Lamarckian factor of the inheritance of acquired characters, has been represented in the time subsequent to Darwin by the two schools of Neo-Darwinians, for whom natural selection is the all-important factor in evolution, and the Neo-Lamarckians, who regard natural selection as of little importance, and lay the chief stress upon the inheritance of acquired characters. This divergence of opinion began in Darwin's own time; A. R. Wallace laying more exclusive stress than did Darwin himself upon Natural Selection, and Herbert Spencer being an advocate of the Lamarckian view. A large part of the work of Biologists since the time of Darwin has been devoted to elucidation of the mode in which variations may be conceived to arise in the organism, and of the conditions under which modifications of the individual organism become hereditary. The question has also arisen, whether the small modifications, the existence of which Darwin presupposed as the basis of his theory, are as important in relation to evolution as larger variations which arise as discontinuous variations, or mutations. The notion of the struggle for existence has been extended to embrace the conception of competition of tissues, cells, and smaller units, within the organism.

The notion that, in addition to the personal selection which depends upon the struggle between an individual and the other members of his race, and also against foes of different race from himself, and against the environment, there is also a struggle of parts and cells within the organism was introduced by Roux in 1881. This histonal struggle, or competition between the tissues for sustenance, may give rise to local modifications which produce acquired adaptations to the environment during the individual lifetime. The question whether such modifications are hereditary, and thus whether they have evolutionary value or not, falls under the general question of the heritability of acquired characters.

The chief inspiration of the Neo-Darwinian school has been derived from the teaching of Weismann, who rejected almost completely the Lamarckian idea of the inheritance of acquired character as a factor in evolution, on the grounds that there has been discovered no sufficient proof that such inheritance in fact takes place, and that, in accordance with his theory of the germ-plasm as the inherited substance, such inheritance is on theoretical grounds impossible. I have in the last lecture given a necessarily summary account of the detailed scheme which was gradually developed by Weismann for the purpose of representing, or as he would have said, of explaining, the facts of inheritance. The chief contribution which Weismann has made, based on his theory of germ-plasm, to evolutionary theory, is the introduction of the idea of germinal selection. This consists in an extension of the notion of natural selection to the sphere of the germ-plasm, with its representative units of ids, determinants, and biophors. This extension is employed to supplement and complete Darwin's theory, by providing a theory of the mode in which the hereditary variations may be conceived to arise. According to Weismann there is not only the struggle between organs, tissues, and cells, going on in the organism, as pointed out by Roux, but also a struggle between the determinants. When the determinants multiply by cleavage, the new determinants differ in size and in power of assimilation of nutrition. Those determinants which have greater power of assimilation become stronger than the others, and increase their superiority over the others in this respect. This struggle is resumed in every successive generation, as each generation receives its germ-plasm with its determinants from the preceding one. The result is that gradually the parts of the organism which are represented by the more vigorous determinants which are successful in the struggle become more strongly marked. This is held to explain the accumulation of small modifications in certain specific directions. The explanation of the fact that useful variations are always to be found, and are increased, is that the corresponding determinants are better nourished and their offspring are stronger than those of other determinants. Whether the original appearance of useful modifications, apart from their increase when already present, can be thus accounted for has been doubted by critics of the theory. Weismann also regards his theory as explaining the correlation of different parts of the organism in the product of adaptations; and also the degeneracy of useless organs he regards as due to the fact that the determinants corresponding to them obtain less nutriment than the others, and have consequently weaker offspring; this process of gradual weakening of the particular kind of determinants leads gradually in the course of many generations to the obliteration of the organ corresponding to them, or to its only remaining as a vestige. The fact of the reappearance of ancestral characters is explained by Weismann as due to a struggle between the various ids, determinants, and biophors, in the fertilized egg-cell. Since these are derived from the germ-plasm of both parents, and that germ-plasm in each case contains ancestral germ-plasm, there is occasion for a contest in which certain selected representative particles become effective in performing the function of forming the embryo, whilst others remain only in germ-plasm which is transmitted unchanged to the offspring. Weismann's theory of germinal selection is intended to explain why variations in a fixed direction take place, as well as why complex organs with many correlated parts appear, whilst these facts remain unexplained by Darwin's theory of selection in its original form. Although Weismann was strongly opposed to the Lamarckian view, he has admitted that in some cases nutritive and other environmental conditions may produce heritable modifications by direct action upon the germ-plasm within the body, but he regards direct action upon the somatic plasm as quite insufficient to produce such heritable variations.

Weismann's theory is a very complicated one which he gradually evolved in various stages; it has elements in common with other theories, but these elements he adapted for his particular purposes. Of all the theories which work with representative bodies in the germ-plasm it is the most complete, and it aims, whether successfully or not is a matter of controversy, at resuming a larger complex of facts than any other such theory. Like every theory dealing with evolution, or with vital phenomena in general, it ends up at a point in which the real difficulties of explanation of the facts of the existence and activities of living organisms are pushed back and concentrated on certain living elementary beings such as biophors, whose existence and activities as living beings are simply postulated without further analysis. This characteristic of such theories is precisely parallel to the analogous case of physical and chemical theories in which the phenomena of inorganic matter are made to depend upon properties assigned to postulated conceptions such as electrons, corpuscles, or atoms, in accordance with certain postulations. In both cases, when a theory has reached at least a provisionally definite form, the theory does not analyse further the ultimate postulations upon which it rests. Thus, in the rigorous sense of the term explanation, such a scientific theory is no nearer an explanation of the phenomena than at its starting point; and yet when the necessary character of this restriction, as inherent in Natural Science, has been grasped, it will be seen that this does not detract from the utility of the theory as a descriptive scheme.

During the time which has elapsed since Darwin's investigations led to the complete acceptance by the Scientific world of the fact of evolution of species, the scrutiny of the factors of evolution has led to an enormous amount of discussion and of detailed work of observation with a view to its elucidation. A large number of theories have been propounded, of which it is impossible for me to give any account. I must confine myself to an indication of the main features of a theory which recognizes the importance of a class of facts which were regarded by Darwin as of little or no importance in relation to evolution. Besides the originally small variations accumulating by slow changes, and consequently known as continuous variations, the rôle of which in Darwin's theory was a fundamental one, there exists another mode of variation which consists of a sudden or discontinuous variation. The characters of a species, or of a variety, sometimes undergo a sudden modification not due to the accumulation of continuous variations. These discontinuous variations, some cases of which were mentioned by Darwin himself, have since been the subject of attention by many naturalists who considered them to be of importance in relation to evolution. A developed theory of evolution regarded as dependent upon these discontinuous variations, or mutations, has been propounded by the Dutch Botanist De Vries, and rests upon the basis of a large number of experiments on transplanted wild plants and on various cultivated plants in a botanical garden at Amsterdam. For the slowly accumulated variations in the Darwinian theory, which may require an enormous amount of time to produce such changes as the evolution of species contemplates, De Vries proposes to substitute periodical, but sudden and quite noticeable steps. He observes that "this assumption only requires a limited number of mutative periods which might well occur within the time allowed by physicists and geologists for the existence of animal and vegetable life on the earth." This theory of periodical mutations is regarded as consonant not only with the fact that species or varieties change, but as reconciling that fact with the constancy of species for long periods of time, subject to individual fluctuations:

> Mutability (he writes) is not a permanent feature but a periodic phenomenon, producing at times new qualities and at other times leaving the plants unchanged during long successions of generations. All lines of the genealogic tree show alternating mutating and constant species. Some lines may be mutating at the present moment; others may momentarily be constant. ... In a complete and systematic enumeration of the real units of nature, the elementary species and varieties are thus observed to be discontinuous and separated by definite gaps. There is no reason to suppose that the world is reaching the end of its development, and so we are to infer that the production of new species and varieties is still going on. In reality, new forms are observed to originate from time to time, both wild and in cultivation, and such facts do not leave any doubt as to their origin from other allied types, and according to natural and general laws.

De Vries, in formulating his theory of the laws of mutation, maintains that new elementary species appear suddenly without intermediate steps, springing laterally from the main stem; and several such new species may arise from the parental form at once, in accordance with his experiments. The new elementary species attain their full constancy at once, and transmit their characters to their progeny independently of any external conditions. Some of the new strains are evidently elementary species, while others are to be considered as retrograde varieties. The term species is used by De Vries in a sense not identical with the Linnaean species; his elementary species is more nearly what is denoted by variety, in the ordinary classification. De Vries attributes an important part in evolution to natural selection, but he regards it as operative between species, and not between individuals of the same species. The external environment he regards as influencing this interspecific selection, and as also probably determining the appearance of a period of mutation, but otherwise of no influence. The origin of the mutations he regards as germinal, congenital, and depending on changes within the sexual cells.

Many cogent criticisms have been made of the theory of De Vries. The importance of the mutations he describes in evolution in general has been the subject of much controversy, but his experiments show clearly that the mutations play some part as a factor in evolution; how great that part is, or how general, remains for determination in the future.

# XIX

# NATURAL SCIENCE AND GENERAL THOUGHT

I HAVE now completed my survey of the methods and implications of Natural Science, as exhibited in its various departments. Necessarily incomplete as this survey has been, I venture to hope that it may serve as a basis for the formation of an estimate of any bearings which Natural Science, as a special domain of Thought, may have upon our more general views of the world, and especially upon the spiritual aspects of experience. In the earlier lectures of the course I stated and discussed in some detail the main conclusions, relating to the true characteristics and the limitations of scope of Natural Science, which I conceive to be borne out and illustrated in the course of the survey of special departments of Science which is made in the later lectures. It seems clear that the position of Natural Science in relation to the more general conceptions of existence which fall within the province of Philosophy and Religion will be largely affected by the nature of our conceptions of the basic characteristics of Natural Science, and of its scope. The nature of any ontological assumptions which may be regarded as necessary for the existence of Natural Science as a systematic scheme of thought, or which may be held to be implicit in that scheme, will be of the highest importance in this regard. The view which I have maintained and illustrated, that the life of Natural Science consists essentially, when it is rightly understood, of an organized attempt, or rather a series of efforts continued through the centuries, to provide conceptual representations of our physical percepts, will, if it be accepted, have a marked effect upon the external relations of this department of thought. This view tends to limit and circumscribe the influence which Natural Science will have upon the wider views of the world with which Philosophy and Religion concern themselves. If it be admitted that Natural Science, when reduced to its essential elements, is independent of any opinions which may be held as regards a reality behind phenomena, and if the notions of final causes and of efficiency be regarded as extraneous to it, it would seem to follow that the existence and the special results of Natural Science cannot be employed in any very direct manner for the purpose of throwing light upon the nature of an assumed reality, or of exercising any decisive influence in the contest between rival views as to the nature of reality.

By adopting phenomenalism as a methodological assumption, independently of any dogmatic assumption that it represents an ultimate philosophical view of existence on its physical side, Natural Science cuts itself off from the possibility of providing criteria which shall be logically effective in relation to metaphysical theories of the nature of the "real," or of "existence." This question of the character of any assumptions about reality which Natural Science may be thought to need is of practical importance because the fact that Natural Science has achieved successes of a kind and degree which cannot be ignored will lead to a presumption, of irresistible weight for many minds, in favour of any existential proposition which Natural Science can show to be really necessary as a foundation for the structure which it has reared. Such existential propositions, if accepted on the authority of Natural Science, and apparently evidenced by its successes, would have a far-reaching and vital effect upon all conceptions, philosophical, or religious, of the nature of reality. As an illustration, we may take the influence, outside

Science itself, of physical realism, frequently supposed to be a necessary assumption for Natural Science.

In one of my earlier lectures I have emphasized the advantages which accrue to Natural Science by taking up a position of complete independence in relation to Philosophical theories as to the origin of the perceptual complex with which Natural Science has to do. These advantages entail the restriction that Natural Science cannot exercise so important an influence in relation to our general attitude towards the world as has sometimes been assigned to it, especially by those who have attempted so to extend its scope that it was developed into a complete and dominant World Philosophy. If the view of the character and scope of Natural Science which I have maintained be correct, Natural Science is circumscribed in its aim and restricted by its method. No doubt the majority of men of Science in all ages have believed that their aim was to penetrate to what they regarded as the reality behind phenomena, and that successful scientific investigations might be expected to give them some detailed information as to the inner relations within that reality. Certainly many, and probably most, men of Science in recent times have been dominated by the conception known as physical realism, in accordance with which the material world not only comes under the category of the real, but is articulated in a manner corresponding closely to the distinctions and constructions which the mind makes, by abstraction and idealization, in the process of analysing and symbolizing its physical percepts. However, the reflections of many of those who have studied the general characteristics of scientific procedure and theories, some of them active investigators in some special department of Science, have led them to the conviction that the adoption of physical realism is an otiose opinion, which, whatever its merits as a philosophical theory may be, is essentially unnecessary as a presupposition of scientific method. I think there is evidence that this emancipatory movement is gaining ground, and is likely in the future to progress much further in the minds of those who find leisure to devote some thought to the underlying assumptions and implications relating to the methodology of Natural Science.

We possess other kinds of knowledge besides that with which Natural Science is *prima facie* alone concerned. There exists a body of conceptual knowledge of the mode of working of the human mind, represented by the results of Psychology and Logic. One of the functions of Psychology is to investigate the process of perception, and to trace out the relations between perception and conception, the bare recognition of which elements, in separation due to abstraction, is sufficient for the purposes of Natural Science, which does not utilize the light thrown by Psychology upon their relations within the fundamental unity of the mind. Psychological investigations appear to fall into two distinct departments. On the one hand, the subject is a purely psychical science in which the various functions and processes in the mind are analysed, and the relations between them investigated. In the other branch, on the border line between Mental Science and Physiology, the functions of the mind are studied by means of their physical manifestations, subject to assumptions as to the correlation of psychical and physical events and processes. On this latter side, Psychology must be regarded as a mixed Science, in the sense in which for the sake of distinction I have already employed the term. Pure logic, the study of the formal side of thought, may be regarded as closely connected with the more abstract side of Psychology. By analogy, Logic may be described as the Grammar of Thought.

Such a Science as Sociology, in its various departments, including Anthropology, Politics, and Economics, may be regarded as a mixed Science, depending, as these departments do, upon investigations and classifications which involve both physical and psychical categories. Psychology and the Sciences of the sociological group have one point in common with Natural Science. They are all concerned not with the individual as such, but with classes of individuals. To all of them, the individual is of concern only so far as it is a member of some more or less extensive class; and only those features of the individual which it has in common with all the members of such a class are relevant to these branches of knowledge. This transcendence of the purely individual is characteristic of all knowledge which, in the widest sense of the term, can be designated as scientific knowledge. Of an individual object, taking the term object in a wide sense, we may have knowledge which, when sufficiently extensive, always differentiates that object from all others, and may be considered collectively as knowledge of the history of that individual object. The history of a particular object, including its present characteristics, is unique, so far as elements in that history are not referred to a system of classification in which the individual is for a specific purpose merged in a class. That history is never identical in all respects with the history of any other object. So far as the individual object is unique, our knowledge of it fails to be subsumed under scientific knowledge, in however general a sense we understand that term; it remains as historical, not as scientific, knowledge. That element in the genesis of an object or complex, in its relations during its existence with its environment, and in its past and present states, which eludes all scientific classification and subsumption under scientific laws and theories, may vary enormously in importance and amount in different cases. No two stones are ever found, upon close examination, to be absolutely indistinguishable from one another in respect of shape, size, and structure. The stories of the past buffetings of the two stones, if it were possible or worth while to obtain knowledge of them, would be full of details unlike in the two cases; this divergence being held to account for their present dissimilarities. The scientific interest in the stones is usually confined to an interest in the heap to which they belong, the individuals being regarded only as samples of that heap; and their irresoluble individual differences are for scientific purposes neglected, unless some specially distinctive features are observed in them which may give rise to some additional scientific inquiry.

When we turn to the case of living organisms, and especially to the case of a particular man, as an historical being, the same considerations hold good, with greatly increased force and import. At each juncture in the life of a man, physical and psychical Science can be applied to give some partial accounts of the physiological processes in his body and nervous system, of his reactions to external stimuli, and of the psychical processes in his mind which are related to the determinations of his will; but these accounts are never complete in all details, and they cannot be welded together into a single coherent whole. Not only his past history, but also the individual peculiarities which distinguish him from every other being of his kind, have some greater or less significance at every juncture of his life. Neither his history nor his character are identical with those of any other being. I have emphasized this fact of the ubiquity of the element of individuality in the perceptual world, because it is one of which the bias of the man of Science, who for his special purpose does not, and cannot, concern himself with the purely individual, except so far as it is assumed to be an instance of the general, tends to minimize the importance in relation to a general view of the world. For the struggles of the man of Science are in the direction of a constant attempt to diminish the importance of the purely individual element, by showing that it is but an instance of the general, and it is in doing this that his successes consist. There remains however in the world the irresoluble element of individuality, of the complete removal of which there is no prospect. The principle of order in the world requires to be supplemented, if not limited, by the principle of individuality. The fact that the history and characteristics of a human being are unique does not however make it impossible for other persons to have a knowledge of that individual, sufficient for the purpose of predicting, with some greater or less degree of assurance and precision, what the actions of that individual will be in given circumstances. This implies the assumption that the determinations of his will are not so wholly irregular and incalculable but that his conduct in a given situation is predictable in some considerable degree. In other words, it is possible to have a knowledge of his character based upon knowledge of his past history, and subject to the assumption that his character has a functional relation with the elements of his past behaviour. The difficult questions how far what is called self-determination goes, and whether indeed it is an ultimately valid conception, including the question of the validity of what is known as psychological determinism, lie quite outside the scope of these lectures, and accordingly cannot be here discussed. The kind of knowledge that we may have of a particular man is essentially knowledge of an individual, not of a class, although a very considerable part of it has no doubt reference to our knowledge of classes to which the individual belongs, and so far is equivalent in kind to scientific knowledge.

Science has however made much use in quite recent times of one method of eliminating the effects of what I have called the principle of individuality. That method is the statistical, by means of which valuable scientific knowledge may be obtained as to the behaviour of groups of individuals; it consists of a process of elimination of the effects of the purely individual characteristics of the individuals of which the groups are composed. By this method, upon the basis of measured facts of observation relating to a large group of individuals, tabulated in numerical form, it is possible to correlate the behaviour of the members of the group with certain kinds of motives, not necessarily conscious, relating to that behaviour, in such wise that, by statistical analysis, a knowledge can be obtained of the *average* effect of the motive upon the members of the group. I have used the term "motive" in a very general sense, to denote any circumstance or set of circumstances which can be correlated with a particular kind of behaviour. Sometimes the statistical method has been applied to detect the existence of such correlation, without previous assumption that a particular set of circumstances was a relevant motive. This method is of special value in the case of a motive of which the intensity is variable; it may then be shown by statistical analysis that the average behaviour, of the particular kind, of the members of a group has a functional relationship with the intensity of the motive. By this method a considerable amount of knowledge may be obtained of how a large group of individuals will behave in certain circumstances, and especially as to the variability of behaviour when those circumstances change in a known manner. If, for example, it has been shown that there is, in a given community, and for a considerable period, a correlation between the number of marriages in a year and the average price of corn in that year, a definite piece of information has been obtained, depending for its establishment upon a process of elimination of most of the many individual characteristics of members of the community which, in any single instance, will be factors in deciding for or against marriage. The effect of all other motives has been eliminated in the statement of the fact of correlation between the number of marriages and the one particular kind of circumstance, the price of corn. A large amount of knowledge of partial correlations, obtained by this method, may afford scientific information about the average behaviour of the members of a community; and this knowledge may have value as a means of forecasting the future. But its value, as enabling us to predict the behaviour of a particular individual, in assigned circumstances, is evanescent, because no particular person is the average individual; neither has he in all respects a close resemblance to the average. For predictions of any value as to the behaviour of a particular person recourse must be had to the purely individual knowledge of those who have sufficient acquaintance with his past history, and his present character and circumstances, to enable them to form an estimate; and such individual knowledge is not in the main scientific knowledge.

Both scientific knowledge and the individual knowledge of which I have spoken have in common the fact that they are of the discursive kind, obtained by an analysis and subsequent synthesis of particular elements. The two are distinguished from one another by the large divergence in the degrees of systematization which they involve; the former is obtained by systematic schematization; the latter by a process of unsystematic synthesis, although it may contain elements in which systematic schematization is not entirely absent. Both these kinds of knowledge are abstract, in very different degrees. But besides these kinds of knowledge, there exists a kind of apprehension which is more immediate and direct, although it is often inextricably combined with knowledge of the other kinds. This is knowledge as given by direct intuition, in which the object in the subject-object relation is apparently apprehended all at once, as a whole, and not by a conscious synthesis of its parts and their relations. This intuitive knowledge, of which the highest example is to be found in the apprehension, in some moments, of the artist, or of the mystic, is transitory and fleeting; for a process of analysis into parts and relations, and resolution into discursive knowledge, is incipient in it. In its purity it is incommunicable; as soon as an attempt is made to describe it in language, the stage of abstraction has already been reached, and the description fails to represent with absolute completeness the unique individuality of the whole, as given by the original intuitive apprehension. In the individual who has been the subject in the intuition, the impression of the whole, as grasped at once intuitively, may remain more or less vividly in the memory, after the inevitable process of dissection has commenced, but only that process gives him the power to communicate what he has experienced to others, and in the act of communication the object, as a unique indivisible whole, becomes merged in systems of classification in which those aspects of it which constitute its uniqueness are largely lost. An exceptional power of obtaining an intuitional grasp of a complex as a whole is an essential element in the mental outfit of a man of Science of the highest order. Such intuitional and imaginative apprehension precedes and conditions any striking success in the process of discovery of the inner relations within the complex.

That element of human experience and life which may be summed up in the word cognition includes common knowledge, which becomes in its developed form scientific knowledge, and also what I have spoken of as individual knowledge and immediate or intuitional knowledge. But when we have taken all these, not completely separable, kinds of knowledge into account, we have still to remember the fact that cognition, however generally the term is understood, is but one among other elements which make up the whole of human experience and life. It would not be necessary for my purpose to give a rigorous analysis of the elements of human experience such as might satisfy the psychologist or philosopher, even if I were competent to attempt such a task. It is sufficient to refer to the elements of feeling and desire which include the fundamental springs of will and activity. These take explicit form in the apprehension of values, spiritual, moral, intellectual, aesthetic, and material. In actual experience, these elements, with the cognitive element, enter as components separable from one another only by abstraction; the fundamental unity of the stream of experience is such that it is not merely a sum of such elements. The qualitative difference between various kinds of experience may be to some extent represented by differences as regards prominence or intensity in these abstractly separable elements.

The main characteristics of the great departments of thought and activity, Religion, Philosophy, Science, and Art, rest upon distinctions in the emphasis which they place, in their aims and procedure, upon the various elements of mental experience to which I have alluded. It may be said that the interests of Philosophy and of Pure Science are cognitive, or as is often said, intellectual; that the object of Philosophy is to obtain systematic knowledge and understanding of experience as a whole, and that the object of Science is to represent certain kinds of experience conceptually. It may also be said that the interests of Religion and of Art are in the main connected with values, spiritual and moral values in the one case, and aesthetic values in the other. But such statements, however correct they may be, if they are understood as applying to ultimate aims and results, cannot be applied without qualification to the processes and activities by means of which the aims are realized and the results obtained. That the aims of Philosophy and of Science are to attain to truth, independently of the specific character of the valuations of that truth when obtained, is doubtless correct; their direct concern is with cognition, and not with valuation. But for a Philosopher or a man of Science, truth is itself a value of the highest kind, even if the truth contain unpleasant features; a recognition of its immediate ideal value, or in some cases of its mediate value as a means for the attainment of practical ends, is an essential spring of action in the mind of the genuine Philosopher or man of Science. The sustained emotion which we call the love of truth as a value is essential to the pursuit of Philosophical and of Scientific knowledge. At every stage in the age-long struggle to reach philosophical or scientific truth the combatants have been animated and sustained by a consciousness of the value of their goal. It is however true that the feeling for values, or rather for specific values, is one which has to be kept in severe restraint by the investigator, in subordination to the cognitive side of his mind, for otherwise it may distort his vision in a manner which may be very detrimental to the attainment of his aims.

That, for the domains of religion and morality, the apprehension of values is the fundamental factor can hardly be denied. In relation to religion this has been formulated by Höffding in the thesis that the fundamental axiom of religion is the conservation of values. But it must nevertheless be recognized that, to found religion exclusively on the basis of feelings involving apprehension of specific values and their conservation, and without any elements of cognition, that is of knowledge, actual or speculative, is an impossibility. Conservation of values implies the existence of forms and modes in which they are conserved; and this fact necessarily brings values and their conservation into connection with conceptions of existence and reality. All valuations arise originally in close connection with ordinary experience of the physical and psychical world; and even in their most clarified form, as they appear in the religious and moral consciousness, they are dependent for their imagery, their concrete expression, and their concepts, upon elements derived from that experience.

Thus the domains of value and of existence are ultimately inseparable. All historical religions have been in close connection with the cognitive side of human experience, and in some periods they have exercised an almost complete domination over that side. In the middle ages Religion represented the unity in which all conceptions not only of the spiritual world, but also of the physical world, were embraced; in which Art and in fact all forms of spiritual activity were contained, and by which all their activities were regulated. All values and all conceptions of existence and reality were bound together in a unified system which, only at the time of the Renaissance, began to dissolve, when Science, Art, and Philosophy set up claims to independence and development on their own lines, unfettered by restrictions imposed upon them by the supposed necessities of the unified system in which they had previously been combined. The advent of this separation of different forms of spiritual life and activity led to the state of tension, manifested in the contest, lasting into our own time, between the conceptions of Natural Science and those older conceptions of the physical world which were for a long time considered to be essentially connected with Religion. The complete cessation of this contest is dependent upon a full recognition of the validity of the claims both of Religion and of Science to independent life and activity, each in its own sphere; and this requires as its condition a more precise formulation of the boundaries of their respective spheres than had formerly been made.

A somewhat similar remark may be made as regards the relations between Religion and Philosophy, but, as I venture to think, the difficulties in this case are essentially greater than in the case of the relations between Religion and Natural Science. Philosophy and Religion are both of them, in different senses, concerned with the whole of human experience. The interests of Philosophy are mainly intellectual, being concerned with the cognitive side of experience; those of Religion are more concerned with the practical side of experience, with the spiritual needs of man, with his hopes and fears, his moral values, and his emotional attitude towards existence. Philosophy, in its eclectic view of human experience, does not, and cannot, leave out of account the nature and specific character of values, and especially of that species of values which we call moral; and here it has a specially close plane of contact with Religion. The fact, to which I have already alluded, that Religion, which is primarily concerned with values, cannot wholly divorce itself from conceptions of existents in which those values are realized, brings it also in this regard into close contact with Philosophical Thought. In point of fact, Religion has at all times found itself impelled, at all events in the minds of more cultivated persons, to ally itself with some form of Philosophy, in which it might find the existential elements which might warrant an ascription of objectivity and of permanence to the carriers of its primary elements of value. The difficulty of combining religious conceptions with a coherent Philosophical scheme has been increased by the historical fact, to which Professor Sorley has recently drawn attention in an emphatic manner, that, in the development of Philosophical systems, the cognitive side of experience has formed almost exclusively the material out of which the systems have been built up; these systems have then been under the necessity of adapting themselves *ex post facto* to include that element of experience which we call the apprehension of values. Although it has been recognized, at least in some quarters, that Religion does not supply a World-Philosophy, and that it should not be expected to fulfil that function, and further that it is not in essence bound up with any philosophical theory of the nature of existence, it remains nevertheless true that, in so far as Religion seeks to relate its concepts of spiritual value with existential forms in which those values are realized, it is impossible to treat Religion as completely independent of philosophical implications. Nearly all actual forms of Religion have been theistic, whether monotheistic or polytheistic, and even Buddhism which, in its esoteric form, is regarded as an exception, is not solely a Religion, but is also a most pronounced form of World-Philosophy. All forms of theism involve ideas, differing widely in their nature, mode of formulation, and degree of precision, which are concerned with the relations between God and the world. Thus the assertion of the existence of God does not simply and solely express a religious belief, or an attitude of faith, but takes also the form of a philosophical proposition, or hypothesis, of an ontological character. This brings with it a series of philosophical questions, some of which at least are of so urgent a character, not merely from a purely cognitive point of view, but in relation to the needs of the religious consciousness, that some answers to them, more or less precise, are an imperative necessity. This is so, not merely or mainly on account of the inherent philosophical importance of these questions, but because the whole character of Religion, and its efficiency in providing for the primary spiritual needs which Religion is to satisfy, are very fundamentally affected by the nature of the answers given to those questions.

Theism has been the fundamental basis of the most multifarious forms of Religion; and these have exhibited, in different ages and in different countries, and even among different groups of persons in one and the same age and country, the most diverse ideas as to the conception of God, and of the place which theism should fill in a general view of existence and life. In fact the particular kind of theism in a religious system is of most fundamental importance in relation to the actual functioning of the creed, as an expression of the spiritual character of its adherents, and as influencing all their activities. There exist, and have always existed, many persons who, whilst dominated in their innermost being by theistic belief, do not feel any need for the kind of support which may be afforded to their belief by philosophical conceptions. Many others, whilst employing such conceptions as an adjunct to their belief, do not feel the need of evidential support of their faith from the side of reasoned Philosophy. To such persons, their theistic belief appears to them in the form of direct intuitive knowledge, obtained in the course of actual experience, and having the character of a certainty which can dispense with any proofs of the sort which reasoned Philosophy can be expected to furnish. Such persons may employ some kind of philosophical or rational scheme, more or less consciously, as a kind of framework into which their faith, resting as it does on intuitive knowledge, may be fitted. This framework is usually supplied by the traditional conceptions of the society in which they live, or of the particular group in which they have been educated, or by which they have been most effectively influenced. The most highly developed form of this attitude of mind towards theistic belief is to be found amongst the mystics of all periods, many of whom may be regarded as specialists in relation to religious experience of the directly intuitional kind. The accounts of this kind of experience given in such a work as William James' *Varieties of Religious Experience* are of extreme interest. It is there shown that mystical phenomena are to be found amongst the adherents of the most diverse creeds, and yet present in all such cases an essential similarity of general character. The accounts given of experiences of this order, by those who have been their subjects, take forms which are coloured in a high degree by their preconceived views; and this appears to indicate the presence of a large element of subjectivity in the interpretation of the actual experience. It is accordingly difficult to assess highly the value of such experience as evidence of the truth of any special cognitive views which may be held by persons who have such experience, and who interpret it in terms of their personal and traditional beliefs. It seems clear that all attempts that may be made to base theistic, or more general, religious belief upon intuitional knowledge obtained directly in actual experience, either of the mystical order or of a more ordinary kind, will ultimately prove insufficient as a basis for such belief amongst a very large number, and probably the majority, of men. The history of religion shows that this is the case. Belief based upon direct intuition will, for those who have it, remain unaffected by discursive thought in relation to Philosophy and Science; and it is not for the sake of such persons that it is necessary to treat of the relations which theistic belief may have with philosophical or scientific views. For the more thoughtful members of the community, these relations have in our time an inestimable importance; and the influence of the views formed of their specific character has an ultimate effect upon the attitude towards theistic belief of multitudes of men who do not consciously concern themselves with such relations on the more theoretical side. Accordingly, the nature and extent of any influence which Natural Science exerts, or ought to exert, upon theistic belief, both in its general and its more specific characters, presents a problem, the importance of which can hardly be overestimated, in view of the effect which solutions of it may have, directly upon the cognitive side, and indirectly upon other sides, of the religious consciousness.

Speaking broadly, the main concern of Religion is with values, and with existence as embodying values, whilst Natural Science, in its results, has no concern with values. On the other hand, Philosophy is concerned with cognition related both to values and to existence. It would then appear that the relation of Natural Science with Theism, so far as such relation exists, is in a sense indirect, as it is through the mediation of Philosophy that it becomes effective. The influence which Natural Science may have upon Theism may be taken to depend, first upon any assumptions of a metaphysical kind which may be held to be necessary for Natural Science to make for its own purposes; and secondly upon the amount and nature of the support which the existence and success of Natural Science may afford to particular theories of a philosophical or metaphysical character. As regards the influence of the first kind, the acceptance of that view of the essential character of Natural Science which I have advocated, and which view I have attempted to establish as correct, leads I think to the inference that, subject to one limiting condition, of which I shall speak presently, Natural Science, taken by itself, does not directly affect the theistic position, either positively by providing support, or negatively by giving rise to objections. In fact, if no philosophical assumptions are made which lie outside the necessities of Natural Science, the position of Natural Science in relation to theism, as in relation to ontological theses generally, is one of neutrality or independence. It is important to emphasize the fact that this position of independence only appertains to Natural Science when all conceptions not strictly necessary for its existence as a schematization of physical percepts are excluded. When Natural Science is taken in combination with metaphysical views which are, in accordance with the opinion here adopted, extraneous to it, its position in relation to theistic, or other ultimate, ideas relating to the nature of reality becomes very different, and its influence upon such conceptions may become of great importance.

It may perhaps seem to be the case that the position of independence here assigned to Natural Science is too absolute; that such a position of apparent isolation in the more general domain of thought and existence is untenable. It may be objected that the fact that it has been found possible to develop such a system as Natural Science, which undoubtedly possesses a certain kind and amount of efficiency in relation to life and activity, must lead to some inferences which show their effect in introducing limitations upon the characteristics of a general philosophy of the world, and in particular upon any theistic, or other, mode in which ultimate reality is conceived. Such an objection would be valid, were it not for one important limitation which the existence of Natural Science, as we know and possess it, places upon any theistic, or other, view of the world. Stated shortly, that limitation consists in the fact that any acceptable view of the world, whether theistic or other, must be such as not to be incompatible with the existence of Natural Science. Any such general Philosophy must provide within itself a place which Natural Science may occupy as an autonomous system. The principle of order in physical phenomena, with the limiting principle of individuality in these phenomena, provide principles with which any Philosophy and any theistic view must not be incompatible, if direct contest between Natural Science and Philosophical or Theistic thought is not to arise. Subject to completely adequate satisfaction of this condition, Natural Science offers no obstacle to the free development of Theistic, or other, Philosophies on their own lines, in the sense that no other purely logical consequences follow from the acceptance of Natural Science, which are effective in the wider domains of Thought. I have already observed that any further influence which Natural Science may have upon general views of the world depends upon the nature of any ontological hypotheses or postulations which may be made in attempts to explore the nature of reality, but which go beyond any assumptions which Natural Science itself needs.

Before discussing the nature of the bearings which

Natural Science may have upon theories of reality, and in particular upon theistic theories, when it is supplemented by ontological postulations which go beyond its own special requirements, I propose to make a few remarks upon what I take to be an essential characteristic of all Philosophical systems. Metaphysical Philosophy is not constructed upon a basis deduced by the canons of pure Logic from any set of axioms, assumptions, or presuppositions, which are accepted by all human minds as possessing self-evidence, or even definiteness of meaning. To the constructor of, and to the adherents of, a particular kind of Philosophy, the assumptions and presuppositions of the system, when indeed they are all explicitly recognized, usually appear to possess apodictic certainty, to be woven in the very web of the mind. To the adherents of a rival system, and to critics in general, these same assumptions or presuppositions may appear to lack this quality of self-evidence, to be only probable in some degree, or in no degree, and often to be either meaningless, or to have only a hazy and ill-defined meaning. In this region of thought, what seems obvious and certain to some minds, appears to be neither obvious nor true, at least without much qualification, to other minds, at the same epoch, or in a different age. If Philosophers were not only logic-machines working in accordance with a single logical canon, but also in possession of a single universally accepted set of premisses or postulations to be employed in the logical processes, the state of philosophical thought would be very different from what it actually is, and always has been. We might then at least hope to attain to a Philosophy which would receive general assent from all those persons who made a sufficient amount of effort, and were possessed of sufficient mental grasp, to enable them to understand it; but this is very far from being the case. The presuppositions which commend themselves to different minds diverge in the widest manner from one another. These presuppositions which serve as premisses, necessary before deductive logic can function, depend upon the widely varying characteristics of particular minds and groups of minds. The causes which make a particular premiss appear to a particular mind, or to a particular class of minds, to have either irresistible cogency, or at least a high degree of probability, are not in their nature purely intellectual. They depend upon the education, traditions, and idiosyncrasies of such individuals or classes; upon irreducible peculiarities of particular minds. The reaction of the individual, as an inseparable whole, to his experience, is involved in their selection. In this whole, understanding, feelings, desires, and tradition, are only in abstraction separable from one another. It is the undifferentiated individuality that is really effective in determining the choice of the axioms and premisses which commend themselves to the individual as the fundamental elements in his Philosophy. The reasoning faculty we distinguish as a separate faculty by abstraction only; its function is coordinative, and it is operative upon data which it does not, and cannot, alone originate.

When an explanation is offered of the fact that there has always existed a large degree of divergence between different philosophical views and creeds, emphasis is often laid upon misunderstandings as to the terminology in which philosophical ideas are expressed. No doubt this factor is a real one in this connection; as language was originally developed chiefly to serve much more purely practical purposes than to provide the means of expressing subtle philosophical distinctions. The process of adaptation of language for the latter purpose is certainly far from complete, and leaves ample scope for shades of variation in the precise meaning that can be attached to the terms employed; and this naturally leads to controversies in which these differences in the interpretation of language are far from negligible. Nevertheless, the reason I have before given to account for the extreme divergences in philosophical opinions is, I think, more fundamental than what is due to the imperfect functioning of language. The particular axioms and postulations, performing the function of premisses, of a particular system of Philosophy are the true characteristics of that system; they form the essential element which distinguishes it from rival views. It is in differences of these characteristic premisses that the everlasting divergence of philosophical views which flourish in the same age, or have been prevalent in different ages, is to be found.

An important consequence of the a-logical character of such a premiss, that is of the fact that it is not a purely logical deduction from other premisses of which the validity is admitted by all normal minds, is that an absolute denial of its truth cannot be refuted by any process of reasoning which rests upon a basis recognized by all men, or even by all Philosophers. Every such presupposition rests in reality upon a judgment of probability, ranging from faith of various degrees up to moral certainty. Some presuppositions of this class are of such a character that they do actually receive the assent of all, or nearly all, men. An example of this is the belief in the existence of other selves besides our own selves. If however a person chooses to take up the position of a solipsist, it would appear that his view cannot be refuted by any process which would show that his opinion leads to logical contradiction. His view appears to us however absurd, and almost insane, because we all have an irresistible belief, that seems to us consonant with our whole experience, that his view is wrong. Although I have described scientific knowledge as essentially public knowledge, it is not necessarily absolutely impossible for a solipsist, if there be such a person, to set up an account of the world, regarded exclusively as a series of psychical happenings in his own consciousness. As regards other ontological hypotheses, there is no such practical unanimity of opinion as in the case of the hypothesis of the existence of other persons with a psychical being resembling our own; and this lack of unanimity exhibits itself in the presence of all the manifold realistic and idealistic species of Philosophy with which the history of Philosophy is conversant.

I have emphasized the fact that the influence of Natural Science upon our general views of the world depends upon the nature of ontological postulations which go beyond any assumptions which Natural Science needs for its own purposes. The complex of physical percepts, taken as an appearance, into the ground of which Natural Science does not need to inquire, cannot be regarded simply as a product of the individual mind, although the activities of the individual are clearly a factor in determining physical presentations. As the fact of the existence of Science as public knowledge testifies, the physical complex contains a large element which appears to be independent of any particular percipient, and in this the objectivity of the physical world for collective mankind consists. For individual minds, this given element can only be taken as a datum which is accepted in perception. The relations between sensations and perception form a domain which the psychologist seeks to describe conceptually, but he, like the worker in Natural Science, accepts as a datum the fact that we have a stream of sensations which we do not appear to originate. The fact that there is, in the complex of percepts, an element independent of any individual mind, of a character consistent with its description and symbolization by rational schemes, to an extent of which we do not know the limits, is the fundamental fact which emerges from the results and history of Natural Science. Scientific laws and theories are the product or creation of mental activity, but are dependent, for the raw material from which they are constructed, upon given characteristics of the perceptual complex. These characteristics are not free creations of the mind, but data without which Science could not even begin to exist. Thus Natural Science exists only in virtue of the fact that the physical domain, the perceptual complex, is of such a character as to render possible, to the extent which we find by experience is actually the case, its conceptual representation by laws and schemes.

When a metaphysical theory or hypothesis is set up as to the fundamental character of this element of the physical complex that is apparently not dependent on the individual mind; that is when some ontological assumption is made having reference to it, there is then provided a bridge by means of which Natural Science is connected with general Philosophy. The precise nature of such an ontological assumption varies widely in different species of metaphysical Philosophy. All forms of realistic Philosophy agree in making the assumption that the element in the physical domain which appears to be independent of the individual mind has a real and independent existence, which does not wholly depend upon its being a co-factor in the subject-object relation. The assumption is that there exists a real complex which is not, as it were, exhausted in the subject-object relation, but can be separated out of that relation not merely in abstraction, so that it has an existence independent of that relation. Different forms of realism vary widely in the conceptions they adopt as to the nature of that reality. In some forms of realism this element is taken to be psychical in its nature, and thus in some degree akin to the human mind; and some systems of thought assume it to consist of a multitude of psychical individuals, or monads. The realism of common sense conceives this element as materialistic, and articulated in the same way as it appears to be in our perceptions, although a distinction between primary and secondary qualities of material objects receives some recognition; the independent real element being more closely related with the former. The ordinary realism of Science, physical realism, regards the real complex as articulated in accordance with the distinctions introduced by scientific theories, and of a non-psychical character; atoms, electrons, etc., are regarded as real. Neither scientific realism, nor that of common sense, is necessarily materialistic in the sense of assuming the psychical to be reducible to, or entirely subordinate to, the non-psychical, although, as we have seen, this assumption has been made by various influential exponents of Natural Science. Some forms of realism have been agnostic, in the sense that, although they have recognized the existence of this real element in the world, they have regarded any precise characterization of it as outside the domain of possible knowledge. All forms of what may be called critical realism, to distinguish it both from physical realism, and naïve realism, assume the existence of a reality which in some sense manifests itself in our sense-perceptions. But critical realism refrains from identifying the real with any of the conceptual constructions of Natural Science. In particular, Monadism refuses to identify Monads with atoms or electrons or biophors. Idealism, even when it does not stop short at purely subjective idealism, does not take the step of separating the object out of the subject-object relation and regarding it as existing independent of that relation. Whilst holding fast to that relation, without which it regards objectivity as meaningless, it usually regards the postulation of a Universal Mind, for which the world is a realm of objectivity, as essential for a Philosophy which shall suffice to give any adequate account of human experience.

In the next lecture I propose to discuss the bearings which Natural Science may have upon the Philosophy of Theism, when some ontological hypothesis is made which will suffice to establish the nexus, of which I have spoken, with an outlook on existence wider than that which Natural Science need adopt for its own special purpose. Such a discussion can naturally deal with but a small part of the whole subject of Theism, but it is concerned with an aspect of the problem of Religion which is of great importance in relation to contemporary thought on the subject.

# XX

## NATURAL SCIENCE AND THEISM

IN describing the subject of which the Gifford Lecturers are to treat, Lord Gifford spoke of it as:

> the true knowledge of God, that is of the Being, Nature, and Attributes of the Infinite, of the All, of the First and the Only Cause, that is, the One and Only Substance and Being, and the true and felt knowledge (not merely nominal knowledge) of the relations of man and the universe to Him, and of the true foundations of all ethics and morals....

Lord Gifford proceeded to make it clear that the most absolute freedom of opinion is to be accorded to the Lecturers; he said that:

> they may be of any religion or way of thinking, or as is sometimes said, they may be of no religion, or they may be so-called sceptics or agnostics or freethinkers, provided only that the "patrons" will use diligence to secure that they be able reverent men, true thinkers, sincere lovers of and earnest enquirers after truth....

He says further:

> I wish the Lecturers to treat their subject as a strictly natural science, the greatest of all possible sciences, indeed, in one sense, the only science, that of Infinite Being, without reference to or reliance upon any supposed exceptional or so-called miraculous revelation.....The lecturers shall be under no restraint whatever in their treatment of their theme.

It is clearly in accord with the broad philosophic spirit in which these words are conceived, and which I have quoted on account of their intrinsic interest, that a particular Lecturer should be free to treat the subject of Theism from a special point of view which may lead up only to one special aspect of Theism. The particular point of view which I have chosen is the one which I conceive to be that of Natural Science. I have already given some indication of the existence of other points of view, which may be of equal, and some of them probably of greater, importance than that of Natural Science, from which the great central problem of reality may be regarded. The path has been prepared for the brief and fragmentary treatment of the subject which I here give by the discussions in the last lecture.

Besides the influence which may accrue to Natural Science in relation to theoretical Theism when existential assumptions are made which are not themselves any part of the necessary basis of Natural Science, there is another kind of influence, more of a practical kind, which Natural Science exerts upon general views of the nature of reality. I have referred to the manifold character of the influences which cause a predisposition to accept or reject ontological hypotheses or assumptions which form the bases of particular views of reality, and have emphasized their a-logical character. The habits of mind induced by the study of Natural Science, and even by acquaintance with the general character of the practical and other results of that study, often play a large part in producing a selective predisposition to accept fundamental postulations of the kind to which I refer. Natural Science has in the past undoubtedly had the effect of producing, in circles wider than those of scientific investigators, a bias in favour of certain forms of realistic philosophy, and even probably of some forms of theism as against other forms. I have maintained throughout that this is a bias, or predisposing influence, and not a logical consequence of the acceptance of Natural Science and its special results as possessing a certain kind of validity. In fact, familiarity with Natural Science may be a cause of belief in the truth of certain premisses employed in philosophical schemes, without providing in any proper sense a reason for such belief.

Theistic Philosophy is dependent upon the two concepts of existence and value, since Theism, in any ordinary sense of the term, is not only concerned with the existence of God, but with His relation to our conceptions of value and its conservation. The estimation of value is, as we have seen, not a matter with which Natural Science has any direct concern; any bearings which Natural Science may be thought to have upon the estimation of values can accordingly be chiefly in and through its bearings upon questions of existence.

Any general theory of reality must be expected to give some account of the origin of the phenomena with which Natural Science has to deal; of the relation of those phenomena with the real, of whatever nature that real may be held to be. Since the real must be such as in some manner to give rise to the phenomena of which Natural Science has to give a conceptual description, it would appear that the kind of knowledge to which Natural Science attains may lead to inferences as to some of the characteristics of the real.

No Theistic theory would be of the slightest value which failed to give some account of the relations of God with finite spirits and with the world of phenomena. Neither on the theoretical nor on the practical side could such a theory serve any useful function; it would set up a conception without any real content. The very various types of theism which exist, and have existed, are mainly differentiated from one another by the character of this relation and the modes in which it is conceived. At one extreme we have Pantheism, in accordance with which God is the only existent, the All, who is identified with the world, or at least with all that can be said to be real in the world. The world contains nothing that is existentially distinct from God, and this statement must include all finite spirits. The great philosophical problem of the One and the Many, pantheism attempts to solve by means of a suppression of the Many. Our actual experience of plurality being reduced to an illusory appearance, the fundamental difficulty of pantheistic or absolutist systems of philosophy consists in their inability to provide a satisfactory account of the origin of this illusion; to account for the apparent differentiation of the One into an apparent multiplicity of phenomenal forms. Pantheistic theories have taken various forms; there exists spiritualistic Pantheism, as for example in the system of Spinoza, and materialistic Pantheism, practically indistinguishable from Atheism; there exists also the absolutism of some forms of Idealistic Philosophy, in which God is the Absolute.

At the other extreme of theistic theory is the view in accordance with which God is purely transcendental; a Being essentially external to, although in relation with, the world of which He may have been the Creator; such Creation being conceived as the calling into existence by His will of a world, external to Himself, existentially separate from Himself, and subject to laws which He has prescribed, but which render the world, at least to a considerable extent, autonomous. In its most pronounced form this view was that of the Deism implicit in the theology of the eighteenth century; it was held not only by some of the writers known as deists, but in at least equal measure by many of their more orthodox opponents. The controversies of that time turned to a large extent upon the degree of autonomy which the Creator had granted to the world, and on questions as to whether or not He had on specific occasions interfered with that autonomy. The analogy of the relation of a watchmaker to the watch he makes, or of an artificer to a machine he constructs, was often uncritically appealed to, as illustrating the relation of the Creator to the world. The fact that the watchmaker, or the artificer, has to deal with materials which he does not make, and with given material properties over which he has no complete control, did not appear to be of sufficient significance to destroy the value of such analogies for exponents of this order of ideas.

Although there still remain in the popular mind distinct traces of this conception of a purely transcendental Deity, conceptions intermediate between the two extremes of which I have spoken are probably dominant in the theistic thought of the present day. God is conceived of as immanent in the world, and more especially in finite spirits, which live and move and have their being in Him. He is also usually conceived to be transcendent, but varying emphasis is placed by different thinkers upon the two elements of immanence and transcendence. All views of this species, however wide their differences may be, agree in the one respect in which they all differ from Pantheism, in refusing to assert that God is the only reality; although they may regard reality other than that of God as derivative, having its ultimate origin in Him, and dependent upon Him for its continued existence, but possessing, at least in some degree, a relative independence. The particular mode in which this kind of Theism is conceived depends to some considerable extent upon whether the general Philosophy adopted is conceived in accordance with a realistic or an idealistic attitude of mind. The theist who combines with his theism some form of realism, naïve, physical, or critical, usually admits real existence to appertain to a domain which is not of a purely psychical character; for him there are real, non-mental, things which may manifest themselves as objects in the subject-object relations of psychical beings, but they exist independently of being mere factors in those relations. For him, there really exists besides God, a world not wholly of a spiritual or mental character; reality consists of God and the world. This is not only the view of the great majority of theists who are not philosophers, but it is also the view of some philosophical theists. On the other hand, for the Idealist, material objects and events have no existence independent of their forming a factor in the subject-object relations of a mind. All reality is taken to be essentially bound up with this form of relation, and consists of minds for which alone objects exist. The existence of God, the Universal Mind, is frequently held to be a necessary inference from this view of the nature of existence. The fact that the objectivity of the world of things is, for the single finite mind, temporary and intermittent, and that we are compelled to look back to a time when minds such as we know did not yet exist, is held to lead to the conclusion that a Universal Mind exists, for whom all objects are eternally present. The conception of the merely potential existence of objects which no one perceives, and at times when there is no sentient being for whom they are objects, as would have been the case when the earth was unfitted for the existence of living beings such as we know, is regarded as inadmissible. It is held that, for the complete objectivity of the world, a Universal Mind is necessary. As Dr Rashdall has written:

> We cannot understand the world of which we form a part except upon this assumption of a Universal Mind for which, and in which, all that is exists. Such is the line of thought which presents itself to some of us as the one absolutely convincing and logically irrefrageable argument for establishing the existence of God.

In accordance with this view, the totality of existence consists of God, the Universal Mind, and derivatively of finite spirits; the world exists solely as eternal object or idea for the Universal Mind, and as, in some partial or fragmentary form, it appears in the subject-object relations of finite minds.

The difficulty of Theism of these types, whether combined with a realistic or an idealistic philosophy, is in some sense the converse of that which presents itself in connection with Pantheism. Commencing with the Many, these types of Theism attempt to provide adequately for the existence of the One, whilst maintaining some degree of real independence for the Many, whereas Pantheism, commencing with the One, finds its crucial difficulty in making any real provision for the appearance of multiplicity. Some of the adherents of Theistic systems which commence with, and maintain, the real existence of the Many do not hesitate to admit that to do this involves the recognition of some limitation in the Being of God. This limitation is often represented as being a willed self-limitation. The position of the world in relation to the Universal Mind is often held to be a necessary one, since a subject without an object is as unthinkable as an object without a subject; it is conceived that it is only in connection with the subject-object relation that either subject or object has a meaning. It was in fact said by T. H. Green that "the world is as necessary to God as God is to the world."

This mode of argument for the existence of God is of a purely metaphysical kind, as indeed are all ontological arguments, whether they rest upon an idealistic or a realistic basis; and this is the case in particular for the traditional "Ontological Proof" of the existence of God. It is clear that Natural Science has no bearings upon arguments of this kind, the validity of which must be estimated upon metaphysical grounds alone. It is only when a theistic theory, based upon such an argument, passes to a later stage, beyond that of the assertion of existence, that the special characteristics of the world of phenomena become relevant in relation to the Nature of God; that the possibility can be contemplated that Natural Science may have a part in the development of Theism.

It may be remarked that, for thinkers of the present day who hold the view that the Divine is immanent in the human spirit, it is not so easy as it appeared to be to the older exponents of Natural Religion to draw a hard and fast line between revealed knowledge and such knowledge as may be obtained by the continued use of the rational faculty of man. For all knowledge obtained by the activities of human spirits in which the Divine Spirit is immanent may be regarded as in some sense a divine revelation; all the activities of finite spirits being conditioned by the immanent presence of the Divine Spirit. The quite sharp distinction between revealed knowledge and other knowledge would thus seem to be characteristic of that conception of a purely transcendental Deity which was commoner amongst the thinkers of the eighteenth century than it is in our time.

The three traditional proofs of the existence of God, the ontological, the cosmological, and the teleological, are, since the famous destructive criticisms of them by Immanuel Kant, no longer regarded as proofs in the sense that they consist of deductions, in accordance with the canons of logic, from premisses which no man can, or does, refuse to admit. They remain, however, in broadened and extended forms, as lines of argument which are still employed, rather inductively than deductively; but the validity of such lines of argument is dependent upon foundations which cannot escape the fullest scrutiny. The moral argument, stated by Kant, has, in modified forms, come to represent the line of thought which, in its insistence upon the fundamental importance of that aspect of existence which we associate with the terms "value" and "valuation," is regarded by most Theists of the present day as outweighing in importance and cogency all other aspects of the subject. Theism is now very frequently regarded as finding its main support in the existence of the domain of moral standards; whereas in much of the thought of the eighteenth century, and even later, this relation of dependence was taken in the reverse order.

The cosmological proof, adopted by Thomas Aquinas

from Aristotle, is founded upon the conception that the world is due to a sequence of causes which may be continued in the backward direction, each cause of the sequence being taken to be the effect of the preceding cause. In accordance with this conception it is argued that we arrive at a first cause, which is taken to be God, for otherwise the causal sequence would constitute an indefinite regress, in which case all explanation of the present existence of the world would be in default. It is accordingly assumed that, in order that the world may be intelligible to us, it must have commenced with a first cause. To this argument there is in point of logic a fatal objection. It is assumed as the law of the sequence that every cause is the effect of a preceding cause, and thus the assumption that, through the sequence, we arrive at a member which is a cause but not an effect is a breach of the law of the sequence. Moreover, the assumption that an indefinite regress is inconceivable, or inconsistent with the existence of an intelligible world, is groundless; in fact it is possible to define such sequences, in accordance with an intelligible law. On these grounds, the proof in its original form has been generally rejected as invalid; but, as we shall see, it is related with a line of argument which is still of importance, and to which the results of Natural Science have made a contribution of great weight. There are two points to notice about the proof, apart from the fatal contradiction which it involves. In the first place, it assumes, as perfectly definite, the conception of causation, without examination of the question whether the causes are to be considered as efficient causes, or whether they are to be considered simply as denoting totalities of conditions which experience shows to precede certain effects. In either case, an examination of the origin of the nexus between cause and effect would appear to be necessary before a proof could be accepted which professes to lead to an explanation of the existence of the world Again it takes, in accordance with an uncritical notion of causation, a temporal succession of causes, and it professes to arrive at an explanation of the whole temporal series by means of a first cause, but not of a law, or final cause, of the whole sequence, which shall be operative or effective at all times. In fact it strongly suggests the conception of a purely transcendental Deity, who as a first cause starts the whole sequence of causes, but then leaves it to itself, not being immanent as the ground of the world. Moreover the proof assumes the validity of the concept of the world as a whole, which in our experience can never be a completed concept.

If we consider the perceptual world from the point of view of Natural Science, we see that one fact about it has been established which may be made the basis of a cosmological argument free from the defects of the proof in its original form, although the fact to which I allude is subject to limitations as to its scope, which can only be removed by means of a postulation or assumption which may commend itself to the mind as appearing probable. The whole history of Natural Science tends to extend the scope of the ascertained fact that the perceptual domain is such that whole tracts of it, and processes in it, are capable of description by rational schemes. Whatever then be taken to be the nature of the reality which the perceptual world manifests, or of the ground of phenomena, that reality or ground must be of such a character that it has some correlation with the rational processes of our minds. This fact may have to be adapted, in the precise form of its interpretation, to whatever philosophical formulation may be adopted of the relation of phenomena with reality or with their ground. Whatever such formulation be employed, whether of a realistic or idealistic type, it will express the fact that the phenomena of perception have as their ground a reality, mental or non-mental, which is an ordered system apprehended by us as rational. This line of thought does not proceed by means of a principle of causation to the recognition of the existence of a first cause, from which the world as we know it has originated, but fixes attention upon the principle of rational order, as in some sense immanent in the world, not as an external cause, but as the ground of the whole life and movement of a reality of which the phenomenal domain is the manifestation. This line of thought has led to the postulation that reality is fundamentally rational and unitary. It is rational, as exhibited in its correlation with human reason, and unitary, as consisting of a completely interconnected system. The coincidence of this conception of a completely unified rational reality with the religious concept of God will require for its realization further elements, outside the purview of Natural Science, in which the conception of values will be the chief factor. It may be held that Natural Science provides a most important part of the justification for the ascription of complete rationality to the real ground of the phenomenal world. But the limitation must be fully recognized, that this postulation of complete rationality of the real goes far beyond anything that has been, or can be, unimpeachably established by Natural Science. Apart altogether from the passage from the phenomenal domain to the real ground, a passage outside the scope of Natural Science itself, the evidence that phenomena can be described by, and correlated with, rational mental processes is incomplete; and thus the complete rationality of the Universe is an hypothesis, not a known fact. It is the methodological axiom of Science that the correlation of phenomena with rational schemes can be carried out to an unlimited extent, but the actual amount of verification which the axiom has received is at all times strictly limited. It cannot be demonstrated that no limits exist, to which the procedure of Natural Science, in accordance with this axiom, may be subject. Accordingly, what Natural Science provides is indicative evidence, and not demonstrative proof, of the unlimited accessibility of natural phenomena to rational schematization. But undoubtedly the actual achievements of Natural Science have been sufficient to cause in many minds a belief in the unlimited rationality of the ground of the real world. With the difficult questions which arise in connection with this line of thought, especially as regards the relation of the wills of sentient beings with the real, conceived as a unitary rational being or principle, I cannot here deal; nor can I discuss the related question which arises as to the significance of what I have called in the last lecture the principle of individuality. I must be content with having briefly indicated a line of thought which has largely supplanted the older mode of thinking embodied in the traditional cosmological proof.

The foregoing discussion of the significance of order or uniformity in Nature, as forming a part of the possible basis of a theistic Philosophy, would be incomplete without some reference to the fact that it is historically far from true that the chief evidence of the existence of God, derived from a contemplation of the phenomena of perception, has always been found in the order and uniformity discernible therein. On the contrary, the religious mind has very frequently fixed its attention, not so much upon that order, as upon the occurrence of supposed breaches of that order, as exhibited in miracles. It is not too much to say that the evidences of divine Power have often been sought rather in miracles, regarded as breaches of natural order, than in the existence of that order itself. As Goethe has said: "Das Wunder ist des Glaubens liebstes Kind." This view of the evidential value of miracles appears to be in close connection with the conception of God as transcendent, as influencing the world not from within but from without. Just as the arbitrary and sporadic acts of an absolute sovereign appeal to many minds as more striking, and more dramatic, evidences of power than are exhibited in the more even course of the orderly working of a constitutional government, the incalculable acts of a transcendental Deity, conceived as a despotic personality, appear to provide more cogent evidence of divine power than does the orderly and ubiquitous working of an immanent Deity. The decay of the belief in miracles which has taken place progressively in modern times is undoubtedly due in large part to the progress of Natural Science, with the emphasis which it places upon order or uniformity in phenomena. It must however be distinctly recognized that there exists, and can exist, no *a priori* proof of the impossibility of what are called miracles. If that impossibility has been sometimes asserted by exponents of Natural Science, the assertion is merely a piece of *a priori* dogmatism, quite incapable of substantiation on scientific grounds. We have no *a priori* knowledge of what can, and what cannot, occur in Nature. We have only presumptions, psychologically explicable as expectations due to habits of thought founded on our past experience. The decay of belief of which I have spoken depends in large part upon a change of attitude towards an unusual or unexpected occurrence, and also consists in large part of a more critical attitude towards the evidence that such alleged events have actually occurred. To the modern man of science, an event which does not appear to happen in accordance with known laws is an occurrence which suggests to him the inadequacy of those laws, or the presence of some disregarded factor. He is incited to attempt, by means of the extension of known laws, to subsume the occurrence under a more complete set of laws, or to determine the character of the particular disturbing factor which had in the first instance been left out of account. What would perhaps formerly have been regarded as a miracle is, for the modern man, not a case of breach of order, but an occasion for extending his knowledge of it. We now know that the evidence of witnesses of miracles which took place in an uncritical age, in which the belief in their actuality was dominant, requires the most rigid scrutiny before it can be accepted as of any value. Such scrutiny is most difficult to apply in the case of events which happened in a remote age and in a distant country. We know how difficult it is to apply crucial tests to the evidence of alleged events of an abnormal kind in our own time. Miracles are always most plentiful at a time, or in a community, in which the belief in their occurrence is most prevalent. Moreover the possibility must be taken into account that superior knowledge of physical laws, or exceptionally great powers, normal in kind, on the part of a supposed worker of miracles, may account for the exceptional occurrences. This may particularly be the case in miracles of healing; in that matter the modern study of Psychotherapeutics may be relevant. There is a further question to be considered in this connection. Supposing it were assumed that a particular occurrence were a breach of natural order, what would the miracle prove, in relation to its assumed agent? Even in the ages of faith, when miracles were commoner than anyone supposes them to be now, this question was one of importance. Some scrutiny of the source of the miraculous intervention was required, as it was by no means always assumed that a miracle was of divine origin. A person who appeared to have powers of influencing natural phenomena which we were wholly unable to bring into line with our scientific knowledge, or with any readily conceivable extension of it, would not now by any means necessarily be accepted as of infallible authority on other matters, such as Philosophy and Theology.

The view of miracles, in their relation to Religion and Natural Science, which is widely held by modern thinkers is aptly and trenchantly expressed by the Danish Philosopher Höffding[1] in the following passage in his *Philosophy of Religion*:

Even if, in spite of all these circumstances, we were to believe in any particular case that we had here before us a real miracle, *i.e.*, a deviation from the law-abiding order of nature, the concept of God which could be based on this fact would necessarily bear the stamp of imperfection; for a miracle is a makeshift, a way out, something which has to make up for a want in the order of nature. The ordering of nature has not been so effected that by it all the divine ends can be attained. God encounters an obstacle within his own order of nature. It is as if there were two gods—one who is active during the ordinary course of things, and another who, in particular cases, corrects the work of the former. Hence the concept of miracle is dangerous from the religious as well as from the scientific standpoint. It is a bastard which neither parent can afford to own. The Church is wise in not acceding to the re-awakened desire for miracles. It is true of increasingly large circles that miracles, which in former times were a proof and support of religion, are now rather a stumbling-block which its apologists have to defend, and which in their hearts they must often wish themselves well rid of. The less we think of the relation between God and the world as a purely external one, analogous to the relation between a clock-maker and his clock, the less there is room for, or possibility of, miracles. The happenings of the world differ widely in value, and excite our admiration in very differing degrees; the highest does not take place every day. But there is nothing to prevent all events being subject to the same great law. It is large enough to embrace an infinite number of things and of problems. May we not assume that that which is of highest value may be reconcilable with the principle of natural causation? The concept of miracle really arises from the negative answer to this question. From whence the right to negate is derived is not easy to discover. The fact that something is of the highest value does not preclude a purely natural origin. The concept of miracle rests on an identification of estimation with explanation, an identification to which is largely due the confusion which at present characterises the religious problem.

[1] English translation, pp. 29–30.

Closely connected with the cosmological argument in favour of a theistic view of the Universe is the teleological argument, which rests upon the fact that there is much in the world of phenomena which has at least the appearance of purposiveness or design. The older forms of this argument have consisted in tracing out in Nature special cases of adaptation, taken as indicating the effects of design on the part of an intelligent Being. By such exponents of the teleological argument as Archdeacon Paley, much emphasis was placed upon the existence, in a great variety of special cases, of intricate arrangements adapted to the attainment of useful ends; it was held that the existence of such contrivances could only be explained as due to intelligent design. The kind of purposiveness taken to be exhibited in such cases was, in accordance with the prevailing view, assumed to be analogous to that of the artificer who constructs a machine which shall fulfil some purpose which he has in view. The argument depended upon the idea that when we find such a machine actually working, so as to attain an end, we are entitled to infer that it has been constructed by an intelligent being who had that end in view. It must be observed that the purposiveness so conceived is not immanent in the machine, but in its designer; the relation of the designer to the machine is conceived as an external relation. The designer works with given materials possessed of given properties; his intelligence is exhibited in the ingenuity with which he works under strictly limited conditions which are for him unalterable; he has to do the best he can with the materials at his disposal. The shortcomings of this line of argument are clear; it might with some plausibility be employed by a polytheist to infer the existence of a number of gods of considerable, but limited, power and intelligence, operating in a world which they did not create. The inference to a God of unlimited power and intelligence would present greater difficulties. The very necessity for contrivance or ingenuity implies limitation of power.

The progress of Natural Science during the time that has elapsed since this form of the teleological argument was commonly employed, and especially the great progress in Biological Science, has resulted in a radical change of the mode in which teleology in relation to natural phenomena is conceived, and in the whole aspect of the place which a teleological view of the world occupies in general Thought. By modern teleologists who regard the matter from the points of view of Natural Science and of Psychology, purposiveness is regarded no longer as indicating a purely external relation, but as in some sense immanent in living organisms, and possibly in a wider sense in the phenomenal world generally. In this connection two questions are of fundamental importance; the question of the existence and scope of natural selection, and the general question of the nature of vital processes and of the mode in which the relation of the physical with the psychical side of the living organism is to be conceived. As I have already pointed out in my lecture on Evolution, the theory of Natural Selection is anti-teleological in its tendency. It is an attempt to give an account of what appears *prima facie* to be progressive adaptation in relation to ends, by means of a scheme in which no teleological factor is admitted. As we have seen, there is at the present time much difference of opinion as to the sufficiency and scope of this theory of Natural Selection, so that it must be regarded as very far from having been established that it gives such an adequate account of the evolution of species, and their adaptations to environment, that a teleological factor can be dispensed with. When the theory is extended to mental evolution, it seems, to say the least, extremely doubtful whether it can, with any plausibility, be appealed to as giving a credible account of the evolution of the higher mental and spiritual faculties of man; in default of any sufficient evidence that these faculties have survival value in the biological sense of the term. The view has been expressed, and is held by many persons, that the development of the higher faculties of man, so far from being an evolution, subject to the law of struggle for existence and the survival of the fittest in the biological sense, involves really a reversal of that law. The difficult question as to the relations between the psychical and physical sides of living organisms I have already spoken of in an earlier lecture. I have contended that there is no sufficient ground for the assertion that what we believe to be our power of purposively influencing natural phenomena is an illusion. If this view be accepted, it is difficult not to recognize a teleological factor as present in all higher organisms at least. But on this matter there is at present much diversity of opinion amongst those who are best able to form a judgment. Natural Science must therefore be taken to speak with an uncertain voice at the present time, as regards the question whether it is compelled or not to recognize a teleological factor, as a supplement to the physico-chemical categories, in its efforts to give an account of vital processes. I think it may fairly be said that there exists no sufficient ground for negativing the hypothesis that something of the nature of entelechy is a necessary factor in all vital processes, even if the positive arguments in its favour that have been given by Driesch and others are regarded as insufficient. But, even on this side, admitting the necessity of conceiving entelechy, or purposive directivity, as exhibited in all vital processes, the teleological argument based upon such admission does not take us very far. It would be a long step to pass on to a unitary purposiveness which should be immanent in, or associated with, all natural phenomena; and indeed, if the argument could be carried so far, it might become pantheistic in its result. The independence of finite spirits which allows their purposiveness to exhibit its effect in the phenomenal domain would appear to be inconsistent with, or at least very difficult to reconcile with, the conception of a completely unified purposiveness which pervaded all natural phenomena. If purposiveness is not to be completely unified, it is difficult, without very arbitrary assumptions, to trace limits to the degree of its diffusion.

I have already referred to the fact that the argument in favour of theism which is in our time generally regarded as the most important and most convincing is the Moral Argument which is based upon the human conception of moral values. In its earliest form, this argument was developed by Kant, who made the belief in the existence of God depend upon a postulation of the practical reason, as distinct from the pure reason which he regarded as impotent to establish the truth of theism. In our day this argument has been developed by various exponents, amongst whom I may allow myself to refer to Professor Sorley, who in his Gifford Lectures on *Moral Values and the Idea of God*, delivered here in Aberdeen, has presented the argument with great force and admirable skill. Any general discussion of this argument would be quite outside the scope of these lectures, even if time allowed me to attempt it. There is however one aspect of the argument to which I must refer, because it is one in which facts of the phenomenal domain, brought to light by Natural Science, have significance. In dealing with the argument from design, the older writers on Natural Religion laid great stress upon the beneficent results for man of many of the contrivances exhibited in Nature. The evidences of design which are discernible in natural phenomena were conceived to afford a proof not only of the intelligence and power of the Designer, but also of His goodness, as exhibited in the benevolence with which Nature has been adapted to serve the needs, and further the well-being, of the human race. The reverse side of this picture of Nature was referred to by Hume, and has been strongly emphasized by later thinkers. J. S. Mill, in his essay on "Nature," framed a terrible indictment against the callousness, cruelty, and injustice exhibited in the ordinary course of natural phenomena.

In sober truth (he writes) nearly all the things which men are hanged or imprisoned for doing to one another, are nature's every day performances....Nature impales men, breaks them as if on the wheel, casts them to be devoured by wild beasts, burns them to death, crushes them with stones like the first Christian martyr, starves them by the quick or slow venom of her exhalations, and has hundreds of other hideous deaths in reserve, such as the ingenious cruelty of a Nabis or a Domitian never surpassed.

Mill drew the conclusion that the God who is responsible for Nature, as he describes it, cannot be a Being both of unlimited power and unlimited goodness. In a later essay, Mill maintained that all the evidences from design, and from the characteristics of Nature, point to a Deity whom he describes as:

A Being of great but limited power, how or by what limited we cannot even conjecture; of great, and perhaps unlimited intelligence but perhaps, also, more narrowly limited than his power; who desires, and pays some regard to, the happiness of his creatures, but who seems to have other motives of action which he cares more for, and who can hardly be supposed to have created the universe for that purpose alone. Such is the Deity whom Natural Religion points to; and any idea of God more captivating than this comes only from human wishes, or from the teaching of either real or imaginary Revelation.

The general picture of organic evolution painted by modern Biology is one in which the struggle for existence and nutriment, involving pain and death, is an important and possibly a dominant feature, although its repulsive aspect may perhaps sometimes have been drawn in too lurid colours. The ruthless sacrifice of multitudes of individuals appears to be a feature of the ordinary course of evolution; it has been said that Nature cares nothing for the individual, but much for the race. The facts brought to light by Bacteriology and Parasitology have disclosed the existence of many organisms, and of what have the appearance of being most ingenious contrivances, the apparent purpose of which is to inflict torture and death upon other, and usually higher, organisms. The result of modern knowledge of this kind has been to heighten the impression produced by the widespread and intricate character of what from our point of view we describe as physical evil. A consistent theist must regard as, in some sense, God's creatures, those living organisms whose presence and activities condition the existence of tetanus, cholera, typhus, and many other diseases. The contemplation of this aspect of phenomena gives rise to a very real problem, not only for Theologians, but for great numbers of thoughtful persons.

Of the problem presented by the existence of moral evil I cannot here speak, but for modern theistic Philosophy, in which the human spirit, with its needs, its rationality, and above all, its moral values, constitutes the basic point of departure, the existence of physical evil gives rise to difficulties which have never been overcome, although many earnest attempts have been made to soften their asperity. Some of these attempts proceed in the direction of indicating the possibility of a resolution of the difficulties, if they could be regarded from a point of view higher than that which the limitations of the human mind, and of its knowledge of the circumstances of the Universe, enable it to occupy. Some expositors lay great stress upon the necessity of an imperfect world as providing a field of activity such as is needed for the progressive perfectibility of free spirits. I must confess that many such attempts to cope with the difficulties of the kind I have indicated impress me as having the unsatisfying character of special pleading.

However this may be, it is, I think, admitted by candid exponents of Philosophical Theism that there do remain very real difficulties in reconciling Theism, of a type such as will completely satisfy the religious consciousness, with some aspects of the actual world of our experience. This does not by any means necessarily entail the consequence that the views of the philosophical theist must be rejected, but it does, I think, necessitate a recognition of the fact that a complete synthesis of the conceptions of existence, and of value, in a unitary theistic view of the Universe, has proved so far to be beyond the reach of the human mind. Many persons think it highly probable, or even morally certain, that such a unified view will prove unattainable so long as the present limitations of the human mind remain.

My main aim has been, by means of a delineation of the domain of Natural Science, to vindicate the perfect freedom of Religious and Philosophical thought from any fear of destructive interference from the side of Natural Science, subject to the sole condition that no encroachment is made upon the autonomy of Natural Science in its own proper domain. It has been no part of my aim in these lectures to indicate the use which Religious and Philosophical Thought may make of this freedom, or to state any results to which I might conceive it to lead. To have attempted to do this would have been to open out a field of discussion of an extent far beyond anything that could possibly have been joined on to the special subject of my course. Successful vindication of this freedom would serve to allay the fears, often perhaps only half conscious, of numbers of thinking men who, intimidated by the striking triumphs of Natural Science in extending our knowledge of the order of phenomena, and in turning that knowledge to practical account, imagined that Natural Science, as a rigidly deterministic world-philosophy, was perhaps destined to exercise a complete domination over the spiritual domain, and to leave no possibility of any real content being assignable to the conception of human freedom. I need hardly emphasize the fact that the removal of the destructive criticism which grounds itself upon Natural Science does not suffice to refute criticism which rests upon philosophical or psychological grounds.

There was a time when Theology claimed to occupy the whole territory of Natural Science, and held it in complete thraldom. The history of the prolonged struggle of Science for autonomy on its own territory has been one in which Theology has lost every battle. Unfortunately, Science has not always remained content with the vindication of its freedom, but has attempted to extend its dominion into territory which is not its own. There are happily at the present time hopeful signs pointing to a cessation, or at least a mitigation, of the conflict. On both sides, the prevailing temper is markedly different from what it has been within living memory. There is greater readiness than formerly to admit that the conditions of life as we experience it are such that different methods are requisite for dealing with differing aspects of our life and experience; and that this involves the necessity of granting freedom to those who pursue the different lines of thought and investigation appropriate to these different parts or aspects of our whole experience. The discursive modes of thinking to which the limitations of the human mind bind us compel us to treat in separation, even on the purely intellectual side, the different aspects of our experience with which Philosophical Theology and Natural Science are concerned. How much more is this seen to be the case when we consider how very different in kind are the needs which are to be satisfied by these two departments of thought. Religion, which is very far from being the same thing as Theology, always sooner or later feels the need of some support, on the intellectual side, from reasoned theological conceptions. Theology, I refer to here on its philosophical side; the philosophical Theologian is simply a Philosopher who pays special attention to those aspects of Philosophy which have a specifically religious bearing. Theology, as distinct from general Philosophy, has as its function the provision of a cognitive basis for Religion, of which the essence is not mainly cognitive, but is mainly concerned with the moral and emotional sides of human nature. If we were in possession of, and able to grasp, a unified view of the Universe, in which all the elements of existence and valuation were completely synthesized, the division of labour of which I have spoken would be unnecessary; we should not require to mark out frontiers between Science and Philosophy or Theology; but of such a synthesis there is not the remotest prospect in view. The secret of the Universe has revealed itself neither to the Theologian nor to the Philosopher. The man of Science, as such, is not even concerned with that secret. The untrammelled freedom which must be allowed to workers in all departments of the great cultural work of humanity, to Philosophers and Theologians, to Historians, to the cultivators not only of Natural Science, but of Science of all kinds, should not however involve the erection of rigid impassable barriers which shall mark off domains which hold no communication with one another. On the contrary, workers in one department will often receive the most valuable enlightenment, and most important suggestions, from quarters outside their own special line.

The summary discussions in the present lecture, and in the one immediately preceding it, deal professedly only with partial aspects of the great central questions which have in all ages been the subject of unending scrutiny on the part of thinkers, great and small; related as these questions are, and always have been, with the deepest thoughts, hopes, and feelings, of multitudes of human beings. The tentative character of much of what I have said in this connection may appear to many persons so devoid of sharply defined results as to be eminently unsatisfactory. I can only plead that anything like a dogmatic statement of personal opinion on most weighty matters, only a few partial aspects of which I have been able to discuss, would have been irrelevant and inappropriate, since it would have been foreign to what I placed before myself as the aim of these lectures.

# INDEX

# SOME DOVER SCIENCE BOOKS

# SOME DOVER SCIENCE BOOKS

COLLEGE ALGEBRA, *H. B. Fine*
Standard college text that gives a systematic and deductive structure to algebra; comprehensive, connected, with emphasis on theory. Discusses the commutative, associative, and distributive laws of number in unusual detail, and goes on with undetermined coefficients, quadratic equations, progressions, logarithms, permutations, probability, power series, and much more. Still most valuable elementary-intermediate text on the science and structure of algebra. Index. 1560 problems, all with answers. x + 631pp. 5⅜ x 8. Paperbound $2.75

HIGHER MATHEMATICS FOR STUDENTS OF CHEMISTRY AND PHYSICS, *J. W. Mellor*
Not abstract, but practical, building its problems out of familiar laboratory material, this covers differential calculus, coordinate, analytical geometry, functions, integral calculus, infinite series, numerical equations, differential equations, Fourier's theorem, probability, theory of errors, calculus of variations, determinants. "If the reader is not familiar with this book, it will repay him to examine it," *Chem. & Engineering News.* 800 problems. 189 figures. Bibliography. xxi + 641pp. 5⅜ x 8. Paperbound $2.50

TRIGONOMETRY REFRESHER FOR TECHNICAL MEN, *A. A. Klaf*
A modern question and answer text on plane and spherical trigonometry. Part I covers plane trigonometry: angles, quadrants, trigonometrical functions, graphical representation, interpolation, equations, logarithms, solution of triangles, slide rules, etc. Part II discusses applications to navigation, surveying, elasticity, architecture, and engineering. Small angles, periodic functions, vectors, polar coordinates, De Moivre's theorem, fully covered. Part III is devoted to spherical trigonometry and the solution of spherical triangles, with applications to terrestrial and astronomical problems. Special time-savers for numerical calculation. 913 questions answered for you! 1738 problems; answers to odd numbers. 494 figures. 14 pages of functions, formulae. Index. x + 629pp. 5⅜ x 8. Paperbound $2.00

CALCULUS REFRESHER FOR TECHNICAL MEN, *A. A. Klaf*
Not an ordinary textbook but a unique refresher for engineers, technicians, and students. An examination of the most important aspects of differential and integral calculus by means of 756 key questions. Part I covers simple differential calculus: constants, variables, functions, increments, derivatives, logarithms, curvature, etc. Part II treats fundamental concepts of integration: inspection, substitution, transformation, reduction, areas and volumes, mean value, successive and partial integration, double and triple integration. Stresses practical aspects! A 50 page section gives applications to civil and nautical engineering, electricity, stress and strain, elasticity, industrial engineering, and similar fields. 756 questions answered. 556 problems; solutions to odd numbers. 36 pages of constants, formulae. Index. v + 431pp. 5⅜ x 8. Paperbound $2.00

INTRODUCTION TO THE THEORY OF GROUPS OF FINITE ORDER, *R. Carmichael*
Examines fundamental theorems and their application. Beginning with sets, systems, permutations, etc., it progresses in easy stages through important types of groups: Abelian, prime power, permutation, etc. Except 1 chapter where matrices are desirable, no higher math needed. 783 exercises, problems. Index. xvi + 447pp. 5⅜ x 8. Paperbound $3.00

## *FIVE VOLUME "THEORY OF FUNCTIONS" SET BY KONRAD KNOPP*

This five-volume set, prepared by Konrad Knopp, provides a complete and readily followed account of theory of functions. Proofs are given concisely, yet without sacrifice of completeness or rigor. These volumes are used as texts by such universities as M.I.T., University of Chicago, N. Y. City College, and many others. "Excellent introduction . . . remarkably readable, concise, clear, rigorous," *Journal of the American Statistical Association.*

ELEMENTS OF THE THEORY OF FUNCTIONS,
*Konrad Knopp*
This book provides the student with background for further volumes in this set, or texts on a similar level. Partial contents: foundations, system of complex numbers and the Gaussian plane of numbers, Riemann sphere of numbers, mapping by linear functions, normal forms, the logarithm, the cyclometric functions and binomial series. "Not only for the young student, but also for the student who knows all about what is in it," *Mathematical Journal.* Bibliography. Index. 140pp. 5⅜ x 8. Paperbound $1.50

THEORY OF FUNCTIONS, PART I,
*Konrad Knopp*
With volume II, this book provides coverage of basic concepts and theorems. Partial contents: numbers and points, functions of a complex variable, integral of a continuous function, Cauchy's integral theorem, Cauchy's integral formulae, series with variable terms, expansion of analytic functions in power series, analytic continuation and complete definition of analytic functions, entire transcendental functions, Laurent expansion, types of singularities. Bibliography. Index. vii + 146pp. 5⅜ x 8. Paperbound $1.35

THEORY OF FUNCTIONS, PART II,
*Konrad Knopp*
Application and further development of general theory, special topics. Single valued functions. Entire, Weierstrass, Meromorphic functions. Riemann surfaces. Algebraic functions. Analytical configuration, Riemann surface. Bibliography. Index. x + 150pp. 5⅜ x 8. Paperbound $1.35

PROBLEM BOOK IN THE THEORY OF FUNCTIONS, VOLUME 1.
*Konrad Knopp*
Problems in elementary theory, for use with Knopp's *Theory of Functions,* or any other text, arranged according to increasing difficulty. Fundamental concepts, sequences of numbers and infinite series, complex variable, integral theorems, development in series, conformal mapping. 182 problems. Answers. viii + 126pp. 5⅜ x 8. Paperbound $1.35

PROBLEM BOOK IN THE THEORY OF FUNCTIONS, VOLUME 2,
*Konrad Knopp*
Advanced theory of functions, to be used either with Knopp's *Theory of Functions,* or any other comparable text. Singularities, entire & meromorphic functions, periodic, analytic, continuation, multiple-valued functions, Riemann surfaces, conformal mapping. Includes a section of additional elementary problems. "The difficult task of selecting from the immense material of the modern theory of functions the problems just within the reach of the beginner is here masterfully accomplished," *Am. Math. Soc.* Answers. 138pp. 5⅜ x 8.
Paperbound $1.50

NUMERICAL SOLUTIONS OF DIFFERENTIAL EQUATIONS,
*H. Levy & E. A. Baggott*

Comprehensive collection of methods for solving ordinary differential equations of first and higher order. All must pass 2 requirements: easy to grasp and practical, more rapid than school methods. Partial contents: graphical integration of differential equations, graphical methods for detailed solution. Numerical solution. Simultaneous equations and equations of 2nd and higher orders. "Should be in the hands of all in research in applied mathematics, teaching," *Nature*. 21 figures. viii + 238pp. 5⅜ x 8. Paperbound $1.85

ELEMENTARY STATISTICS, WITH APPLICATIONS IN MEDICINE AND THE BIOLOGICAL SCIENCES, *F. E. Croxton*

A sound introduction to statistics for anyone in the physical sciences, assuming no prior acquaintance and requiring only a modest knowledge of math. All basic formulas carefully explained and illustrated; all necessary reference tables included. From basic terms and concepts, the study proceeds to frequency distribution, linear, non-linear, and multiple correlation, skewness, kurtosis, etc. A large section deals with reliability and significance of statistical methods. Containing concrete examples from medicine and biology, this book will prove unusually helpful to workers in those fields who increasingly must evaluate, check, and interpret statistics. Formerly titled "Elementary Statistics with Applications in Medicine." 101 charts. 57 tables. 14 appendices. Index. vi + 376pp. 5⅜ x 8. Paperbound $2.00

INTRODUCTION TO SYMBOLIC LOGIC,
*S. Langer*

No special knowledge of math required — probably the clearest book ever written on symbolic logic, suitable for the layman, general scientist, and philosopher. You start with simple symbols and advance to a knowledge of the Boole-Schroeder and Russell-Whitehead systems. Forms, logical structure, classes, the calculus of propositions, logic of the syllogism, etc. are all covered. "One of the clearest and simplest introductions," *Mathematics Gazette*. Second enlarged, revised edition. 368pp. 5⅜ x 8. Paperbound $2.00

A SHORT ACCOUNT OF THE HISTORY OF MATHEMATICS,
*W. W. R. Ball*

Most readable non-technical history of mathematics treats lives, discoveries of every important figure from Egyptian, Phoenician, mathematicians to late 19th century. Discusses schools of Ionia, Pythagoras, Athens, Cyzicus, Alexandria, Byzantium, systems of numeration; primitive arithmetic; Middle Ages, Renaissance, including Arabs, Bacon, Regiomontanus, Tartaglia, Cardan, Stevinus, Galileo, Kepler; modern mathematics of Descartes, Pascal, Wallis, Huygens, Newton, Leibnitz, d'Alembert, Euler, Lambert, Laplace, Legendre, Gauss, Hermite, Weierstrass, scores more. Index. 25 figures. 546pp. 5⅜ x 8. Paperbound $2.25

INTRODUCTION TO NONLINEAR DIFFERENTIAL AND INTEGRAL EQUATIONS,
*Harold T. Davis*

Aspects of the problem of nonlinear equations, transformations that lead to equations solvable by classical means, results in special cases, and useful generalizations. Thorough, but easily followed by mathematically sophisticated reader who knows little about non-linear equations. 137 problems for student to solve. xv + 566pp. 5⅜ x 8½. Paperbound $2.00

AN INTRODUCTION TO THE GEOMETRY OF N DIMENSIONS, *D. H. Y. Sommerville*

An introduction presupposing no prior knowledge of the field, the only book in English devoted exclusively to higher dimensional geometry. Discusses fundamental ideas of incidence, parallelism, perpendicularity, angles between linear space; enumerative geometry; analytical geometry from projective and metric points of view; polytopes; elementary ideas in analysis situs; content of hyper-spacial figures. Bibliography. Index. 60 diagrams. 196pp. 5⅜ x 8.
Paperbound $1.50

ELEMENTARY CONCEPTS OF TOPOLOGY, *P. Alexandroff*

First English translation of the famous brief introduction to topology for the beginner or for the mathematician not undertaking extensive study. This unusually useful intuitive approach deals primarily with the concepts of complex, cycle, and homology, and is wholly consistent with current investigations. Ranges from basic concepts of set-theoretic topology to the concept of Betti groups. "Glowing example of harmony between intuition and thought," David Hilbert. Translated by A. E. Farley. Introduction by D. Hilbert. Index. 25 figures. 73pp. 5⅜ x 8.
Paperbound $1.00

ELEMENTS OF NON-EUCLIDEAN GEOMETRY, *D. M. Y. Sommerville*

Unique in proceeding step-by-step, in the manner of traditional geometry. Enables the student with only a good knowledge of high school algebra and geometry to grasp elementary hyperbolic, elliptic, analytic non-Euclidean geometries; space curvature and its philosophical implications; theory of radical axes; homothetic centres and systems of circles; parataxy and parallelism; absolute measure; Gauss' proof of the defect area theorem; geodesic representation; much more, all with exceptional clarity. 126 problems at chapter endings provide progressive practice and familiarity. 133 figures. Index. xvi + 274pp. 5⅜ x 8.
Paperbound $2.00

INTRODUCTION TO THE THEORY OF NUMBERS, *L. E. Dickson*

Thorough, comprehensive approach with adequate coverage of classical literature, an introductory volume beginners can follow. Chapters on divisibility, congruences, quadratic residues & reciprocity. Diophantine equations, etc. Full treatment of binary quadratic forms without usual restriction to integral coefficients. Covers infinitude of primes, least residues. Fermat's theorem. Euler's phi function, Legendre's symbol, Gauss's lemma, automorphs, reduced forms, recent theorems of Thue & Siegel, many more. Much material not readily available elsewhere. 239 problems. Index. I figure. viii + 183pp. 5⅜ x 8.
Paperbound $1.75

MATHEMATICAL TABLES AND FORMULAS, *compiled by Robert D. Carmichael and Edwin R. Smith*

Valuable collection for students, etc. Contains all tables necessary in college algebra and trigonometry, such as five-place common logarithms, logarithmic sines and tangents of small angles, logarithmic trigonometric functions, natural trigonometric functions, four-place antilogarithms, tables for changing from sexagesimal to circular and from circular to sexagesimal measure of angles, etc. Also many tables and formulas not ordinarily accessible, including powers, roots, and reciprocals, exponential and hyperbolic functions, ten-place logarithms of prime numbers, and formulas and theorems from analytical and elementary geometry and from calculus. Explanatory introduction. viii + 269pp. 5⅜ x 8½.
Paperbound $1.25

A Source Book in Mathematics,
*D. E. Smith*
Great discoveries in math, from Renaissance to end of 19th century, in English translation. Read announcements by Dedekind, Gauss, Delamain, Pascal, Fermat, Newton, Abel, Lobachevsky, Bolyai, Riemann, De Moivre, Legendre, Laplace, others of discoveries about imaginary numbers, number congruence, slide rule, equations, symbolism, cubic algebraic equations, non-Euclidean forms of geometry, calculus, function theory, quaternions, etc. Succinct selections from 125 different treatises, articles, most unavailable elsewhere in English. Each article preceded by biographical introduction. Vol. I: Fields of Number, Algebra. Index. 32 illus. 338pp. 5⅜ x 8. Vol. II: Fields of Geometry, Probability, Calculus, Functions, Quaternions. 83 illus. 432pp. 5⅜ x 8.
Vol. 1 Paperbound $2.00, Vol. 2 Paperbound $2.00,
The set $4.00

Foundations of Physics,
*R. B. Lindsay & H. Margenau*
Excellent bridge between semi-popular works & technical treatises. A discussion of methods of physical description, construction of theory; valuable for physicist with elementary calculus who is interested in ideas that give meaning to data, tools of modern physics. Contents include symbolism; mathematical equations; space & time foundations of mechanics; probability; physics & continua; electron theory; special & general relativity; quantum mechanics; causality. "Thorough and yet not overdetailed. Unreservedly recommended," *Nature* (London). Unabridged, corrected edition. List of recommended readings. 35 illustrations. xi + 537pp. 5⅜ x 8. Paperbound $3.00

Fundamental Formulas of Physics,
*ed. by D. H. Menzel*
High useful, full, inexpensive reference and study text, ranging from simple to highly sophisticated operations. Mathematics integrated into text—each chapter stands as short textbook of field represented. Vol. 1: Statistics, Physical Constants, Special Theory of Relativity, Hydrodynamics, Aerodynamics, Boundary Value Problems in Math, Physics, Viscosity, Electromagnetic Theory, etc. Vol. 2: Sound, Acoustics, Geometrical Optics, Electron Optics, High-Energy Phenomena, Magnetism, Biophysics, much more. Index. Total of 800pp. 5⅜ x 8.
Vol. 1 Paperbound $2.25, Vol. 2 Paperbound $2.25,
The set $4.50

Theoretical Physics,
*A. S. Kompaneyets*
One of the very few thorough studies of the subject in this price range. Provides advanced students with a comprehensive theoretical background. Especially strong on recent experimentation and developments in quantum theory. Contents: Mechanics (Generalized Coordinates, Lagrange's Equation, Collision of Particles, etc.), Electrodynamics (Vector Analysis, Maxwell's equations, Transmission of Signals, Theory of Relativity, etc.), Quantum Mechanics (the Inadequacy of Classical Mechanics, the Wave Equation, Motion in a Central Field, Quantum Theory of Radiation, Quantum Theories of Dispersion and Scattering, etc.), and Statistical Physics (Equilibrium Distribution of Molecules in an Ideal Gas, Boltzmann Statistics, Bose and Fermi Distribution. Thermodynamic Quantities, etc.). Revised to 1961. Translated by George Yankovsky, authorized by Kompaneyets. 137 exercises. 56 figures. 529pp. 5⅜ x 8½.
Paperbound $2.50

MATHEMATICAL PHYSICS, *D. H. Menzel*
Thorough one-volume treatment of the mathematical techniques vital for classical mechanics, electromagnetic theory, quantum theory, and relativity. Written by the Harvard Professor of Astrophysics for junior, senior, and graduate courses, it gives clear explanations of all those aspects of function theory, vectors, matrices, dyadics, tensors, partial differential equations, etc., necessary for the understanding of the various physical theories. Electron theory, relativity, and other topics seldom presented appear here in considerable detail. Scores of definition, conversion factors, dimensional constants, etc. "More detailed than normal for an advanced text . . . excellent set of sections on Dyadics, Matrices, and Tensors," *Journal of the Franklin Institute*. Index. 193 problems, with answers. x + 412pp. 5⅜ x 8. Paperbound $2.50

THE THEORY OF SOUND, *Lord Rayleigh*
Most vibrating systems likely to be encountered in practice can be tackled successfully by the methods set forth by the great Nobel laureate, Lord Rayleigh. Complete coverage of experimental, mathematical aspects of sound theory. Partial contents: Harmonic motions, vibrating systems in general, lateral vibrations of bars, curved plates or shells, applications of Laplace's functions to acoustical problems, fluid friction, plane vortex-sheet, vibrations of solid bodies, etc. This is the first inexpensive edition of this great reference and study work. Bibliography, Historical introduction by R. B. Lindsay. Total of 1040pp. 97 figures. 5⅜ x 8. Vol. 1 Paperbound $2.50, Vol. 2 Paperbound $2.50, The set $5.00

HYDRODYNAMICS, *Horace Lamb*
Internationally famous complete coverage of standard reference work on dynamics of liquids & gases. Fundamental theorems, equations, methods, solutions, background, for classical hydrodynamics. Chapters include Equations of Motion, Integration of Equations in Special Gases, Irrotational Motion, Motion of Liquid in 2 Dimensions, Motion of Solids through Liquid-Dynamical Theory, Vortex Motion, Tidal Waves, Surface Waves, Waves of Expansion, Viscosity, Rotating Masses of Liquids. Excellently planned, arranged; clear, lucid presentation. 6th enlarged, revised edition. Index. Over 900 footnotes, mostly bibliographical. 119 figures. xv + 738pp. 6⅛ x 9¼. Paperbound $4.00

DYNAMICAL THEORY OF GASES, *James Jeans*
Divided into mathematical and physical chapters for the convenience of those not expert in mathematics, this volume discusses the mathematical theory of gas in a steady state, thermodynamics, Boltzmann and Maxwell, kinetic theory, quantum theory, exponentials, etc. 4th enlarged edition, with new material on quantum theory, quantum dynamics, etc. Indexes. 28 figures. 444pp. 6⅛ x 9¼. Paperbound $2.75

THERMODYNAMICS, *Enrico Fermi*
Unabridged reproduction of 1937 edition. Elementary in treatment; remarkable for clarity, organization. Requires no knowledge of advanced math beyond calculus, only familiarity with fundamentals of thermometry, calorimetry. Partial Contents: Thermodynamic systems; First & Second laws of thermodynamics; Entropy; Thermodynamic potentials: phase rule, reversible electric cell; Gaseous reactions: van't Hoff reaction box, principle of LeChatelier; Thermodynamics of dilute solutions: osmotic & vapor pressures, boiling & freezing points; Entropy constant. Index. 25 problems. 24 illustrations. x + 160pp. 5⅜ x 8. Paperbound $1.75

PRINCIPLES OF STRATIGRAPHY,
*A. W. Grabau*

Classic of 20th century geology, unmatched in scope and comprehensiveness. Nearly 600 pages cover the structure and origins of every kind of sedimentary, hydrogenic, oceanic, pyroclastic, atmoclastic, hydroclastic, marine hydroclastic, and bioclastic rock; metamorphism; erosion; etc. Includes also the constitution of the atmosphere; morphology of oceans, rivers, glaciers; volcanic activities; faults and earthquakes; and fundamental principles of paleontology (nearly 200 pages). New introduction by Prof. M. Kay, Columbia U. 1277 bibliographical entries. 264 diagrams. Tables, maps, etc. Two volume set. Total of xxxii + 1185pp. 5⅜ x 8. Vol. 1 Paperbound $2.50, Vol. 2 Paperbound $2.50,
The set $5.00

SNOW CRYSTALS, *W. A. Bentley and W. J. Humphreys*

Over 200 pages of Bentley's famous microphotographs of snow flakes—the product of painstaking, methodical work at his Jericho, Vermont studio. The pictures, which also include plates of frost, glaze and dew on vegetation, spider webs, windowpanes; sleet; graupel or soft hail, were chosen both for their scientific interest and their aesthetic qualities. The wonder of nature's diversity is exhibited in the intricate, beautiful patterns of the snow flakes. Introductory text by W. J. Humphreys. Selected bibliography. 2,453 illustrations. 224pp. 8 x 10¼. Paperbound $3.25

THE BIRTH AND DEVELOPMENT OF THE GEOLOGICAL SCIENCES,
*F. D. Adams*

Most thorough history of the earth sciences ever written. Geological thought from earliest times to the end of the 19th century, covering over 300 early thinkers & systems: fossils & their explanation, vulcanists vs. neptunists, figured stones & paleontology, generation of stones, dozens of similar topics. 91 illustrations, including medieval, renaissance woodcuts, etc. Index. 632 footnotes, mostly bibliographical. 511pp. 5⅜ x 8. Paperbound $2.75

ORGANIC CHEMISTRY, *F. C. Whitmore*

The entire subject of organic chemistry for the practicing chemist and the advanced student. Storehouse of facts, theories, processes found elsewhere only in specialized journals. Covers aliphatic compounds (500 pages on the properties and synthetic preparation of hydrocarbons, halides, proteins, ketones, etc.), alicyclic compounds, aromatic compounds, heterocyclic compounds, organophosphorus and organometallic compounds. Methods of synthetic preparation analyzed critically throughout. Includes much of biochemical interest. "The scope of this volume is astonishing," *Industrial and Engineering Chemistry*. 12,000-reference index. 2387-item bibliography. Total of x + 1005pp. 5⅜ x 8. Two volume set, paperbound $4.50

THE PHASE RULE AND ITS APPLICATION,
*Alexander Findlay*

Covering chemical phenomena of 1, 2, 3, 4, and multiple component systems, this "standard work on the subject" (*Nature*, London), has been completely revised and brought up to date by A. N. Campbell and N. O. Smith. Brand new material has been added on such matters as binary, tertiary liquid equilibria, solid solutions in ternary systems, quinary systems of salts and water. Completely revised to triangular coordinates in ternary systems, clarified graphic representation, solid models, etc. 9th revised edition. Author, subject indexes. 236 figures. 505 footnotes, mostly bibliographic. xii + 494pp. 5⅜ x 8.
Paperbound $2.75

A HISTORY OF PHYSICS: IN ITS ELEMENTARY BRANCHES (THROUGH 1925), INCLUDING THE EVOLUTION OF PHYSICAL LABORATORIES,
*Florian Cajori*
Revised and enlarged edition. The only first-rate brief history of physics. Still the best entry for a student or teacher into the antecedents of modern theories of physics. A clear, non-mathematical, handy reference work which traces in critical fashion the developments of ideas, theories, techniques, and apparatus from the Greeks to the 1920's. Within each period he analyzes the basic topics of mechanics, light, electricity and magnetism, sound, atomic theory and structure of matter, radioactivity, etc. A chapter on modern research: Curie, Kelvin, Planck's quantum theory, thermodynamics, Fitzgerald and Lorentz, special and general relativity, J. J. Thomson's model of an atom, Bohr's discoveries and later results, wave mechanics, and many other matters. Much bibliographic detail in footnotes. Index. 16 figures. xv + 424pp. 5⅜ x 8.
Paperbound $2.50

OPTICKS, *Sir Isaac Newton*
In its discussions of light, reflection, color, refraction, the wave theory and the corpuscular theory of light, this work is packed with scores of insights and discoveries. In its precise and practical discussion of construction of optical apparatus, contemporary understandings of phenomena it is truly fascinating to modern physicists, astronomers, mathematicians. Foreword by Albert Einstein. Preface by I. B. Cohen of Harvard University. 7 pages of portraits, facsimile pages, letters, etc. cxvi + 414pp. 5⅜ x 8. Paperbound $2.25

DIALOGUES CONCERNING TWO NEW SCIENCES,
*Galileo Galilei*
This classic of experimental science, mechanics, engineering, is as enjoyable as it is important. A great historical document giving insights into one of the world's most original thinkers, it is based on 30 years' experimentation. It offers a lively exposition of dynamics, elasticity, sound, ballistics, strength of materials, the scientific method. "Superior to everything else of mine," Galileo. Trans. by H. Crew, A. Salvio. 126 diagrams. Index xxi + 288pp. 5⅜ x 8.
Paperbound $1.75

TREATISE ON ELECTRICITY AND MAGNETISM,
*James Clerk Maxwell*
For more than 80 years a seemingly inexhaustible source of leads for physicists, mathematicians, engineers. Total of 1082pp. on such topics as Measurement of Quantities, Electrostatics, Elementary Mathematical Theory of Electricity, Electrical Work and Energy in a System of Conductors, General Theorems, Theory of Electrical Images, Electrolysis, Conduction, Polarization, Dielectrics, Resistance, etc. "The greatest mathematical physicist since Newton," Sir James Jeans. 3rd edition. 107 figures, 21 plates, 1082pp. 5⅜ x 8.
Vol. 1 Paperbound $2.50, Vol. 2 Paperbound $2.50
The set $5.00

*Prices subject to change without notice.*

Available at your book dealer or write for free catalogue to Dept. Adsci, Dover Publications, Inc., 180 Varick St., N.Y., N.Y. 10014. Dover publishes more than 150 books each year on science, elementary and advanced mathematics, biology, music, art, literary history, social sciences and other areas.